International Trade and Food Security

The Future of Indian Agriculture

FSC
www.fsc.org
MIX
Paper from responsible sources
FSC® C013604

International Trade and Food Security

The Future of Indian Agriculture

Edited by

Floor Brouwer

LEI Wageningen UR, The Netherlands

and

P.K. Joshi

International Food Policy Research Institute, India

CABI

CABI is a trading name of CAB International

CABI
Nosworthy Way
Wallingford
Oxfordshire OX10 8DE
UK

Tel: +44 (0)1491 832111
Fax: +44 (0)1491 833508
E-mail: info@cabi.org
Website: www.cabi.org

CABI
745 Atlantic Avenue
8th Floor
Boston, MA 02111
USA

Tel: +1 (617) 682 9015
E-mail: cabi-nao@cabi.org

A catalogue record for this book is available from the British Library, London, UK.

Library of Congress Cataloging-in-Publication Data

Names: Brouwer, Floor, editor. | Joshi, P. K., editor.
Title: International trade and food security : the future of Indian agriculture / editors, Floor Brouwer, P.K. Joshi.
Other titles: Future of Indian agriculture
Description: Boston, MA : CAB International, [2016] | Includes bibliographical references and index.
Identifiers: LCCN 2015046431 | ISBN 9781780642826 (hbk : alk. paper) | ISBN 9781780648866 (epub)
Subjects: LCSH: Agriculture--Economic aspects--India--Forecasting. | Food supply--India--Forecasting. | International trade.
Classification: LCC HD2072 .I685 2016 | DDC 382/.410954--dc23 LC record available at http://lccn.loc.gov/2015046431

ISBN-13: 978 1 78064 282 6

Commissioning editor: Alex Hollingsworth
Editorial assistant: Emma McCann
Production editor: Shankari Wilford

Typeset by SPi, Pondicherry, India
Printed and bound in the UK by CPI Group (UK) Ltd, Croydon, CR0 4YY, UK

Contents

Contributors

Brouwer, Floor, Researcher, LEI Wageningen UR, PO Box 29703, 2502LS The Hague, The Netherlands; e-mail: Floor.Brouwer@wur.nl

Ganguly, Kavery, Director, Confederation of Indian Industry (CII), India Habitat Centre, Core 4A, Ground Floor, Lodi Road, New Delhi-110 003, India; e-mail: Kavery.ganguly@gmail.com

Joshi, P.K., Director-South Asia, International Food Policy Research Institute, NASC Complex, CG Block, Dev Prakash Shastri Marg, Pusa, New Delhi-110 012, India; e-mail: p.joshi@cgiar.org

Joshi, Shwetima, University of Texas El Paso, 500 W University Avenue, El Paso, TX 79968, USA; e-mail: shwetimajoshi87@gmail.com

Karandikar, Bhushana, Fellow, Indian School of Political Economy, Arthbodh, S.B. Road, Pune, 411016, India; e-mail: kbhushana@gmail.com

Kumar, Praduman, Former Professor, Division of Agricultural Economics, Indian Agricultural Research Institute, New Delhi-110 012, India; e-mail: pkumariari@gmail.com

Mehta, Rajesh, formerly of Research and Information System for Developing Countries, Zone IV-B, Fourth Floor, India Habitat Centre, Lodhi Road, New Delhi-110 003, India. Rajesh Mehta passed away at the age of 65 years on 12 May 2015, during the process of editing this book. We deeply regret the loss to his colleagues, friends and family.

Meijerink, Gerdien, Program manager, CPB Netherlands Bureau for Economic Policy Analysis, The Hague, The Netherlands; e-mail: g.w.meijerink@cpb.nl

Mythili, G., Professor, Indira Gandhi Institute of Development Research (IGIDR), Gen. A.K. Vaidya Marg, Goregaon (E), Mumbai-400 065, India; e-mail: mythili@igidr.ac.in

Nambiar, R.G., Professor-Economics, FLAME University, 401, Phoenix Complex, Bund Garden Road, Opp. Residency Club, Pune-411 001, Maharashtra, India; e-mail: nambiar@flame.edu.in

Pandey, Vijay Laxmi, Associate Professor, Indira Gandhi Institute of Development Research (IGIDR), Gen. A.K. Vaidya Marg, Goregaon (E), Mumbai-400 065, India; e-mail: vijay@igidr.ac.in

Roy, Devesh, Research Fellow, International Food Policy Research Institute, NASC Complex, CG Block, Dev Prakash Shastri Marg, Pusa, New Delhi-110012, India; e-mail: d.roy@cgiar.org

Rutten, Martine, DLO Researcher, LEI Wageningen UR, PO Box 29703, 2502LS The Hague, The Netherlands; e-mail: martine.rutten@wur.nl

Smeets, Edward, Researcher, LEI Wageningen UR, PO Box 29703, 2502LS The Hague, The Netherlands; e-mail: Edward.Smeets@wur.nl

Squicciarini, Mara P., LICOS – Center for Institutions and Economic Performance, KU Leuven, Waaistraat 6, 3000 Leuven, Belgium; e-mail: mara.squicciarini@kuleuven.be

Swinnen, Johan F.M., LICOS – Center for Institutions and Economic Performance, KU Leuven, Waaistraat 6, 3000 Leuven, Belgium; e-mail: jo.swinnen@kuleuven.be

Vandeplas, Anneleen, LICOS – Center for Institutions and Economic Performance, KU Leuven, Waaistraat 6, 3000 Leuven, Belgium and European Commission; e-mail: anneleen.vandeplas@kuleuven.be

Woltjer, Geert, DLO Researcher, LEI Wageningen UR, PO Box 29703, 2502LS The Hague, The Netherlands; e-mail: Geert.Woltjer@wur.nl

Foreword

Global food demand will increase in coming decades, mainly in response to changing global diets and rapidly growing middle-income populations in emerging economies. To ensure food and nutrition security, this future demand must be met at affordable prices. Because international trade will necessarily play a significant role in balancing food demand and supply, its potential for improving global food security needs to be better understood. India provides a critical case for investigating the links between trade and food security. It is one of the major emerging economies, and has experienced a population increase of some 100 million over the last decade. The proportion of undernourished people is high and the population is young, with 40 per cent between the ages of 10 and 30, and highly rural, with only 30 per cent living in urban areas.

To date, the importance of the international trade of Indian agricultural products in securing global and national food supplies has not been properly addressed. The current volume fills this gap. It provides an in-depth understanding of the driving role of food security in Indian debates about opening up to international markets for food products, and explores the potential benefits and risks of international trade in food commodities. A mix of global, national, and regional assessments, complemented with qualitative approaches, include demand–supply projections under different scenarios and modelling of the impacts of different trade regimes on agricultural growth and food security. The role of price support systems, input subsidies, and government programmes in food security are also covered.

We welcome the insights provided by *International Trade and Food Security: The Future of Indian Agriculture*, which are the product of a fruitful collaboration between the International Food Policy Research Institute (Washington DC, USA), LEI Wageningen UR (the Netherlands), Indira Gandhi Institute of Development Research (India), KU Leuven (Belgium), IAMO (Germany), and CRPA (Italy). We compliment the efforts of Floor Brouwer and P.K. Joshi in compiling the studies and bringing out this volume. This work will undoubtedly generate discussion and contribute to policy formulation related to domestic policies, international trade, and food security.

Shenggen Fan, Director General, International Food Policy Research Institute (IFPRI), Washington, DC, USA
Jack van der Vorst, Managing Director Social Sciences Group, Wageningen UR, Netherlands

Preface

India has been successful in recovering from the global financial crisis and food price spikes. However, poverty and food insecurity always remain a high priority in the development agenda. A large majority of the Indian population dependent on agriculture is poor and food insecure. Therefore, to improve the livelihood of those dependent on agriculture, the aim is to achieve a 4% annual growth rate in agriculture during the 12th Five-Year Plan (2012–2017). This plan also targets a 9% economic growth rate for generating employment opportunities in the non-farm sector. These are ambitious targets and call for a paradigm shift in policies and institutional arrangements. To evolve effective policies, a clear perspective and empirical evidence on supply, demand and trade of agri-food commodities is required in the medium term. The question often raised is whether 'business-as-usual' will meet the targets in a scenario of increasing income, rising food demand and growing trade. There are also apprehensions among policy makers on agricultural trade liberalization that this would lead to the domestic market being flooded by imports that may adversely affect farmers.

This book presents recent work on demand and supply projections for food grains and high-value agri-food commodities in India. It intends to decipher the curiosity by projecting implications of policies and trade on farm and non-farm incomes, poverty and income distribution in rural and urban areas. It also offers the future potential for international trade of agricultural commodities and its implications on food security and poverty.

This title is the product of an international collaboration among researchers from India and Europe. The work presented in this volume has been (co-)funded by the project 'Trade, Agricultural Policies and Structural Changes in India's Agri-food System: Implications for National and Global Markets (TAPSIM)' under the Seventh Framework Programme (Grant agreement no: 212617). The financial support from the Strategic Research Programme 'Food Security, Scarcity and Transition', which is funded by the Ministry of Economic Affairs in The Netherlands, is also acknowledged for enabling the work on this volume. In addition to the chapters based on the research conducted under this project, we have solicited additional chapters to enrich the range of perspectives.

The book was not possible without the cooperation of all the authors. Their deep understanding and empirical analysis make the book a rich source of information on prospects for food supply, demand and trade under different scenarios. We appreciate their immense cooperation in undertaking the studies, preparing reports and finalizing chapters.

The contents of all the chapters were presented in two policy workshops in India to validate the findings of the research studies. We express our sincere thanks to: R.B. Singh, K.L. Chadha, Mahendra Dev, A.K. Srivastava, Ramesh Chand and Kirit Parikh for their invaluable contributions in improving the analysis. We also benefited from very useful feedback on different chapters from a number of professionals including: A. Ganesh-Kumar, A.K. Bandyopadhyay, Anjani Kumar, K.S. Ramchandra, Manoj Panda, Meetu Kapur, Mruthyunjaya, Pratap S. Birthal, Purvi Mehta, Shashanka Bhide, Sukhpal Singh, Surabhi Mittal, T. Haque and Tushar Pandey. We are grateful to all participants for lively discussions and invaluable inputs. We also acknowledge the support received from B.S. Agarwal for carefully and patiently editing the chapters.

We received incredible support and motivation from Shenggen Fan, Director-General, International Food Policy Research Institute, and Jack van der Vorst, Managing Director, LEI Wageningen UR, for undertaking the studies and completing the book.

One of the contributors to this book, Rajesh Mehta, passed away at the age of 65 years on 12 May 2015, during the process of editing the manuscript. We deeply regret the loss to his colleagues, friends and family.

The content of this book does not represent the official position of the European Commission and is entirely the responsibility of the authors. We are grateful to the European Commission for their financial support for this research.

Floor Brouwer and P.K. Joshi
October 2015

Acronyms

ADF	Augmented Dickey-Fuller
BAU	Business-as-usual
BIS	Bureau of Indian Standards
BPL	Below poverty line
CAC	Codex Alimentarius Commission
CACP	Commission for Agricultural Costs and Prices
CES	Constant elasticity of substitution
CGE	Computable general equilibrium
CSO	Central Statistical Organization
DES	Directorate of Economics and Statistics
EC	European Commission
EIC	Export Inspection Council of India
EU	European Union
FAO	Food and Agriculture Organization of United Nations
FCI	Food Corporation of India
FCR	Feed conversion ratio
FDA	Food and Drug Administration
FDI	Foreign direct investment
FMD	Foot and mouth disease
FSSA	Food Safety and Standards Act
FSSAI	Food Safety and Standards Authority of India
FTA	Free trade agreement
GCF	Gross Capital Formation
GCFA	Gross Capital Formation in Agriculture
GDP	Gross domestic product
GDPA	Gross domestic product of agriculture
GoI	Government of India
GTAP	Global Trade Analysis Project
ha	hectare
HACCP	Hazards analysis and critical control points
IARI	Indian Agricultural Research Institute
IEA	International Energy Agency

IFAD	International Fund for Agricultural Development
IFC	International Finance Corporation
IFPRI	International Food Policy Research Institute
IMAGE	Integrated Model to Assess the Global Environment
IMF	International Monetary Fund
ISO	International Organization for Standardization
ITC	International Trade Centre
LOU	Livestock output units
Mha	million hectares
MAGNET	Modular Applied GeNeral Equilibrium Tool
MGNREGA	Mahatma Gandhi National Rural Employment Guarantee Act
MoA	Ministry of Agriculture
MoC	Ministry of Commerce
MRL	Maximum residue levels
MSP	Minimum support price
Mt	million tons
NDDB	National Dairy Development board
NFSM	National Food Security Mission
NHM	National Horticulture Mission
NSSO	National Sample Survey Organization
NTM	Non-tariff measures
OBC	Other Backward Community
OECD	Organization for Economic Cooperation and Development
PDS	Public distribution system
PP	Phillips-Perron
RKVY	Rashtriya Krishi Vikas Yojna (National Agriculture Development Plan)
Rs	Rupees (Indian)
SAM	Social Accounting Matrix
SC	Scheduled Caste
SMEs	Small manufacturing enterprises
SNF	Solid Not Fat (non-fat solid)
SPSS	Sanitary and Phytosanitary Standards
ST	Scheduled Tribe
t	ton
TFP	Total factor productivity
TPDS	Targeted Public Distribution System
US$	Dollars (USA)
UT	Union Territory
VECM	Vector Error Correction Model
WTO	World Trade Organization

1 Introduction

Floor Brouwer[1]* **and P.K. Joshi**[2]
[1]*LEI Wageningen UR, the Hague, The Netherlands;* [2]*International Food Policy Research Institute, Pusa, New Delhi, India*

India has made considerable progress on the overall macro-economic indicators since independence in 1947. The country showed resilience to economic shocks, which was evident during the global food, fuel and financial crisis of 2008–2009. However, agriculture has remained lagging behind the other sectors of the economy and is facing a higher degree of volatility. In the first decade of 21st century, average economic growth was more than 7%, while growth was less than 4% in the agriculture sector on which more than half of the population depends for its livelihood. Although the Government of India has launched several programmes and reformed policies to increase agricultural production, the target to achieve 4% annual growth rate could not be realized. The main reasons for relatively poor growth in the agriculture sector were falling size and fragmenting of landholdings, near stagnating public investment, increasing pressure of farm subsidies, slowing irrigation expansion, hindering access to credit, marginalizing agricultural labour, and growing environmental stresses. Such factors remain a major challenge to the public sector for accelerating agricultural performance. This needs to create greater space for the private sector engagement, from seeds to storage, to processing and retailing that can help lift the overall growth in the agriculture sector and ensure food and nutritional security for the masses.

The present volume has a forward-looking approach, exploring structural changes in India's agrifood system in the coming 10 to 20 years. The dynamics in the agrifood sector are explored in the context of the overall economy, taking into account agricultural and trade policies and their impacts on national and global markets, and assessing their implications on food security and poverty alleviation. The book draws from qualitative and quantitative approaches, using a national model – to focus on urban–rural relations and income distribution – and an international model – to focus on patterns of economic growth and international trade.

The Organization of the Book

Following the Introduction, Part 1 presents the main features of Indian agriculture in the changing global context. Chapter 2 documents the transformation of Indian agriculture following economic liberalization. Kavery Ganguly and Vijay Laxmi Pandey present the

* Corresponding author, e-mail: floor.brouwer@wur.nl

main economic trends in India since the early 1990s, targeted towards greater economic integration, both domestically and internationally. Policies are designed for high growth rates in agriculture (4% during the 12th Five-Year Plan: 2012–2017) aiming at poverty alleviation and considering food security concerns. Agricultural trade policies remain subservient to food security concerns, which is particularly true with respect to grains. The export of high-value agricultural commodities has increased over time, but India remains a small player in the global market. However, huge investments would be needed to develop adequate infrastructure to enable large-scale agricultural exports. A free trade agreement between India and the European Union (EU) would be beneficial for unskilled labour in India, since their wages would increase.

A range of demand–supply projections for food commodities is presented in Part 2 of this volume. Chapter 3 presents the impacts of rural employment policies on food consumption patterns and nutritional security among rural households. Praduman Kumar and P.K. Joshi use household unit data on dietary patterns and employment collected at the national level. Implementation of the Mahatma Gandhi National Rural Employment Guarantee Act (MGNREGA) stipulates a guarantee of providing 100 days of wage employment in a financial year to adult members of any rural household willing to do unskilled manual work. The programme increases income of the rural poor, especially in the economically weaker states of the country which have implemented this programme more vigorously. Moreover, the programme has also supported diversification in the dietary pattern of households.

Food demand and supply projections for India are essential from the perspective of food security and are presented in Chapter 4. Praduman Kumar and P.K. Joshi project the demand to 2030 for food grains, as well as horticultural, livestock and fisheries products. The study is aware of the diversification and structural changes in the food basket of India and due importance is given to the future demand for high-value food commodities. Future demands for rice and wheat are likely to be met by domestic supplies, but the authors indicate that demand for pulses, edible oils and sugar would be greater than supply.

Patterns of economic growth and agreements on international trade both impact the economy, nationally and globally. Chapter 5, co-authored by Geert Woltjer and Martine Rutten, compares alternative patterns of economic growth with a bilateral trade agreement between India and the EU and a multilateral trade agreement in the context of the World Trade Organization (WTO). The authors conclude that rising patterns of economic growth in India would be beneficial for the country with gaining importance as a net-exporter of industrial products and a net-importer of services. With respect to agriculture, the net-imports of crop products would increase. Meanwhile, the increasing prices of farm land would lead to intensification in Indian agriculture.

Chapter 6 by G. Mythili addresses whether poverty has been reduced in both rural and urban areas. The author indicates that India had a share in global GDP of 6% in 2002, which is foreseen to rise to 11% by 2025. The share of agriculture in GDP is on the decline. High growth rates (over 8% per annum) seem to be achieved largely in the industrial and service sectors, and are benefiting mainly the high-income population groups in urban areas. The study also concludes that growth patterns exceeding 8% would reduce the share of real income in rural areas, largely because agricultural growth patterns remain lower than the rest of the economy.

The emergence of the livestock sector is presented in Part 3 of the volume. Chapter 7 reviews the relevance of food safety and quality standards for India, for both domestic and global markets. Anneleen Vandeplas and Pasquamaria Squicciarini address the importance of food quality standards in international trade, highlighting that sanitary and phytosanitary (SPS) measures are often significant barriers to trade between countries. The authors focus on the dairy sector in India and conclude that public food quality and safety measures remain addressed to a limited extent in the bulk of marketed milk. This is largely because the dairy sector is a rather unorganized sector. However, the increasing awareness of consumers about food quality has spurred the emergence of private food standards and quality standards among dairy processors. The authors conclude that SPS measures might be disregarded in trade modelling assessments, but their implications for international trade can be substantial and therefore should not be ignored.

The trade prospects of India's poultry sector are presented in Chapter 8. Rajesh Mehta, R.G. Nambiar and P.K. Joshi present the notion that the poultry industry is price competitive, and facing major non tariff barriers for international trade. The authors identify the trade opportunities for India's poultry sector. Since India is price competitive in eggs, but not in poultry meat, the authors conclude that the country may trade in eggs in the world market. Drastically reducing import tariffs of poultry meat could also result in large imports, and may adversely affect the domestic poultry industry.

Part 4 of the volume addresses a couple of policies and their implications on food consumption and supply. Chapter 9 examines the income distribution effects of the Mahatma Gandhi National Rural Employment Guarantee Act (MGNREGA). In doing so, the chapter builds on Chapter 3. So far, the implications of the programme over time, using a general equilibrium modelling approach, have been ignored and G. Mythili does fill this gap. The author indicates that the poorer households do not benefit from the fiscal measures to support the economy, and may even be worse off. The MGNREGA is implemented to benefit the rural poor, enhancing their real incomes. The analysis concludes that the MGNREGA is likely to have a negative impact on the agriculture sector in the long-term. It contributes to the growth of the industry sector, the labour and capital-intensive manufacturing industry as well as the construction industry. Market wages of unskilled labour do not rise due to the programme. Per capita income of urban poor is supported from the programme in the long-term, but not of the rural poor, as was intended originally.

Agricultural price policy in India is targeted towards assuring remunerative prices to farmers and providing subsidized food grains to the poor at reasonable prices. Chapter 10, co-authored by Gerdien Meijerink and P.K. Joshi, examines the impact of global food prices on the Indian price policy. Although the price policy has shielded domestic rice and wheat prices from global volatility, this led to increase in their minimum support prices and rising stocks. The study also clarifies the trade-offs involved with the multiple policy targets.

Chapter 11 examines the international trade implications of biofuel commitments in India and of biofuel policies in other parts of the world. Geert Woltjer and Edward Smeets use a general equilibrium model framework and show that biofuel policies outside India will not reduce poverty in India. While the urban poor will face higher food prices, the effect for the rural poor will be dampened because they would benefit from increased wages in agriculture. The national biofuel policy in India would increase global production of sugarcane by almost 20% and sugarcane prices by about a quarter of present values. The welfare effects for India are negative, because biofuel production is (implicitly or explicitly) subsidized.

The development of Indian agriculture is driven by input subsidies and farm technology. Therefore Chapter 12 examines which of these is more important for agricultural development. Praduman Kumar and P.K. Joshi evaluate the effects of price and non-price factors on factor demand, output supply and demand, as well as prices and farmers' income. The authors conclude that technology has a substantial impact on food supply. Although the input subsidy has a positive effect on input use, crop supply and farm income, the authors conclude that technology shifters have a strong influence on commodity supply and a negative effect on farm income because of the decline in market price in the absence of minimum support policy.

Part 5 of the volume focuses on the importance of high-value commodities and the rise of modern markets. Chapter 13 studies the extent to which dairy production has the potential to act as a motor of pro-poor growth in India. In order to achieve this, Anneleen Vandeplas, Mara P. Squicciarini and Johan F.M. Swinnen study whether the poorest rural households are effectively involved in dairy production. The analysis draws from a survey of 1000 rural households on dairy production in the Indian state of Andhra Pradesh. The study shows that dairy production has increased, but mainly through a larger number of households engaging in dairy rather than in scaling-up the operations. The productivity remains low compared to other dairy-producing countries. The study concludes that the rural poor are less likely to be dairy producers than wealthier households, which might be largely due to their constrained access to land.

Chapter 14 analyses whether the growth in supermarkets is associated with greater demand for product features, such as food safety and customization in consumer needs. Devesh Roy, Shwetima Joshi, P.K. Joshi and Bhushana Karandikar, using a site survey in three Indian cities (Mumbai and Pune, Maharashtra state; and Mysore, Karnataka state), notice that the demand for such features remains low. However, a market segment does care for such attributes and looks for imported products to satisfy its demand. The authors find that the retail sector so far has limited incentives for investment in the development of back end activities, the input service system (lime, seed, fertilizer, credit, etc.). A command approach by the government to enforce back end activities would require enforcement to be sufficiently strong. The authors argue that this might not be likely the case in India.

Finally, the editors present the way forward, including policies for changing business-as-usual. P.K. Joshi and Floor Brouwer focus on measures for accelerating agricultural growth, reforming policies for developing markets, and promoting agricultural trade for increasing farm incomes and reducing poverty. Recommendations are made to strengthen the agricultural sector. Involvement in international markets and securing national food supplies remain a challenge for Indian agriculture in the coming decades. Meanwhile, the country is well placed with the available human capital and the young society benefiting from rising welfare.

2 Transformation of Indian Agriculture Following Economic Liberalization

Kavery Ganguly[1]* and Vijay Laxmi Pandey[2]

[1]*Confederation of Indian Industry (CII), New Delhi;* [2]*Indira Gandhi Institute of Development Research, Mumbai, India*

Indian Economic and Agricultural Performance

Agriculture is critical for India, not just from the growth objective, given that it supports a huge agriculture-dependent industry, but because of its pivotal role in ensuring food security of the masses through its larger livelihood opportunities. Hence, it is a tough balancing act for the government and policy makers to design a high growth path for agriculture without neglecting the food security concerns. Since the 1950s, i.e. the post-Independence era, the socio-economic scenario has changed favourably in India, with higher economic growth, savings and investment patterns, rising foreign exchange reserves, increasing food production, rising income and reducing poverty levels.[1] This is not to deny that there have been rough patches that need strategic intervention and efforts are underway to address and contain the rising adversities. Despite higher economic growth, issues related to malnutrition, declining yet high poverty rates, rising subsidies and inadequate incentives for investments continue to be the key challenges. The agriculture and allied sector has been subject to piecemeal reforms governed largely by food security concerns and is yet to witness a major breakthrough.

In this backdrop of changing economic environment, it is interesting to understand how far we have come on the agricultural growth path and what more is to be done to ensure that agricultural growth becomes sustainable and has a positive impact on the socio-economic conditions of the people dependent on this sector.

Post-1991, the era of economic liberalization, India has undergone significant macroeconomic modifications to realize the growth potential and step towards greater economic integration, both domestically and globally. The Indian economy grew at an average annual rate of 3.6% during the period 1950/51 to 1980/81. The GDP growth accelerated to 5.6% (average annual) during the 1980s, but following the balance of payments crisis, it plunged to 1.4% in 1991/92. In response to the economic reforms initiated in the early 1990s, the economy recovered, clocking a growth of 5.4% in 1992/93 to as high as 8% in 1996/97. But thereafter, the overall GDP hovered between 4 and 6% per year during 1997/98 to 2002/03, a period of East-Asian crisis. The economic growth gained momentum during 2003/04 to 2007/08 at

* Corresponding author, e-mail: Kavery.ganguly@gmail.com

8.9% per year, giving a per-capita GDP growth of 7.2% per annum between 2003/04 and 2007/08. This period seems to be the 'golden period' in the economic history of India. This gave a new confidence and hope to Indian policy makers, and even when the global recession hit in 2008/09 and overall GDP in India slid to 6.7%, it quickly recovered to 7.2% in 2009/10 and was growing at 6.2% in 2010/11 (CSO, various years).

While overall economy has grown steadily from 5.6% during the 1980s and 5.8% during the 1990s to over 7% during the 2000s (until 2011/12), agricultural growth has slipped from more than 4.0% in the 1980s to 3.2% in the 1990s and 3.0% in the 2000s (until 2011/12). The period 2007/08 to 2011/12 witnessed a higher agricultural growth of 3.7% (although less than the targeted rate of 4%) (Fig. 2.1).

Nevertheless, agricultural performance has become less volatile during the decade of 2000 and has turned resilient, as observed from the year-to-year growth in the face of fluctuations in monsoons and its adverse impact on production systems. The state-wise performance of agriculture has been quite heterogeneous. The states that championed the success of the first green revolution in the 1960s (notably Punjab and Haryana) have been showing signs of stagnation or deceleration in the post-reform period, as has also been observed in Tamil Nadu and Andhra Pradesh. Some states, such as Gujarat and Madhya Pradesh, have sustained a continuous growth and Bihar has shown potential for a higher growth in agriculture. While high volatility in growth patterns has been a concern, investment in infrastructure (roads, markets, etc.) and irrigation, among others, has enabled a sustained high growth rate.

Although the share of the agriculture and allied sector (which includes crops, livestock and forestry) in the gross domestic product (GDP) has declined steadily from 38.8% in 1980/81 to 13.9% in 2011/12, it continues to be the major source of livelihood for 55% of the workforce (MoF, 2012; GoI, 2013a). For India, the agricultural sector and its employment opportunities are critical to sustained poverty reduction with nearly 27.5% of the population living below the poverty line[2] (as in 2004/05) (Planning Commission, 2012a).

The agricultural production basket has diversified considerably from traditional grains to high-value products and food grains account for less than 25% of the total value of output. Growth in production of food grains has declined over time and some grains, such as rice and wheat, and pulses have witnessed declining or stagnating yields. As the food grains are centric to the food security issue, efforts have been underway in the form of large flagship programmes and interventions to boost productivity of these crops. While the high-value agricultural

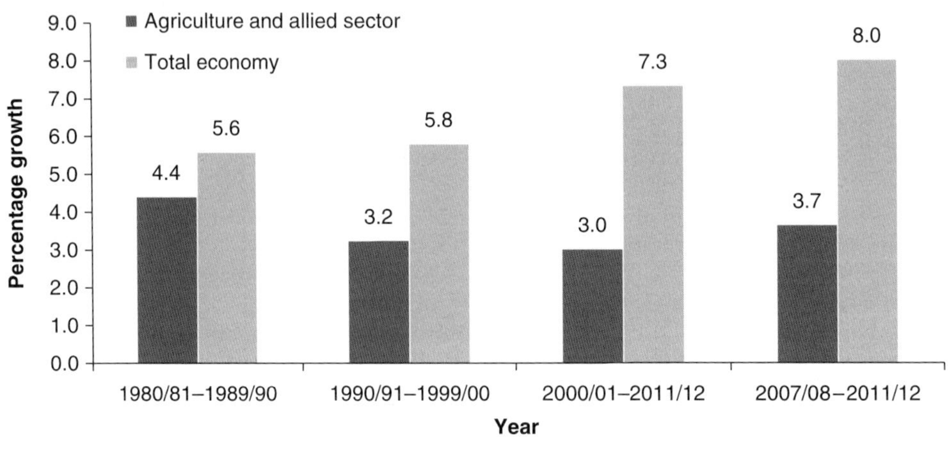

Fig. 2.1. Growth of the agriculture and allied sector and overall economy (Planning Commission, 2012b).

commodities, particularly fruits, vegetables, milk, fish, meat and eggs, are increasingly contributing to the value of agricultural output, production of these commodities has been growing at a faster rate compared to food grains. The increasing demand for these high-value commodities will require further growth in production, much of which is likely to come from productivity gains.

Despite the increasing resilience of the agricultural sector to adverse conditions and a not-so-poor growth performance, issues related to higher and sustainable growth continue to be of prime consideration.

It is often cited that parts of northern India are no longer suited to growing paddy or other water-intensive crops given the fast depleting groundwater table, degrading soil and water quality due to excessive use of fertilizers and chemicals. The impacts of climate change also threaten sustainable agricultural growth. It is estimated that due to global warming by 1°C, India will have to suffer yield losses in production of wheat, soybean, mustard, groundnut and potato by 3–7% (Aggarwal, 2009). Addressing some of these issues in the medium to long term will be critical to ensure that agricultural growth is sustainable and also inclusive as a large number of farmers in India operate on less than 2 ha of land. The scope of technology and innovation to overcome the natural and man-made challenges need to reach the smallholders (having less than 2 ha of land) and be able to address the challenges that vary across the geographies in India.

Evolution of Agricultural Trade

Agricultural trade in India has been growing with time and during 1990/91 to 2011/12 (Provisional; P) the net agricultural exports have been positive and have increased from US$2.7 billion in 1990/91 to US$23 billion in 2011/12 (P). India emerged as the largest exporter of rice with exports of nearly 10 Mt soon after the lifting of the export ban in late 2011. While exports and imports (in particular) have grown at different rates, in value terms, the agricultural exports are much higher than imports. In 2011/12, the composite agricultural export basket was worth US$39.1 billion with exports of fruit and vegetables of US$1.7 billion, cereals of US$6.4 billion, oil meals of US$2.5 billion, tea, coffee and spices of US$4.6 billion and marine products of US$3.5 billion. Since 1994/95, marine products have been the biggest export item, worth more than US$1 billion, but were overtaken by oil meals and cotton in 2007/08. Cotton exports were as high as US$1.9 billion in 2007/08. Export of cotton has skyrocketed in the recent past owing to the technological breakthrough brought about by use of the Bt variety that provided a major boost to production and thereby generated export surpluses, unprecedented in India. Cotton production increased from 15.8 million bales (of 170 kg each) in 1997/98 to 31.5 million bales in 2007/08 and simultaneously exports of cotton increased to 5.8 million bales in 2006/07 and further to 8.5 million bales in 2007/08, which is nearly a 47% increase in a single year. Agricultural imports have also grown over a period of time and reached US$16.1 billion in 2011/12 (P). The key import items were edible oils, which have increased manifold and posed a bill of US$10 billion, and pulses costing US$2 billion in 2011/12. India being a significant consumer of edible oils and pulses will have to augment its domestic supply to reduce its dependence on imports.[3]

Structure of global agricultural trade

While agricultural exports and imports have increased considerably over a period of time, India's share in global trade is negligible: it was 0.81% in world agricultural imports and 11.7% in global exports in 2004 (FAOSTAT, 2009). The biggest increase in Indian exports has been in the share of rice, which rose from 6.4% in 1990 to 18% in 2004, although this declined to 14% in 2006. All other agri-products have registered only a marginal increase in world share, while that of sugar and spices has been quite impressive. One positive feature of India's exports is that its export markets are well diversified with the group of export destination countries not having changed much during the period of study. In 2005, Indian agricultural exports were going mainly to the Asian countries; the value of exports has been increasing over time (almost doubling its quantity). However, the share in total agricultural exports from India has not changed that much, accounting for 39% in 1995 and 40% in 2005. Europe is the second biggest destination for Indian agricultural exports.

On the other hand, most of the agricultural imports come from the Asian countries. The value of agricultural imports from Asia has more than doubled, but its share in total agricultural imports has not changed much; their share was about 51% in 1995 and 49.6% in 2005. In 1995, Africa was the second main origin of Indian imports with a share of 14.3%, which declined to 12% in 2005. The share of imports from Europe has also declined over time, 6.5% in 1995 to 4.5% in 2005, although the value of imports to India has increased.

India–European Union trade in agricultural commodities

An analysis of the flow of trade between India and the European Union (EU) is important due to current negotiations of a possible Free Trade Agreement between them. A look at the trade relations between India and the EU has revealed that the EU is a more important partner of trade for India than India is for the EU. Indian exports to the EU are greater than its imports from the EU. However, both these shares have been declining for the past decade or so. The decline can be explained in terms of the different trade agreements that India has been signing with other countries and this might have played a role in changing the directions of trade towards these new signatory countries. Agricultural exports from India to the EU increased from US$1.5 million in 1996/97 to US$2.1 million in 2007/08. Agricultural imports from the EU have also increased over time, from US$167.40 million in 1996/97 to US$643.78 million in 2007/08. While both exports and imports of agricultural commodities have increased over time, the trends have been somewhat fluctuating. Fish and crustaceans (HS code 03) are the biggest agricultural export items from India to the EU and their exports have grown significantly since 1996/97, followed by coffee, tea and mate (HS code 09). The EU's agricultural exports to India have been quite a mixed bag over the years. In 1998/99, India imported fats and oils of animal and vegetable origin (HS code 15) worth US$152.19 million, highest ever, which gradually declined with time. In 2006/07, wheat and meslin worth US$224.18 million (HS code 1001) were imported by India (MoC, 2009). However, India is not one of the main trading partners of the EU for its agricultural trade as it accounts for only 0.58% of the EU total agricultural imports and 0.08% of total agricultural exports (Sequeros, 2008). These figures increased to 0.63% and 0.12%, respectively, in 2006, indicating a rise in the volume of trade between India and the EU.

Challenges and Opportunities Thus Far

While agricultural performance has not been all that bad and certain sectors have done well, there is no doubt that much needs to be done to realize the full potential of the sector in terms of its contribution to food security, overall economic growth and also as a source of livelihood to millions of farmers. The key areas of concern that pose both a challenge and opportunity to overcome the barriers to higher growth and more income for the farmers include near stagnation of public investment, increasing pressure of subsidies, limited access to formal credit, slowing of irrigation expansion and a limited role of the private sector. The increasing demand for food, particularly for high-value, protein-rich food, will serve as the key pull factor for the agricultural sector in India.

Boosting investments and rationalizing subsidies

Both public and private sector investments are important for agricultural growth. In the early 1980s, the share of public sector investment and private sector investment (including household sector) in gross capital formation in agriculture was almost equal, but by the early 2000s the share of private sector investment became much higher than that of the public sector (CSO, various years). However, about 90% of the public sector investment in agriculture has been for irrigation. The ratio of gross capital formation in agriculture (GCFA) to gross domestic product of agriculture (GDPA), which was declining in the 1980s and continued to decline even after the economic reforms, took a big leap after 1998/99. The ratio, which was 11.7% in 1980/81, fell to 7.6% in 1998/99 and started rising thereafter to reach 14.2% in 2007/08 (CSO, various years). Public investment, which declined continuously in real terms until the 1990s, was showing signs of improvement in the 2000s. The share of public investment in total agricultural investment improved to about 33% in 2007/08. However, input subsidies have depicted a rising trend and are linked to the excessive use of resources leading to environmental problems. Hence, there is a need to rationalize the subsidies and enhance the investment further.

Agricultural credit – extending the outreach of formal credit

The access to formal credit is critical for farmers to boost productivity and net returns that have a positive impact on the agricultural sector. While the flow of agricultural credit has increased over time (from Rs2546 billion in 2007/08 to Rs4468 billion in 2010/11) and more importantly, in this flow the formal sources have overtaken the informal sources, there are concerns related to the regional disparity and access to formal credit by small and marginal farmers (with <2 ha operated land) (Fig. 2.2). Despite institutional changes, small and marginal farmers, the proportions of which continue to increase in the total number of farmers, depend on informal credit sources at the cost of high interest rates due to their credit unworthiness. Lack of assets and land titles that are popular collaterals, restrict small farmers from accessing formal credit.

Irrigation – expanding infrastructure and ensuring better management of water resources

Irrigation is critical to Indian agriculture and improvement in irrigation systems will play an important role in enhancing agricultural productivity. There are concerns around water management in India, which are being addressed through participatory efforts. Although the net irrigated area increased from 38.7 Mha in 1980/81 to about 63 Mha in 2009/10, the average increase in

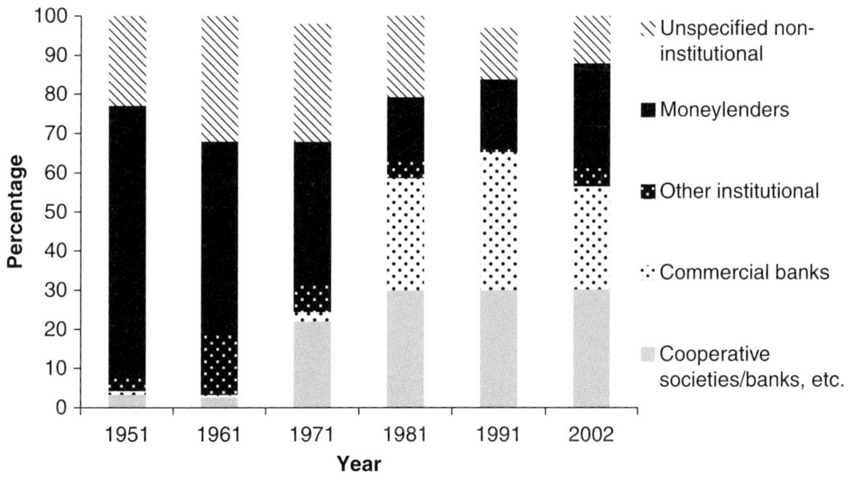

Fig. 2.2. Share of farm households in credit from different sources, 1951–2002 (Planning Commission, 2011).

the net irrigated area was less than 1 Mha/year. However, 45% of the total area under crops in India is irrigated. Considering the importance of irrigation for agricultural growth and the potential available, the central gross budgetary support for development of water resources (including through the Accelerated Irrigation Benefit Programme; AIBP) has been increased to Rs1095.5 billion during the 12th Five-Year Plan (2012–2017), up from Rs414.3 billion in the 11th Five-Year Plan (2007–2012). The increasing gap between the irrigation potential created and utilized is a big concern (2.7 Mha out of 9.5 Mha was utilized during the 11th Five-Year Plan) (Planning Commission, 2012b). Also, completion of the ongoing irrigation projects, particularly major irrigation projects which have a long gestation period, is critical to control cost escalation and also to ensure availability of services. Irrigation through groundwater exploitation has resulted in rapid depletion of water resources in certain states in north India. Studies suggest that groundwater level has been declining annually by about 4 cm during the past decade. The average stage of groundwater development in India is 61% and for states like Punjab, Haryana, Rajasthan and Delhi, it is more than 100%, indicating that utilization is far in excess of recharge. This has raised concerns about the sustainability of growing water-intensive crops in the northern region of the country (CGB, 2012).

Emerging private sector participation

Today, it is being envisaged that the private sector will play a major role in boosting agricultural growth through investments in agricultural research, technology and infrastructure. The role of the private sector in agriculture has been evident from the success of Bt cotton and the seed sector in India. The contribution of private companies to seed production has increased over the years, from 49% in 2004 to 58% in 2006. Introduction and adoption of Bt cotton, India's first transgenic crop, is one such success story. After the adoption of genetically modified (GM) Bt cotton in fields (officially released in 2002), the production and productivity of cotton have almost doubled. In the 7 years to 2009, more than 80% cotton area has come under Bt cotton at an all-India level, while in states such as Maharashtra and Gujarat, more than 90% of the cotton area is under Bt cotton. Adoption of Bt seeds resulted in a breakthrough in cotton yield that helped increase production. GM technology in agriculture is being widely debated in India and there is an increasing need to bring in institutions and regulatory mechanisms that are science-based and not driven by popular perceptions. There is a need for wider scientific consultation and awareness generation about the advantages of adopting GM technologies that help address the productivity challenges confronting Indian agriculture. The Seed Bill 2004, which is yet to be passed through the legislative process, has been subject to debate and does not create a platform for a level playing field for the private and the public sectors. Moreover, it is not in the interest of the seed sector, given that the private sector is better positioned to play a greater role in ensuring availability of quality seeds and proper utilization of such seeds in the farmers' fields through their extension and crop advisory services.

Diversifying consumption patterns – inducing supply response

Changes in the consumption basket are driving changes in production patterns. While an average Indian spends about 50% of the monthly expenditure on food, the bottom 30% of the population spend more than 60% on food. Within food, there is a shift in consumption preference toward high-value commodities such as fruit, vegetables, dairy products, meat, fish and eggs. The demand for cereals has decreased, from 14 kg per capita per month in 1983 to 11.1 kg per capita per month in 2007/08. Strong economic growth (7.2% GDP growth in 2009/10), rising income levels, increasing numbers of young and working population together with expanding urbanization and changing lifestyles of the people are driving these changes. India's urban population is projected to grow to 590 million by 2030 from 340 million in 2008, accounting for 40% of India's population by 2030 and also likely to generate 70% of the country's GDP in 2030 against 58% in 2008. The number of middle-class households (earning Rs200,000 to Rs 1 million per year, i.e. approximately US$4380 to US$21,890) is projected to increase from 5% of the population in 2005 to 41% of the population by 2025 (Mckinsey Global Institute, 2010).

The supply patterns are also responding to the changing demand patterns, but at a less desirable pace, as reflected in price inflation of high-value commodities in general. Agricultural production is diversifying and high-value commodities account for 47.4% of the value of output of crop, livestock and fisheries combined. However, there is a big gap between the fresh and the processed food baskets and a host of factors ranging from lack of availability of processable varieties to inadequate infrastructural facilities, which result in postharvest losses, are held accountable. The export of horticulture produce from India is considered to be quite uncompetitive, the higher transportation costs being one of the key reasons (Mattoo *et al.*, 2007). The factors related to food safety and standards and adherence to the stringent norms in the importing countries affect the export potential of India.

Key agricultural policies and initiatives – implications for growth and food security

The issue of food security has been very high on the policy agenda of India and remains critical, given a large population base and socio-economic and human development concerns. The food security concerns can be addressed through increasing domestic production or by import of food and making it accessible to the people at affordable prices. Therefore, concerned with the underperformance of agriculture during the past decade and the incidence of farmers' suicides in certain parts of the country, the Government of India has initiated several measures for improving the food security and livelihood of the people engaged in this sector. These programmes and policies can be broadly classified into three categories: (i) increase productivity of food grain/cereal crops; (ii) improve accessibility to food through economic empowerment; and (iii) boost agricultural diversification towards high-value commodities.

Addressing productivity issues

The National Food Security Mission (NFSM) and *Rashtriya Krishi Vikas Yojana* (RKVY; or the National Agriculture Development Plan) are the major government programmes aimed at increasing production through area and productivity improvements of key food crops. Given the scale and scope of these programmes, positive results in terms of increase in productivity of targeted crops in specific areas have been realized. However, it is difficult to ascertain the impact of these flagship programmes on controlling for other factors. There is a need for independent impact evaluation studies to understand better their progress. In general, there are issues related to implementation, accountability and transparency of these programmes, which have been often debated and need to be addressed.

NFSM, a centrally sponsored scheme (Fig. 2.3), was launched in 2007/08 to enhance the production of rice, wheat and pulses by 10 Mt, 8 Mt and 2 Mt, respectively, by the end of the 11th Five-Year Plan (2007–2012) through: area expansion and productivity enhancement; restoring soil fertility and productivity; creating employment opportunities; and enhancing farm-level economy to restore the confidence of farmers across targeted districts in the country.

The productivity gains in rice, wheat and pulses during the period of 2007/08 to 2012/13 have been attributed to the successful implementation of the NFSM programme together with other initiatives such as increasing support price and effective

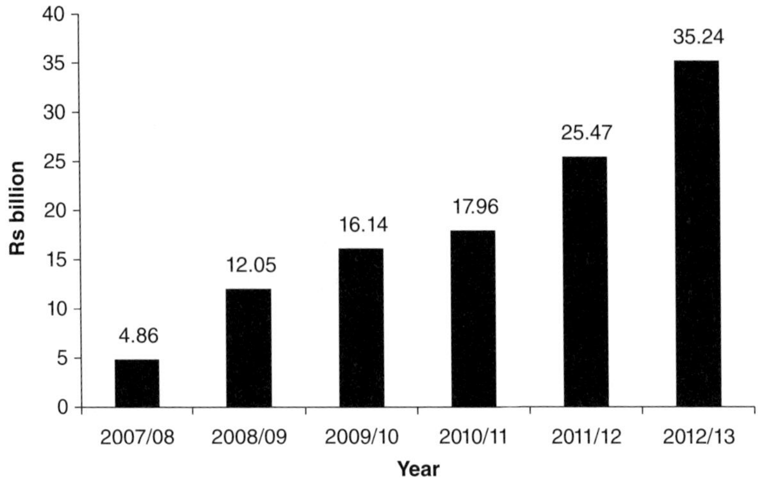

Fig. 2.3. Budgetary allocation to National Food Security Mission (NFSM) from 2007/08 to 2012/13 (GoI, 2013b). The budgetary allocation to NFSM increased from Rs4.9 billion in 2007/08 to Rs35.2 billion in 2012/13.

procurement mechanisms (Planning Commission, 2012b). Cereals production increased by 37 Mt (8 Mt coarse cereals, 11 Mt rice and 18 Mt wheat) between 2006/07 and 2011/12. All-India average growth in wheat yields was negligible during the 9th and 10th Five-Year Plans, but increased to 3% in the 11th Five-Year Plan. The 12th Five-Year Plan approach towards NFSM will focus on strategic area development in addition to ensuring that the productivity gains achieved during the 11th Five-Year Plan are sustained and the programme is extended to cropping systems rather than individual crops.

The RKVY was introduced in 2007/08 as an additional central assistance scheme to incentivize states to draw up plans, down to the district levels, for comprehensive development of agriculture (including crops, livestock, fisheries), taking agro-climatic conditions, natural resource issues and technology into account, and integrating livestock, poultry and fisheries more holistically. With an outlay of Rs250 billion in the 11th Five-Year Plan (2007/08 to 2011/12), the programme aims at incentivizing states to enhance public investment. The RKVY format permits taking up national priorities as sub-schemes, allowing the states to have flexibility in project selection and implementation. The annual allocation of funds for RKVY normal and sub-schemes increased from Rs15 billion in 2007/08 to Rs99.5 billion in 2013/14 (Fig. 2.4).

The budgetary allocation under RKVY for 2012/13 was Rs92 billion linking 50% of central assistance to those states that have stepped up the percentage of state plan expenditure on the agriculture and allied sector. The state governments, keeping in view their priorities, have approved the project proposals for implementation under RKVY in wide-ranging sectors, which include crops, horticulture, organic farming, farm mechanization, micro-irrigation, watershed development, marketing, storage, dairy development, fisheries, etc. The critical infrastructure, such as state seed farms, soil and fertilizers testing labs etc., have received substantial support under RKVY.

The state plan expenditures (excluding RKVY receipts) as percentage of GDP in the agricultural and allied sector increased from 1.0% in the 10th Five-Year Plan to 1.4% in the 11th Five-Year Plan. The state plan expenditures on agriculture and allied sector (excluding RKVY) have also increased as a percentage of total plan spending by states, from about 5% during the 10th Five-Year Plan to over 6% during the 11th Five-Year Plan, indicating some success in motivating states to pay greater attention to agriculture.

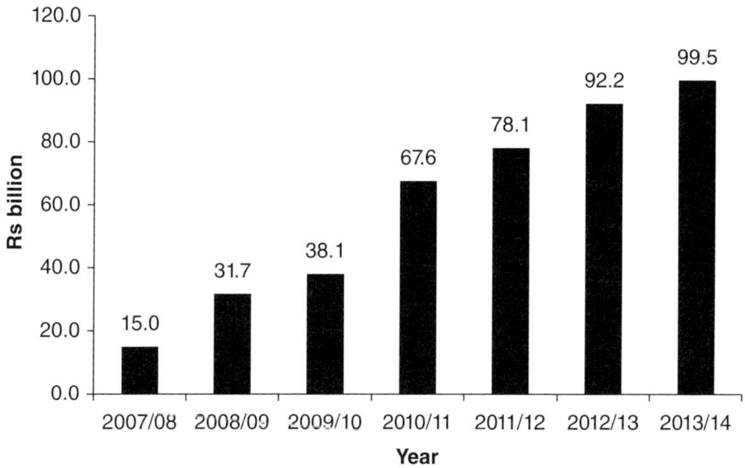

Fig. 2.4. Budgetary allocation to *Rastriya Krishi Vikas Yojana* (RKVY) normal and sub-schemes from 2007/08 to 2013/14 (GoI, 2013c).

Going forward, during the 12th Five-Year Plan, several improvements have been proposed in the area of setting priorities for agricultural development, capacity building of the implementing wing of the government, and leveraging the presence of private corporate players through public–private partnerships, among others.

As a sub-segment of RKVY, 'Bringing Green Revolution to Eastern India', initiated in 2010/11, intends to address the constraints limiting the productivity of 'rice-based cropping systems' in eastern India comprising seven states, viz. Assam, Bihar, Chhattisgarh, Jharkhand, Odisha, eastern Uttar Pradesh and West Bengal. A sum of Rs4 billion each was allocated for the programme during 2010/11 and 2011/12 and of Rs10 billion during 2012/13. Taking forward the objective of ensuring sustainable agricultural practices, the government has made efforts to shift the grain basket to the resource-abundant eastern region and release the burden on groundwater, soil health and the environment that has been rapidly increasing owing to intensive grain cultivation in parts of northern India. It is well understood that while the eastern region is well suited for the production of grains, particularly paddy, there is a need for better infrastructure (roads, rural electrification) and incentives (moving public procurement of grains to the eastern states) and investments in the region.

Improving economic access to food through improved livelihood options

The Mahatma Gandhi National Rural Employment Guarantee Act (MGNREGA) (implemented in 2006/07) aims at enhancing the livelihood security of people in rural areas by guaranteeing 100 days of wage-employment in a financial year to a rural household whose adult members volunteer to do unskilled manual work. With a budgetary outlay of Rs113 billion and covering 21 million households spread across 200 districts of India in 2006/07, the allocation to the programme was increased to Rs401 billion in 2011/12, covering 50 million households spread across 626 districts. Thus, the annual budgetary outlay was increased from Rs113 billion in 2007/08 to Rs400 billion in 2012/13 and Rs330 billion in 2013/14 (Fig. 2.5).

The average wage under MGNREGA was increased from Rs65 per person-day in 2006 to Rs115 per person-day in 2012 (GoI, 2013d). Since wages under MGNREGA do not differ for male and female workers, another positive impact of this programme has been on the women participation rate that ranged between 40 and 48% of the total

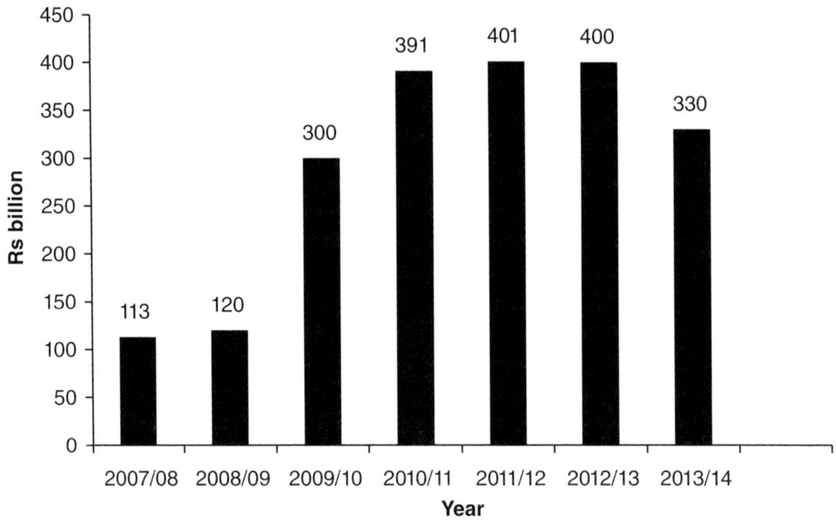

Fig. 2.5. Annual budgetary outlay under Mahatma Gandhi National Rural Employment Guarantee Act (MGNREGA) during 2007/08 to 2013/14 (GoI, 2013d).

person-days' work generated (considerably higher than the statutory requirement of 33%). It is also reported by independent research studies that MGNREGA has had significant impact on positive spending trends in the rural areas, which for the first time in 25 years outpaced trends in urban consumption in India (between 2009/10 and 2011/12) and contributed to enhancing rural food security.

While there are reports about a large number of people benefiting from the work opportunities, and reductions in rural to urban migration in search of jobs, there are issues related to the execution of the programme at the village level, the method, amount and timeliness of payment to the job seekers. The expansion of MGNREGA has resulted in labour shortage in the agricultural sector, resulting in an increase in agricultural wages. Although Gulati *et al.* (2013) have argued that MGNREGA is not a real game changer and has not resulted in an increase in farm wages, there are arguments in favour of the positive impact of MGNREGS in raising farm wages. Between 2008 and 2010, agricultural wages shot up by 106.5% in Andhra Pradesh, 84.4% in Punjab, 74.7% in Haryana and 73.6% in Tamil Nadu. Among the economically weaker states, the wages rose by 58.3% in Bihar, 56.3% in Madhya Pradesh, 62.8% in Odisha and 62.3% in Uttar Pradesh (Aiyar, 2011).

Improving economic access to food through support price and procurement policies

In order to ensure a sustainable supply of food grains and a supportive price policy aimed at guaranteeing remunerative prices to the farmers, the government has launched an elaborate system of food management consisting of procurement, storage and public distribution of food grains so as to give price insurance to farmers and food security to the poor consumers. Many studies have suggested that the price policy in India was fairly successful in price insurance through a minimum support price (MSP) to the farmers (Acharya, 2001; Reddy and Reddy, 2003; Bhalla, 2007) and also price stabilization during the pre-reform period. An analysis of price behaviour of rice and wheat has shown that the procurement prices were mostly around or above the MSP. This was observed even in the areas of large surpluses. Historically, the MSP for wheat and rice were increased at a rate below inflation and remained well under import parity prices. However, MSP has lost connection with both domestic as well as world prices as a substantial rise in the MSP of wheat and rice was observed after 1997/98, despite falling world prices (Landes and Gulati, 2004).

The procurement of food grains is undertaken to ensure that the market prices do not fall below the MSP. However, effectiveness of MSP is observed only in a few states (Punjab, Haryana, Uttar Pradesh and Andhra Pradesh) and for few crops (rice and wheat). Therefore, there is a need to shift the procurement to other regions as well, where government procurement is almost absent and water stress is less, such as the eastern states. There is also a need to extend the effective procurement of other cereals and pulses. At present, debate is on going with regard to de-linking the procurement price from the MSP, which is *de facto* the same. It is being suggested that the MSP should be there to protect the minimum prices to the farmers. However, the procurement price should be an incentive price over and above the MSP and should depend upon the prevailing market conditions and in competition to the private trade.

Promoting agricultural diversification

The diversification of agriculture towards non-food grains and high-value commodities (HVCs) has potential for income augmentation, employment generation, poverty alleviation and export promotion. Hence, it offers an opportunity to the very large number of smallholders to utilize their surplus labour resource and augment their incomes. In this connection, livestock, poultry, fisheries and horticulture have greater potential for supporting farm incomes as employment elasticities for these activities are very high. Nevertheless, there are obstacles towards this endeavour as farmers may lack skills on production methods, and require initial investment, access to credit, and risk-bearing ability. Recognizing this, the Government of India launched National Horticulture Mission (NHM) in 2005/06. The objectives of the mission are to enhance the production of horticulture crops and improve nutritional security and income support to farm households and others through area-based regionally differentiated strategies. Crops such as fruits, spices, flowers, medicinal and aromatic plants and plantation crops of cashew and cocoa are included for area expansion, whereas vegetables are covered through seed production, protected cultivation, integrated nutrient management, integrated pest management and organic farming.

Presently, 344 districts have been included under NHM. The physical target and achievement show an impressive overall performance of this programme. An additional area of 2.7 Mha has been put under various horticultural crops till 2010/11. About 300,000 ha of old and degraded plantations have been rejuvenated. In fact the achievement of physical targets was met in almost all components from 2007/08 onwards, except for the number of markets. Thus, the achievement of both financial and physical targets under NHM has helped in increasing both the area and production of horticultural crops in the country.

Agricultural trade policies

Indian agriculture had been characterized by relative dis-protection and was discriminated relative to the manufacturing sector during much of the period from 1970 to 1990 (Hoda and Gulati, 2008; Pursell *et al.*, 2009), resulting from the inbuilt urban bias that prioritized industry over agriculture. High levels of protection were applied to the manufacturing sector through a combination of tariff and non-tariff barriers and also an overvalued exchange rate. This scenario resulted in prices of essential farm inputs and machineries being inflated artificially which together with an overvalued exchange rate and export restrictions, dampened the export competitiveness of agricultural commodities. The implicit protection on agriculture has gradually improved since 1981, the trend being in sync with fluctuations in world prices of key agricultural commodities (in a countercyclical manner).

In July 1991, India chose a definite directional change in its economic policies.

The set of policies, often termed as the 'economic reforms package', was to have a long-lasting impact on the structure and performance of the economy. This package was comprised primarily of four policy changes:

1. Major correction in the exchange rate.[4]
2. Restructuring trade policies with a view to expose the domestic economy to global competition, and promote trade.[5]
3. De-licensing of a substantial part of the industrial sector, thereby abolishing much of the 'licensing system'.
4. Fiscal and monetary corrections that could promote growth and contain inflation.

However, unlike reforms carried out in the manufacturing sector, those in the agricultural sector were carried out in a piecemeal manner and quite often with hiccups and even reversals.

Self-reliance in staples has been an overarching objective whereby, time and again, the country has resorted to a 'stop-go' trade policy approach. Agriculture in India is critically linked to food security and serves as an important source of livelihood for millions, dominated by smallholders. In 2008, rising food prices the world over resulted in India further liberalizing imports of several key agricultural products (such as staples, edible oils, sugar, etc.) by lowering tariffs to very low levels, almost zero in many cases, and also imposing export restrictions on many of these commodities, especially rice and wheat, which has been the subject of much criticism in the global arena.

Agricultural trade policy reforms introduced since 1994 can be categorized as: exchange rate policies; import policy; export policy; and domestic policies. The exchange rate policies were very instrumental in changing the agricultural and manufacturing trade scenario in the country. The real rupee devaluation was large during the second half of the 1980s, about 62% between 1985 and 1990, and it was around 145% over the entire period to 1993.

Agricultural imports were earlier restricted through quantitative restrictions (QRs), even after the signing of the Uruguay Round of Agricultural Agreement (URAA) in 1995, due to the balance of payments cover under Article 18-B. But finally, India agreed to remove QRs in 1997 in response to WTO ruling, except for a few sensitive commodities. Starting in 1998, the general import licensing system was gradually dismantled, and on 1 April 2001, the last 715 of 2714 tariff lines (which included nearly all the agricultural tariff lines) were removed and the system itself was abolished. In 2006/07, un-weighted average tariffs protecting these sectors (HS 01-24) were about 40%, almost four times the level of average industrial tariffs. As judged by this criterion, India's agricultural sector appears to be one of the most protected in the world. However, in reality these tariffs 'contained a lot of water for facilitating flexibility in negotiations at the WTO' in case the negotiations to cut tariffs started from applied tariffs and not bound tariffs (Pursell *et al.*, 2009).

During much of 1980s there existed an anti-export bias and export of agricultural commodities was subject to restrictions, licences, quotas, controls and minimum export prices. Quantitative restrictions were administered through the state trading enterprises. However, with the economic liberalization in 1991, the policy of cash incentives was abolished but the income tax exemption continued. Ad hoc export subsidies were provided to compensate for the poor infrastructure (freight cost and stock holding cost), which were used periodically to dispose of agricultural surpluses (e.g. wheat and sugar). Under the *Vishesh Krishi Upaj Yojana* (VKUY) (special agricultural production scheme) introduced in 2004, exporters of selected commodities such as fruits and vegetables, dairy products and poultry, among others, were provided with import duty credit. Furthermore, agricultural export zones (AEZs) have been established to further encourage exports of value-added agricultural products. The domestic price support policy (price support scheme) has not been much affected by the economic reforms of 1991.

The Way Forward

Improving agricultural performance has a greater scope for poverty alleviation and addressing the food security concerns as well as increasing the footprint of the sector in the global food markets. Business-as-usual efforts have very little to offer and, particularly in the context of emerging structural changes in consumption, diversifying production and the dynamics of climate change, the volatile global markets cannot be wished away. The government intervention through programmes like RKVY, NFSM and the new initiatives for a second green revolution are highly laudable. A similar initiative, the NHM, has been operational in the horticulture sector and there is a need for better programmes in other high-value sectors such as livestock and fisheries. Despite all the controversies shrouding the success of Bt cotton in India, it is proven that a technology that benefits farmers is well taken and fetches high returns. Technology has played an important role in Indian agriculture, beginning with the success of high-yielding varieties from the 1960s. However, while public sector action was instrumental in pushing across the first Green Revolution, the current technology-based innovations and research are increasingly being taken ahead by the private sector. There is a need to recognize the role of private participation and encourage their presence in other segments of the agricultural sector. Investments are the key to improving agriculture in a sustainable manner, be it in irrigation, infrastructure (roads, markets), agricultural R&D, extension services, etc. Subsidies have outlived their significance, and it is time to rationalize them to boost investments. Agricultural input subsidies have increased manifold and, in certain cases, have even proved to be environmentally damaging, as observed in the case of fertilizer and power. These subsidies often also do not reach the targeted beneficiaries, particularly the marginal and small farmers who dominate the agricultural sector. Investments to enhance quality and delivery of public goods will have to come from the public sector.

Agricultural trade policy in India will remain subservient to food-security concerns. This is true particularly with respect to grains in the country. Despite large reserves of foreign exchange and the ability to play world markets, agricultural trade policies are driven by food-security concerns and often trigger knee-jerk reactions. Liberalization of agricultural trade had aroused apprehensions in the minds of the policy makers that the domestic market would be flooded by imports, but such was not the case. Agricultural production is diversifying and the share of high-value commodities such as horticulture, livestock and marine products is increasing and this provides a boost to the export of these items. The export of high-value commodities has increased over a period of time, but India is still a very small player in the global market and herein lies the scope to expand further. One of the key challenges confronting the agricultural sector is the lack of world class physical infrastructure, which has an adverse impact on agricultural exports. There is a need for large investments to build adequate infrastructure and bring in the right technology, but it will be possible only when subsidies give way to investments.

Notes

[1] The overall trend has been quite promising, except for the slowdown in 2013 and its adverse impact on the overall economic environment.

[2] Based largely on economic development, the country as well as each state declare for rural and urban households separately an income level which is termed as the 'poverty line' (PL), and the households having income less than this level are termed as 'below the poverty line' (BPL) households.

[3] All data related to value or quantity of import or export of agricultural commodities have been sourced from MoA (2012).

[4] The rupee was depreciated in 1991 and then again in 1992 with the Rp/US$ exchange rate becoming nearly 70% lower in 1992/93 than in 1990/91.

[5] Much of the export subsidies and licensing of imports were abolished. They were replaced by import entitlements (Exim scripts) linked to export earnings, etc. Import tariffs on industrial goods were substantially lowered in stages.

References

Acharya, S.S. (2001) Domestic agricultural marketing policies, incentives and integration. In: Acharya, S.S. and Chaudhri, D.P. (eds) *Indian Agricultural Policy at the Crossroads: Priorities and Agenda*. Rawat Publications, New Delhi.

Aggarwal, P.K. (2009) *Vulnerability of Indian Agriculture to Climate Change: Current State of Knowledge*. Indian Agricultural Research Institute, New Delhi. Available at: http://agricoop.nic.in/Climatechange/ccr/files/Vulnerability%20of%20Indian%20Agriculture,%20Climate%20Chnage%20PK%20Aggarwal.pdf (accessed 20 August 2013).

Aiyar, S.S.A. (2011) Agricultural wages have skyrocketed; poors have benefited from GDP growth. Available at: http://articles.economictimes.indiatimes.com/2011-07-07/news/29747766_1_wage-rate-gdp-growth-monsoon (accessed 10 July 2013).

Bhalla, G.S. (2007) *Indian Agriculture since Independence*, 1st edn. NBT Publications, New Delhi.

CGB (Central Groundwater Board) (2012) *Groundwater Yearbook India*. Ministry of Water Resources, Government of India, New Delhi.

CSO (Central Statistical Organization) (Various years) *National Accounts*. Government of India, New Delhi.

FAOSTAT (2009) *Trade Data*. Available at: http://faostat.fao.org (accessed 25 March 2009).

GoI (Government of India) (2013a) *Primary Census Data Highlights - India. Census 2011*. New Delhi.

GoI (Government of India) (2013b) *Base Data on Funds Allocated under National Food Security Mission (NFSM)*. Available at: http://nfsm.gov.in/nfsmmis/login.aspx (accessed 14 July 2013).

GoI (Government of India) (2013c) *Data on Allocations under Rashtriya Krishi Vikas Yojana (RKVY)*. Available at: http://rkvy.nic.in (accessed 14 July 2013).

GoI (Government of India) (2013d) Report to the People. Available at: http://nrega.nic.in/netnrega/WriteReaddata/circulars/Report_to_the_people_English2013.pdf (accessed 3 July 2013).

Gulati, A., Jain, S. and Satija, N. (2013) *Rising Farm Wages in India: The 'Pull' and 'Push' Factors*. Discussion Paper 05. Commission for Agricultural Costs and Prices. Department of Agriculture and Cooperation, Ministry of Agriculture, Government of India, New Delhi.

Hoda, A. and Gulati, A. (2008) *WTO Negotiations on Agriculture and Developing Countries*. Oxford University Press, New Delhi.

Landes, M. and Gulati, A. (2004) Farm sector performance and reform agenda. *Economic and Political Weekly* 39(32), 3611–3619.

Mattoo, A., Mishra, D. and Narain, A. (2007) *From Competition at Home to Competing Abroad: A Case Study of India's Horticulture*. Oxford University Press, New Delhi.

Mckinsey Global Institute (2010) *India's Urban Awakening: Building Inclusive Cities, Sustaining Economic Growth*. Mckinsey & Company, New Delhi.

MoA (Ministry of Agriculture) (2012) *Agricultural Statistics at a Glance 2012*. Directorate of Economics and Statistics, Department of Agriculture and Cooperation, Government of India, New Delhi.

MoC (Ministry of Commerce) (2009) *Trade Data*. Government of India, New Delhi. Available at: http://commerce.nic.in/eidb/default.asp (accessed 4 March 2009).

MoF (Ministry of Finance) (2012) *Economic Survey 2011–12*. Government of India, New Delhi.

Planning Commission (2011) *Report of the Working Group on Outreach of Institutional Finance, Cooperatives and Risk Management for the 12th Five Year Plan (2012–17)*. Government of India, New Delhi.

Planning Commission (2012a) *Press Note on Poverty Estimates 2009–10*. Government of India, New Delhi.

Planning Commission (2012b) *Twelfth Five Year Plan Document*. Government of India, New Delhi.

Pursell, G., Gulati, A. and Gupta, K. (2009) Distortions to agricultural incentives in India. In: Anderson, K. and Martin, W. (eds) *Distortions to Agricultural Incentives in Asia*. World Bank Publications, Washington, DC, pp. 339–377.

Reddy, V.R. and Reddy, P.P. (2003) Domestic price policies in the context of trade liberalization. *Indian Journal of Agricultural Economics* 58(3), 333–352.

Sequeros, T. (2008) Agricultural Trade in India. MSc thesis. Wageningen University, Wageningen, The Netherlands.

3 Food Consumption Pattern and Nutritional Security among Rural Households in India: Impact of Cross-cutting Rural Employment Policies

Praduman Kumar[1]* and P.K. Joshi[2]

[1]Indian Agricultural Research Institute, New Delhi; [2]International Food Policy Research Institute, New Delhi, India

Introduction[1]

The Government of India has launched various programmes from time to time in order to alleviate poverty in rural areas. These include: Integrated Rural Development Program (IRDP), Employment Assurance Scheme (EAS), *Pradhan Mantri Rojgar Yojna* (Prime Minister Job Scheme), *Swaranjayanti Gram Swarojgar Yojna* (Golden Jubilee Rural Self-Employment Scheme) and *Pradhan Mantri Gramodaya Yojna* (Prime Minister Rural Development Scheme). The latest programme, covering all the earlier poverty alleviation schemes, was implemented by the Government of India through the legislation entitled the 'National Rural Employment Guarantee Act' (NREGA). This is the largest employment-providing programme in the world started by a country for the development of its rural areas. The Act was later renamed as the Mahatma Gandhi National Rural Employment Guarantee Act (MGNREGA). This cross-cutting scheme stipulates a legal guarantee of providing 100 days of wage employment in a financial year to adult members of any rural household willing to do unskilled manual work at the statutory minimum wage. In 2009, these wages were Rs120 (US$2.39) per day (GoI, 2005). The wages paid under MGNREGA correspond to the minimum wages stated by the central government but vary across states. In 2014/15, the per day wages varied from Rs154 in Himachal Pradesh to Rs236 in Haryana (GoI, 2014).

For operation of a scheme under MGNREGA, the Government of India meets the costs towards payment of statutory wages, three-fourths of the material cost and some percentage of the administrative cost. The Act has a provision of payment of unemployment allowance also to a job-seeker who is not provided employment within 15 days of his/her request date. However, this unemployment allowance is to be met by the state governments along with one-fourth of the material cost and the remaining administrative cost. The implementation of MGNREGA was started with an initial outlay of Rs125 billion in the year 2006/07. It was enhanced to Rs440 billion in 2010/11. The outlay was Rs400 billion in 2011/12 and Rs340 billion in 2014/15.

A scheme under MGNREGA adopts a direct and most effective way of reducing

* Corresponding author, e-mail: pkumariari@gmail.com

poverty by providing (extra) wage employment to the rural poor, whether they belong to the below poverty line category or not. Contrary to the traditional practice, a scheme under MGNREGA provides equal wages to both men and women workers. It is open to all the rural households including those of scheduled castes (SCs), scheduled tribes (STs) and other backward classes (OBCs).

MGNREGA aims to achieve the twin objectives of providing rural employment and undertaking rural development simultaneously. The works under MGNREGA must be targeted towards a set of stipulated rural development activities like water and soil conservation, afforestation, flood control, watershed development, road connectivity, construction and repair of embankments, digging of new tanks/ponds, construction of percolation tanks, check dams, etc. Up to the end of 2010/11, various schemes under MGNREGA have provided employment to 25.7 million rural households with around 12,054 million person-day's work.

The landless and small farm (<1 ha land) households constitute more than 50% of India's population and account for more than half of the poor people in the country. This scenario of prevalence of wide poverty on one side and the implementation of a massive livelihood security-providing act, MGNREGA, on the other, raises some fundamental questions, such as: (i) Is the small farm size of the majority of rural households the main cause of perpetuating poverty and undernourishment in the country? (ii) Is there any prospect of liberating these farm households from poverty and undernourishment? (iii) How can the scope of MGNREGA be enhanced in empowering the poor to combat poverty and raise nutritional status? and (iv) Will MGNREGA help in uplifting the socio-economic status of poor households? Some of these questions have been addressed in this study by examining the dynamics of rural households through geographic and socio-economic dimensions across states and regions of India, and by finding the dietary pattern, nutritional status, and expenditure on food and non-food commodities by these households. The study has been conducted with the specific objectives of: (i) examining the changes in food consumption and nutritional security of rural poor households; and (ii) assessing the impact of MGNREGA on food consumption and food security.

Components of MGNREGA

Some significant components of MGNREGA include the following:

- It is a cross-cutting programme of reducing poverty by providing employment and income to the poorest in the rural areas.
- It provides legal rights to wage employment.
- It has the provision of 'unemployment allowance' in case employment is not provided within 15 days of demand.
- It provides equal wages to men and women and thus empowers women socially and economically.
- It is open to all rural households, irrespective of their farm size, household type, caste and religion.
- It has enhanced the bargaining power of poor men and women in the labour market by providing statutory minimum wages.
- It provides work site facilities such as drinking water, first-aid, crèches, etc.

Categorization of Rural Households by Region, Farm size and Income Level

The study has used Indian household unit data on dietary patterns and employment collected at the national level by a survey method based on the 66th round of the National Sample Survey (NSS) Organization and pertaining to the year 2009 (GoI, 2009a, b). For analysis, the sample rural households were grouped into six regions: Eastern states, Western states, Northern states, Southern states, Hill states and North-East states; six land classes: landless, sub-marginal (<0.5 ha), marginal (0.5–1.0 ha), small (1.0–2.0 ha), medium (2.0–4.0 ha) and large (>4.0 ha);

five household types: self-employed in non-agricultural sector, agricultural labour, non-agricultural labour, self-employed in agricultural sector and others; and four income groups based on the poverty line (PL): very poor (75% below PL), poor (on PL), middle-income (PL to <150% above PL) and high-income (>150% PL). The PL depicts a specific income level of a household; the households having income less than this level are termed as BPL (below poverty line) households and greater than this level are called APL (above poverty line) households. The PL is adopted by the Planning Commission of the Government of India for rural households by provinces (states of India). The PL values for the year 2009 were used to classify households into different income groups (Fig. 3.1).

The calories and protein intakes in the study refer to their respective consumption through different food commodities, calculated using the conversion factors provided by the NSSO (1996). The minimum (threshold) food-energy requirement was taken as 1800 kcal/person/day, which is 75% of the recommended energy requirement of 2400 kcal/person/day for rural households. An intake less than this threshold is considered as not sufficient to maintain health and body mass, or to support physical activity. The threshold level of food protein intake was used as 48 g/person/day (which is also 75% of the recommended level of protein for an average rural Indian). The households whose average intake of calories and protein was below these recommended threshold levels were categorized as 'undernourished' and 'malnourished', respectively.

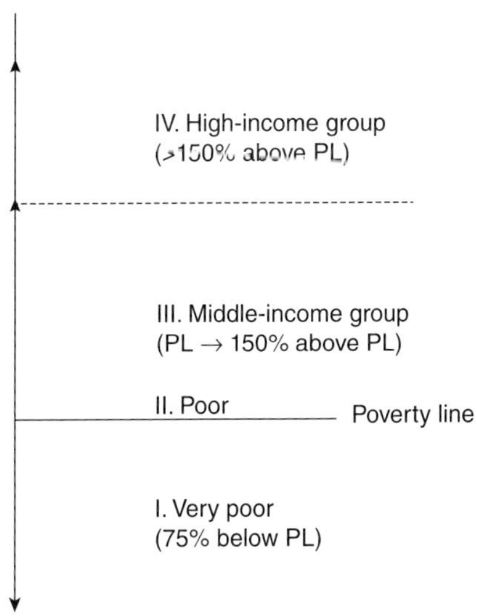

Fig. 3.1. Classification of households into different income groups. Poverty Line (PL) is the threshold income level for a household, declared by the Government of India, and households having income less than this are called Below Poverty Line (BPL) households.

Rural Households' Linkage with MGNREGA

In rural areas, there are some households that are in need of a job, while some others are either comfortable with their income or are not interested to do the type of work offered under a MGNREGA scheme (Fig. 3.2). For seeking employment under a MGNREGA scheme, adult members of a rural household, willing to do unskilled manual work, are required to obtain a 'job card' from the local Gram Panchayat after registering with it. These persons were classified as 'job card holders', while the non-applicants were termed as 'non-job card holders' and were not of relevance for the present study. The job card holders were classified into 'job seekers' and 'non-job seekers'. The non-job seekers were those who were not serious about obtaining employment under a MGNREGA scheme but had the job card issued to be used under emergency or as a trump card for getting higher wages from the present employer.

The job seekers had two types of members. Those who asked for employment and were able to get it were termed 'beneficiaries' and those who asked for employment but were not provided because of work-shortage or for some other reasons were termed as 'non-beneficiaries'. Of course, these non-beneficiaries were entitled to daily unemployment allowance, the liability of payment of unemployment allowance being the state's.

The impact of MGNREGA in providing nutritional security and energy security to

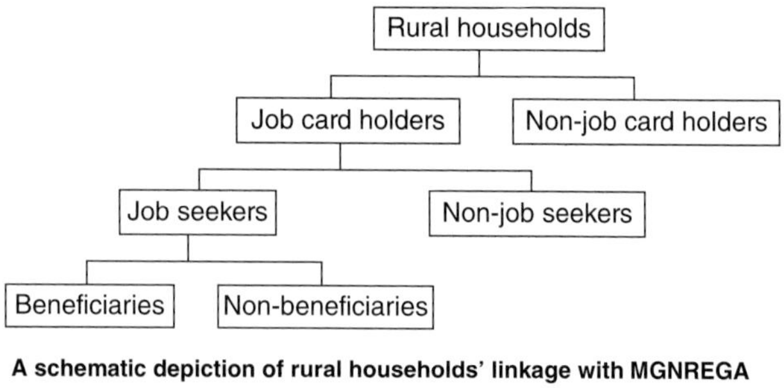

Fig. 3.2. A schematic depiction of rural households' linkage with MGNREGA.

the rural households has been studied by comparing the consumption pattern of beneficiary and non-beneficiary households.

Dynamics of MGNREGA Job Card Holders

The dynamics of MGNREGA job card holders has been studied across geographic regions of the country, land class, household type and income group. The dynamics of MGNREGA job card holders by socio-economic dimensions have also been studied across different states of India.

Job card holding

Of the total all-India sample of rural households, only about one-third (32.1%) had registered for seeking employment under MGNREGA, i.e. were job card holders. Across different regions of the country (eastern, western, northern, southern, hills and north-eastern), the number of MGNREGA job card holders in total sample households was a maximum (55.2%) in the north-eastern region, followed by the western region (38.6%), hill region (30.6%) and the eastern region (29.7%), and was lowest in the northern region (17.6%). The trend in seeking job cards across regions clearly depicts that registration for employment was maximal in the economically weaker regions of the country.

Across income groups, as expected, the maximum percentage of job card holders was of very poor (45.2%) and poor (41.9%) households and the least percentage was of high-income (21.3%) households. It shows that MGNREGA has been successful in the first step of its aim of providing employment to the poor. Land class-wise also, the percentage of job card holders was high across landless (29.2%), sub-marginal (37.5%) and marginal (30.8%) households and was least but still substantial for large households (19.7%). In household types, agricultural labour and non-agricultural labour households were far ahead in getting job cards than self-employed and other types of households.

Employment seeking

Table 3.1 depicts the number (and percentage) of job card holders who sought employment in the total job card holders (columns 5 and 6, respectively). Overall, 84.0% of the total job card holders sought employment in the states of India. Region-wise, the number of job seekers was maximum in the north-eastern region (95.7%), followed by northern (90.2%) and eastern (85.1%) regions. Thus, in the north-eastern region, not only was the number of job card holder households the highest in the total sample households, but the number of job seekers was also highest, showing the incidence of extreme poverty in the region.

Across income groups, it was surprising to see that not all the job card holders in the poor and very poor household categories sought employment. As far as the trend is concerned, it was the same as observed in

Table 3.1. Dynamics of MGNREGA households by geographic and socio-economic dimensions in India, 2009 (authors' computation).

Category of households	No. of sample rural households	MGNREGA job card holders: No. of households	MGNREGA job card holders: % of households	MGNREGA job seekers: No. of households	MGNREGA job seekers: % of households	MGNREGA beneficiaries: No. of households	MGNREGA beneficiaries: % of job seekers	MGNREGA beneficiaries: Employment (number of days in year)	MGNREGA beneficiaries: % of sample rural households
Region									
Eastern region	14,227	4,219	29.7	3,591	85.1	2,491	69.4	22.5	17.5
Western region	12,870	4,966	38.6	3,689	74.3	2,950	80.0	49.6	22.9
Northern region	10,042	1,770	17.6	1,596	90.2	1,342	84.1	30.0	13.4
Southern region	12,344	3,490	28.3	2,894	82.9	2,567	88.7	42.9	20.8
Hills region	3,108	952	30.6	748	78.6	609	81.4	45.4	19.6
North-eastern region	6,538	3,612	55.2	3,456	95.7	3,367	97.4	57.6	51.5
India (All states)	59,129	19,009	32.1	15,974	84.0	13,326	83.4	43.1	22.5
Income group									
Very poor	5,289	2,393	45.2	2,106	88.0	1,659	78.8	37.1	31.4
Poor	9,965	4,173	41.9	3,621	86.8	3,032	83.7	41.3	30.4
Middle income	20,447	7,453	36.5	6,381	85.6	5,395	84.5	45.1	26.4
High income	23,428	4,990	21.3	3,866	77.5	3,240	83.8	44.6	13.8
Land class									
Landless	25,087	7,314	29.2	6,243	85.4	5,153	82.6	41.3	20.6
Sub-marginal (<0.5 ha)	21,266	7,974	37.5	6,905	86.6	5,797	84.0	42.9	27.3
Marginal (0.5–1.0 ha)	8,158	2,514	30.8	2,040	81.1	1,734	85.0	45.7	21.3
Small (1.0–2.0 ha)	2,816	735	26.1	516	70.2	415	80.4	51.3	14.7
Medium (2.0–4.0 ha)	1,330	379	28.5	225	59.4	185	82.2	52.4	13.9
Large (>4.0 ha)	472	93	19.7	45	48.4	39	86.7	67.2	8.3
Household type									
Self-employed in non-agricultural sector	14,330	4,094	28.6	3,278	80.1	2,665	81.3	40.6	18.6
Agricultural labour	6,453	3,209	49.7	2,891	90.1	2,395	82.8	35.4	37.1
Non-agricultural labour	10,215	4,515	44.2	4,029	89.2	3,394	84.2	44.5	33.2
Self-employed in agricultural sector	16,837	5,497	32.6	4,487	81.6	3,775	84.1	47.4	22.4
Others	11,294	1,694	15.0	1,289	76.1	1,097	85.1	46.9	9.7

job card holding, i.e. very poor (88.0%), followed by poor (86.8%), middle income (85.6%) and high income (77.5%) households. Land class-wise, employment was sought by sub-marginal (86.6%), landless (85.4%) and marginal (81.1%) households. Although the job seeking was least among large households at 48.4%, it was still substantial. Across household types, agricultural labour and non-agricultural labour households were far ahead in seeking jobs under a MGNREGA scheme.

MGNREGA beneficiaries

These were those job card holders who sought employment under a MGNREGA scheme and were successful in getting employment. It is significant, because some job card holders who sought employment were not given employment because of shortage of work, more seekers than the quantity of work available, faulty planning, etc. The analysis is based on the number (and percentage) of beneficiaries in the total number of job seekers; these values are given in columns 7 and 8 of Table 3.1.

Region-wise, it was the north-eastern region that provided maximum employment (97.4%) to its MGNREGA job seekers, followed by the southern (88.7%) and northern (84.1%) regions. Overall, 83.4% MGNREGA job seekers were successful in getting jobs under a MGNREGA scheme. The success rate in getting employment across income group-wise, land class-wise as well as household type-wise was quite high and varied between 80 and 85% of total MGNREGA job seekers.

Duration of employment

This is an important aspect of MGNREGA and is the basis of rural poverty reduction. The Act has a provision of providing 100 days of wage employment in a year, but it was found that no rural household could get employment for 100 days in a year. The details about days of work are given in column 9 of Table 3.1. On an average, employment for 43.1 person days was provided in 2009, with the maximum (57.6 person days) in the north-eastern region and minimum in the eastern region (22.5 person days). Across income groups, APL households received employment for a higher number of days (45 person days) than BPL households (34–41 person days). Similarly, large households could manage to get more work (67 person days) than landless and small households, who could get employment for only 41–46 person days. Household type-wise, agricultural labour obtained employment for the minimum duration, only for 35.4 days. It was surprising to note that resource-poor households could get employment for a smaller number of days than resource-rich households.

The benefits of MGNREGA have reached 22.5% of the rural households at the national level. About 30% BPL households, 37% agricultural labour, 27% sub-marginal farmers and 21% landless households have benefited from the launching of MGNREGA schemes. However, during the study year 2009, none of the socio-economic groups got employment for 100 days, as stipulated in the Act.

State-wise economic dynamics of rural households

The state-wise socio-economic dynamics of sample rural households are depicted in Table 3.2. It is seen that more than two-thirds of households received job cards under MGNREGA in the states of Rajasthan, Manipur, Mizoram and Tripura, while in the states of Arunachal Pradesh, Assam, Bihar, Chhattisgarh, Goa and Daman, Gujarat, Haryana, Jammu and Kashmir, Karnataka, Kerala, Lakshadweep, Maharashtra, Punjab and Uttar Pradesh, not even one-third of rural households had obtained job cards. In some states like Punjab, Haryana and Goa and Daman, this percentage had not even touched a two-digit figure. However, the percentage of job seekers among the registered job card holders was very high in almost all states, with states like Andhra Pradesh, Jharkhand, Manipur, Meghalaya, Nagaland, Sikkim, Tripura, Pondicherry, Arunachal

Table 3.2. State-wise dynamics of MGNREGA households by socio-economic dimension in India, 2009 (authors' computation).

State	No. of sample rural households	MGNREGA job card holders: No. of households	MGNREGA job card holders: % of households	MGNREGA job seekers: No. of households	MGNREGA job seekers: % of households	MGNREGA beneficiaries: No. of households	MGNREGA beneficiaries: % of job seekers	MGNREGA beneficiaries: Employment (number of days in year)	MGNREGA beneficiaries: % of sample rural households
Andaman and Nicobar Islands	271	96	35.4	64	66.7	33	51.6	34.1	12.2
Andhra Pradesh	3926	1557	39.7	1331	85.5	1203	90.4	50.1	30.6
Arunachal Pradesh	1042	228	21.9	214	93.9	188	87.9	48.4	18.0
Assam	2616	591	22.6	520	88.0	354	68.1	30.5	13.5
Bihar	3300	457	13.8	376	82.3	229	60.9	26.0	6.9
Chhattisgarh	1759	464	26.4	395	85.1	243	61.5	24.1	13.8
Goa and Daman	224	22	9.8	13	59.1	7	53.8	13.6	3.1
Gujarat	1721	426	24.8	300	70.4	212	70.7	30.3	12.3
Haryana	1440	79	5.5	74	93.7	60	81.1	30.8	4.2
Himachal Pradesh	1660	719	43.3	581	80.8	506	87.1	48.2	30.5
Jammu and Kashmir	1448	233	16.1	167	71.7	103	61.7	31.3	7.1
Jharkhand	1495	655	43.8	575	87.8	514	89.4	34.2	34.4
Karnataka	2038	233	11.4	148	63.5	103	69.6	31.8	5.1
Kerala	2606	433	16.6	310	71.6	246	79.4	25.8	9.4
Lakshadweep	56	15	26.8	15	100.0	12	80.0	46.0	21.4
Madhya Pradesh	2735	1690	61.8	983	58.2	697	70.9	29.1	25.5
Maharashtra	4017	452	11.3	297	65.7	126	42.4	34.0	3.1
Manipur	1376	923	67.1	922	99.9	922	100.0	56.1	67.0
Meghalaya	864	375	43.4	326	86.9	315	96.6	51.1	36.5
Mizoram	632	508	80.4	497	97.8	496	99.8	76.3	78.5
Nagaland	704	384	54.5	367	95.6	327	89.1	38.8	46.4
Odisha	2976	1033	34.7	813	78.7	520	64.0	27.4	17.5
Pondicherry	128	47	36.7	46	97.9	40	87.0	14.5	31.3
Punjab	1560	107	6.9	92	86.0	63	68.5	27.6	4.0
Rajasthan	2582	1714	66.4	1514	88.3	1387	91.6	70.3	53.7
Sikkim	608	229	37.7	224	97.8	220	98.2	58.5	36.2
Tamil Nadu	3319	1109	33.4	980	88.4	930	94.9	40.7	28.0
Tripura	1312	965	73.6	906	93.9	899	99.2	59.7	68.5
Uttar Pradesh	5903	1080	18.3	960	88.9	832	86.7	32.8	14.1
Uttarakhand	1048	504	48.1	470	93.3	387	82.3	24.3	36.9
West Bengal	3576	1674	46.8	1487	88.8	1145	77.0	16.8	32.0

Pradesh, Rajasthan, Tamil Nadu and Uttarakhand experiencing demand of jobs from about 90% MGNREGA-registered households. In terms of provision of work, some states like Andhra Pradesh, Rajasthan and north-eastern states of Manipur, Meghalaya, Sikkim, Tripura and Mizoram, could provide work for 50 days or more in different schemes under MGNREGA in 2009.

In terms of benefits of MGNREGA schemes, it was found that the percentage of beneficiary households in the total sample households remained less than 10% in Bihar, Goa and Daman, Haryana, Jammu and Kashmir, Karnataka, Kerala, Maharashtra and Punjab, while it passed the 50% mark in Rajasthan, Manipur, Mizoram and Tripura. The overall performance of Rajasthan and north-eastern states has been commendable in terms of the high number of job card holders, job seekers, and the number of days' work provided in various schemes under MGNREGA.

Dietary pattern and nutritional status

MGNREGA was introduced with the aim of improving the purchasing power of rural people in India. Table 3.3 provides a comparison of the dietary pattern and nutritional status of MGNREGA job card holders versus non-job card holders, job-seekers versus non-job seekers and beneficiaries versus non-beneficiaries among rural households in India. It was found that job card holders under MGNREGA spent a higher percentage (57%) of their income on food commodities compared to non-job card holders (51%). The expenditure share on food decreased with increase in income level. Among food commodities, the share of cereals dominated the food expenditure accounting for about 28–32%. With improvement in the purchasing power, the dietary pattern diversified towards non-cereal high-value commodities, accounting for 68–72% of the total food expenditure.

MGNREGA job card holders are economically more weak than non-job card holders. The non-job card holders spent Rs37/capita/day on food and non-food commodities while job card holders spent only Rs29/person/day on these items. Non-job seekers among the job card holders were richer than the job-seekers. The MGNREGA targeting the weaker sections of the rural households contributed towards providing additional income through employment for 43.1 days (average) in a year. It facilitated buying more food and consumption of more calories and protein by the beneficiaries vis-à-vis non-beneficiaries of MGNREGA.

Cereals accounted for 55–61% of the total calories intake and 44–52% of total protein intake by the households across various MGNREGA groups formed for the analysis of data. The households who desired to seek employment but failed to obtain any (non-beneficiaries) had a lower energy intake (2199 kcal/capita/day) as well as lower protein intake (67.6 g/capita/day) compared to the households who received employment (beneficiaries) (2332 kcal/capita/day and 81.4 g/capita/day, respectively).

This comparison of beneficiaries and non-beneficiaries has clearly revealed that MGNREGA is attaining its aim of providing nutritional security to the weaker sections of rural households. The calorie intake has increased from 2199 kcal/capita/day to 2332 kcal/capita/day, and protein intake has increased from 67.6 g/capita/day to 81.4 g/capita/day. About 4.3% households will be lifted above the poverty line. The number of nutrition-deficit households has reduced by nearly 8%, from 44.2% to 36.3%, and the number of undernourished (deficit in protein) households has come down by 9%, from 26.9% to 17.7% at the country level.

Non-food expenditure pattern

The study has revealed that among non-food commodities, the expenditure on fuel and light takes the major share, followed by clothing, transport, healthcare and education (Table 3.4). The MGNREGA beneficiary households spent more on non-food items as compared to non-beneficiary households. The beneficiary households spent more on fuel and light, transport, clothing and other non-food items but less on education and healthcare and medicines than the non-beneficiary households.

Table 3.3. Dietary pattern and nutritional status of rural households in India, 2009 (authors' calculations).

Socio-economic dimension	MGNREGA job card holders	MGNREGA non-job card holders	MGNREGA job card holders: Job seekers: Beneficiaries	MGNREGA job card holders: Job seekers: Non-beneficiaries	MGNREGA job card holders: Non-job seekers
Share of food expenditure in total expenditure (%)					
Food commodities	57	51	58	58	54
Non-food commodities	43	49	42	42	46
Expenditure (Rs/capita/day)	28.80	37.20	28.60	26.60	31.60
Dietary pattern (annual per capita consumption in kg)					
Cereals	146.7	143.5	148.4	141.9	143.2
Pulses	7.0	8.7	6.8	7.4	7.8
Milk	43.9	66.2	41.6	39.3	58.0
Edible oils	6.1	7.4	5.9	6.3	7.1
Vegetables	57.7	63.1	58.1	57.0	56.2
Fruits	7.3	10.8	6.9	7.0	9.1
Meat, eggs and fish	5.8	5.9	6.1	5.4	5.0
Sugar	7.7	9.5	7.5	7.6	9.0
Food budget shares, % of total food expenditure					
Cereals	31.9	27.8	32.6	32.4	28.5
Non-cereals	68.1	72.2	67.4	67.7	71.6
Food expenditure (Rs/capita/year)	6025	6925	6050	5591	6277
Sources of calories, % of total calories intake					
Cereals	59.9	55.5	60.1	60.9	58.3
Non-cereals	40.1	44.6	39.9	39.1	41.7
Calorie intake (kcal/capita/day)	2311	2440	2332	2199	2317
Sources of protein, % of total protein intake					
Cereals	46.0	46.8	44.1	52.2	50.2
Non-cereals	54.0	53.2	55.9	47.8	49.9
Protein intake (g/capita/day)	78.4	77.5	81.4	67.6	74.5
Percentage of households					
Below poverty line	39.9	26.4	40.6	44.9	32.4
Nutritional deficit	37.6	33.8	36.3	44.2	37.1
Undernourished	19.6	18.0	17.7	26.9	21.3
Employment under MGNREGA schemes (person days/year)	30.2	0.0	43.1	0	0

Conclusions

The study has revealed that implementation of MGNREGA is a direct way of increasing income of the rural poor. It has benefited 22.5% of the rural households by providing wage employment for 43 days, on average. MGNREGA has been successful in reducing the poverty level by 4%. It has provided almost equal employment benefits to all the categories of farm sizes, household types and income groups. The state-wise study has

Table 3.4. Expenditure on major non-food items by MGNREGA beneficiaries and non-beneficiaries in India, 2009 (authors' calculations).

	Expenditure (Rs/capita/year)		Percentage share in total non-food expenditure	
Non-food items	MGNREGA beneficiaries	MGNREGA non-beneficiaries	MGNREGA beneficiaries	MGNREGA non-beneficiaries
Fuel and light	965	915	22.0	22.2
Education	383	402	8.7	9.8
Transport	419	347	12.3	15.0
Medicine and health care	540	617	9.6	8.4
Clothing	583	558	13.3	13.6
Other non-food items	1493	1276	34.1	31.0
Total non-food expenditure	4384	4115	100	100

revealed that all states have benefited from MGNREGA, but with wide variations. It is observed that the economically weaker states of the country have benefited more and have implemented MGNREGA more vigorously.

The study has shown that the raise in income has led to an increase in food consumption level – of both cereals and non-cereals – by all the categories of rural households. A diversification in the dietary pattern of different households has also been observed, which is again a strong indicator of better food consumption. These have resulted in a substantial increase in calorie intake as well as protein intake by different categories of households, leading to a decrease in undernourished and nutrition-deficit households by 8–9%.

Overall, the impact of MGNREGA has been positive and effective in increasing household food consumption, changing dietary pattern and providing nutritional food security to the poor rural households of India.

Note

[1] The authors thank the Agricultural Economics Research Association (India), New Delhi, India for permission to reproduce their article on which this chapter is based: Kumar, P. and Joshi, P.K. (2013) Household Consumption Pattern and Nutritional Security among Poor Rural Household: Impact of MGNREGA. *Agricultural Economics Research Review* 26(1), January–June 2013, 73–82.

References

GoI (Government of India) (2005) Mahatma Gandhi National Rural Employment Guarantee Act, 2005. Available at: http://www.nrega.nic.in (accessed 16 December 2015).

GoI (Government of India) (2009a) *Socio-Economic Survey: Employment and Unemployment*. Sixty-sixth Round: July 2009–June 2010. National Sample Survey Organization, New Delhi.

GoI (Government of India) (2009b) *Socio-Economic Survey: Consumer Expenditure*. Sixty-sixth Round: July 2009–June 2010. National Sample Survey Organization, New Delhi.

GoI (Government of India) (2014) *State-wise Wage Rate for Unskilled Manual Workers*. Ministry of Rural Development, New Delhi.

Kumar, P. and Joshi, P.K. (2013) Household Consumption Pattern and Nutritional Security among Poor Rural Household: Impact of MGNREGA. *Agricultural Economics Research Review* 26(1), 73–82.

NSSO (National Sample Survey Office) (1996) *Report No. 405*. Government of India, New Delhi.

4 Food Demand and Supply Projections to 2030: India

Praduman Kumar[1]* and P.K. Joshi[2]

[1]*Division of Agricultural Economics, Indian Agricultural Research Institute, New Delhi;* [2]*International Food Policy Research Institute, New Delhi, India*

Introduction

Achieving food self-sufficiency has always been the primary objective of agricultural policy in India. Driven by rising population, growing economy, expanding urbanization and changing tastes and preferences, the demand for food is continuously increasing in the country. On the other side, India is facing the problems of plateauing productivity, decreasing farm sizes and diminishing natural resources. This scenario has raised questions such as: 'Will India be able to produce enough to meet its growing food demand or will it be open for imports of food commodities by 2030? What would be the likely trends of future demand for various food commodities? Will the supply of key food commodities continue to keep pace with their demand?' These questions need to be answered in order to evolve an appropriate strategy for meeting the future demand for food commodities in India. In this chapter, an attempt has been made to project the demand for and supply of key food commodities by 2020 and 2030. It also assesses their trade potential by computing demand–supply gaps. This information will help to evolve appropriate medium- and long-term strategies in the Indian food sector.

A review of past studies has revealed wide variations in food demand projections due to their dependence on the type of data used and magnitudes of demand elasticities, income distribution, regional dietary pattern and dietary diversification. These estimates for food demand also have some limitations: (i) the model specification ignores theoretical restrictions of demand relationship; (ii) aggregate analysis is done at the national level, ignoring the effect of structural changes on economy such as urbanization and regional variations; (iii) national income growth assumption is superimposed on the regions and income groups; (iv) per capita income growth is used, which ignores the population growth in the projected years and underestimates the income effect on demand because of declining population growth; and (v) they ignore the surge caused in 'home-away demand' for food by the sustained rise in per capita income, fast growing urban population and increasing employment opportunities for urban women. In the present study, these deficiencies are addressed while projecting the demand to 2030 for food grains, and horticultural, livestock and fisheries products at the disaggregated level. In most previous studies, the main attention has been on demand predictions for cereals

* Corresponding author, e-mail: pkumariari@gmail.com

and pulses, but in the present study, due importance has been given to the demand for high-value commodities, considering the diversification and structural changes in the food basket.

Categorization of Households

The study is based on the household data on dietary pattern and consumer expenditures provided through regular surveys conducted by the National Sample Survey Organization (NSSO), Government of India. The household data collected under major rounds of the National Sample Survey (NSS) covering the years 1983 and 2004–2005 pertaining to 38th and 61st rounds, respectively, were used for assessing the changes in dietary pattern and estimating the food demand elasticities. The per capita expenditure was considered as a proxy for per capita income and therefore these have been used interchangeably in the study. The sample households were categorized into three expenditure/income groups: poor group, middle-income group and high-income group (Fig. 4.1). The 'poor-income group' comprised households that had income levels below the poverty line (PL).[1] The 'middle-income group' households had income levels between PL and 150% of PL. The 'high-income group' comprised those households whose per capita income was above the 150% level of PL.

Food Demand Projections

The demand for food comprises its direct demand and indirect demand. The direct demand consists of food consumption at home and outside the home. The indirect demand includes its use as seed and feed, for industrial uses and loss as wastages. In this study, an attempt has been made to provide credible estimates of future demand for food grains and other food commodities by estimating their demand at the disaggregated level, in terms of income levels, rural and urban households and states/union territories (UTs) of India, and their summation to derive the national estimates. To capture their effects, we classified rural/urban households of 35 states/UTs of India into three income groups.

The base line consumption, demand elasticity, income growth and population are the important factors in demand projections. The growth rates in per capita income were obtained by subtracting population growth rate from economic (gross domestic product; GDP) growth rate and were used in predicting per capita consumption. The estimated per capita consumption was multiplied by the projected population, and aggregated by state/UTs, income groups and rural/urban households to obtain the household demand at the national level.

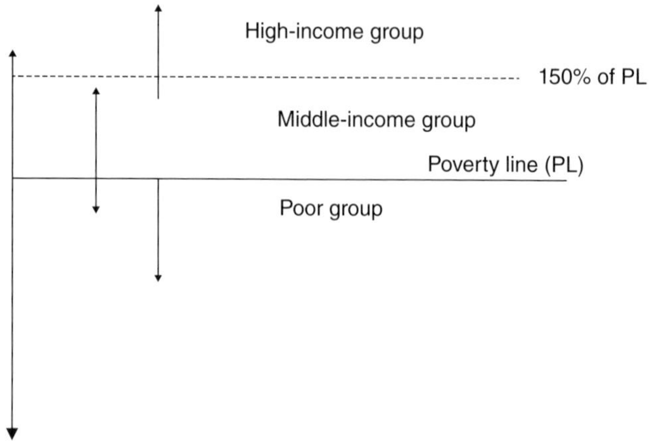

Fig. 4.1. Household categories based on income level.

Analytical Approach

Food demand at home

The per capita food demand at house level is predicted as:

$$d_{ijkt} = d_{ijkt-1}\left(1 + y_{ijt}.e_{ijk}\left(1-s\right)\right) \quad (4.1)$$

where d_{ijkt} is the per capita consumption for the subgroup '*i*' rural/urban households in the '*j*' state/UT of '*k*' income group in year '*t*', d_{ijkt-1} is the per capita consumption for the subgroup '*i*' rural/urban households in the '*j*' state/UT of '*k*' income group in year '*t*–1', y_{ijt} is the per capita GDP growth for the subgroup '*i*' rural/urban households in the '*j*' state/UT in the year '*t*', e_{ijk} is the expenditure elasticity for the subgroup '*i*' rural/urban households in the '*j*' state/UT of '*k*' income group, and 's' is the saving rate assumed at 36%, as estimated by the Central Statistical Organization (CSO), Government of India.

The total home demand is obtained by multiplying the per capita food demand by the population:

$$D_{ijkt} = d_{ijkt}.N_{ijkt} \quad (4.2)$$

where D_{ijkt} is the total household demand for a commodity of the subgroup '*i*' rural/urban households in the '*j*' state/UT of '*k*' income group in year '*t*', and N_{ijkt} is the population in the year '*t*' belonging to '*i*' rural/urban households in the '*j*' state/UT of '*k*' income group.

The aggregate food demand at home is obtained by the summation of food demands across income groups of rural/urban households in each state/UT as:

$$D_t = \Sigma\Sigma\Sigma D_{ijkt} \quad (4.3)$$

where D_t is the total household demand for a commodity in the year '*t*'.

The expenditure elasticities were estimated at the regional levels using the Food Characteristic Demand System following Bouis and Haddad (1992). These regional expenditure elasticities were superimposed on the corresponding state/UT. The aggregate household human demand for the '*j*' state/UT (*j* =1,... 35) in the year '*t*' was computed by summation of '*i*' rural/urban households (*i*=1, 2) and '*k*' income group (*k*=1,... 3). The summation over the states/UTs gave the household demand at the national level for a commodity in the year '*t*'.

Home–away food demand projections

The food demand of a household away from home was estimated based on the FAO Food Balance Sheet. In this approach, the total food consumption C was obtained as: C = (Q+M+S) – E – (Seed + Feed + Wastages + Industrial uses) using data on domestic food production (Q), imports (M), stock change (S), export (E) and indirect requirement for seed and feed, wastages and industrial use. The C included food consumption at home (H) and outside home (OH). The NSS survey data on household consumption was used to estimate H and the food consumption outside home (OH) was obtained by subtracting H from C (see Appendix 4.1). The baseline per capita food consumption outside home (oh_{ijko}) for subgroup '*i*' rural/urban households in the '*j*' state of '*k*' income group in the base year was computed as:

$$oh_{ijko} = d_{ijko} \times (OH/H) \quad (4.4)$$

where, d_{ijko} is the per capita food consumption for the subgroup '*i*' rural/urban households in '*j*' state of '*k*' income group in the base year 2004.

Food demand projections outside home

The per capita food demand outside home is predicted as:

$$oh_{ijkt} = oh_{ijkt-1}\left(1 + y_{ijt}.fe_{ijk(1-s)}\right) \quad (4.5)$$

where oh_{ijkt} is the per capita consumption outside home for the subgroup '*i*' rural/urban households in '*j*' state/UT of '*k*' income group in year '*t*', oh_{ijkt-1} is the per capita consumption of food outside home for the subgroup 'i' rural/urban households in the 'j' state/UT of 'k' income group in year 't–1', y_{ijt} is the per capita GDP growth for the subgroup 'i' rural/urban households in 'j' state/UT in year 't' and fe_{ijk} is the expenditure demand elasticity for food for the subgroup 'i' rural/urban household in the 'j' state/UT of 'k' income group. It was computed as the weighted average of demand elasticities for different food commodities. The weights were the individual shares in total food expenditure on the respective commodities; s is the saving rate assumed at 36%.

The total demand for food outside home is obtained as:

$$OH_{ijkt} = oh_{ijkt} . N_{ijkt} \quad (4.6)$$

where OH_{ijkt} is total demand for food outside home for the subgroup 'i' rural/urban households in 'j' state/UT of 'k' income group in year 't', oh_{ijkt} is the per capita consumption outside home for the subgroup of 'i' rural/urban household in 'j' state/UT of 'k' income group in year 't' and N_{ijkt} is the population of subgroup 'i' rural/urban households in 'j' state/UT of 'k' income group in year 't'.

The national demand for food outside home in the year 't' (OH_t) can be obtained by summation of all the disaggregated demands for food outside the home for the subgroup 'i' rural/urban households in 'j' state/UT of 'k' income group in year 't' as:

$$OH_t = \sum\sum\sum OH_{ijkt} \quad (4.7)$$

Total household food demand

The total food demand is obtained by summation of food demands at home and outside home:

$$FD_t = D_t + OH_t \quad (4.8)$$

where, FD_t is the total household food demand in the year 't', D_t is the food demand at home in the year 't' and OH_t is the food demand outside home in the year 't'.

Population projections

The Registrar-General of Census, Government of India, provided the numbers of rural and urban population by state and UT of India with projections to 2030 (Registrar General, 1996). The rural and urban population was further categorized into three income groups for each state/UT by using the weights derived from the sample households of the 61st NSS round.

Income growth

The Central Statistical Organisation (CSO), Government of India, provides data on the GDP at factor cost from the agricultural and allied activities sector and national economy at 1999 prices. From these data series, 5-year moving growths for agricultural, non-agricultural and total economic activities were computed up to the year 2009. The slowdown of the economy in the year 2008 and the start of its picking-up in the year 2009 were assumed with recovery by 25 percentage point growth in the years 2010 and 2011 and the economic growth was assumed to be constant till the projected year 2030. The agricultural GDP growth was assumed as the income for rural households and the non-agricultural GDP growth was assumed as the income for urban households.

Indirect demand for food

The indirect demand for food grains, constituting their use as seed and feed, loss as wastages (SFW) and for industrial uses was estimated as follows.

1. Seed: The seed requirement was estimated on the basis of projected area under a crop, and the application of seed rate.

2. Feed: The demand for feed grains for livestock consumption was computed using their demand for livestock products in terms of livestock output units (LOU) and the average feeding ratio (that is the quantity of feed required per unit of livestock product). The LOU was worked out by adding the required quantities of meat and eggs and one-tenth of milk. Looking at the importance of aquaculture, one-tenth of fish production was also included in LOU. The feed demand was estimated by multiplying the LOU with the feeding ratio. The feed requirement is largely met by oilcakes, cotton seed and food grains. The share of different food grains and other feeds is as follows: rice, 2.6%; wheat, 12.7%; coarse grains, 30.6%; pulses, 5.1%; oilcakes, 33%; cotton seed, 11%; and other concentrates (salt, *gur*, *methi*, oil, etc.), 5%. These estimates were used in deriving the requirement of rice, wheat, coarse grains and pulses as feed.

3. Industrial uses: An industrial use allowance of 5 Mt of food grains was provided by the National Commission on Agriculture in the year 2000. In the total industrial use, the allowances are 40% each for rice and wheat, 13% for coarse grains and 7% for pulses. The industrial use has been projected assuming the growth rates of 2.13% for rice, 1.5% for wheat and 3.0% each for coarse grains and pulses.

4. Wastages: The wastages in food grains have been derived as 1.1% of rice production, 3.0% of wheat production, 4.6% of coarse grains production and 2.2% of pulses production, as assumed by the Directorate of Economics and Statistics (DES), Ministry of Agriculture, Government of India. The wastage allowances also included grains not fit for human consumption and used as feed. To overcome the problem of double accounting, only half of these allowances have been accounted towards the feed.

In the present study, the share of indirect demand (seed, feed, wastages and industrial use) in total demand has been estimated to be 5.1–5.6% for rice, 11.2–11.4% for wheat, 30.6–40.5% for coarse cereals, 11.6–13.7% for total cereals, 18.6–19.5% for pulses and 12.2–14.1% for food grains.

Looking at the FAO Food Balance Sheet for commodity, the share of indirect demand in total supply (production + import + change in stocks) was assessed as 7.52% for rice, 9.52% for wheat, 32.45% for coarse cereals, 12.50% for cereals, 16.39% for pulses and 12.80% for total food grains. An IARI study (Kumar, 1998) has estimated the share of indirect demand in total demand as 4.7–5.0% for rice, 10.9–11.4% for wheat, 32.3–40.3% for coarse grains, 11.5–12.7% for total cereals, 15.2–15.5% for pulses and 11.8–12.9% for food grains. The DES has taken into account the share of seed, feed and wastages in production as 7.6% for rice, 12.5% for wheat, 26.5% for coarse grains and 12.5% for pulses while computing the availability of cereals and pulses (DES, 2009). The indirect estimates based on various studies are quite close.

Based on FAO Food Balance Sheet, the indirect demand was computed as 17.2% for edible oils, 12.8% for sugar, 11.3% for vegetables, 21.0% for fruits, 2.4% for milk, 7.8% for fish, 13.7% for eggs and 1.14% for meat and poultry and was used for computing the total household demand.

Different Demand Scenarios

The annual per capita consumption and total food demand have been projected under three scenarios: (i) current GDP growth scenario (S1); (ii) 25% lower GDP growth scenario (S2); and (iii) 25% higher GDP growth scenario (S3).

Food basket in India: the changing trends

The food basket is more diversified today in both rural and urban India. The consumption of cereals as food is declining while that of non-cereals, such as horticultural, livestock and fisheries products, is increasing. During the two-decade period of 1983–2004, the per capita annual consumption of cereals declined from 181 kg to 149 kg in the rural areas and from 142 kg to 128 kg in the urban areas (Table 4.1). A declining trend in sugar consumption is also observed in both rural and urban households. On the other hand, the per capita annual consumption of fruits and vegetables has increased from 49 kg to 76 kg in rural and from 55 kg to 81 kg in urban areas. The annual milk

Table 4.1. Structural changes in food consumption at rural, urban and national levels in India, 1983–2004 (authors' computations).

	Rural			Urban			India		
Food commodity	1983	2004	Change	1983	2004	Change	1983	2004	Change
Annual per capita food consumption (kg)									
Cereals	181.4	149.1	–32.3	142.0	127.8	–14.2	167.5	141.9	–25.6
Pulses	11.3	8.6	–2.7	12.4	9.5	–2.9	11.7	8.9	–2.8
Edible oils	3.5	5.9	2.4	6.1	7.5	1.4	4.4	6.4	2.0
Sugar	10.7	9.3	–1.4	11.7	9.9	–1.8	11.1	9.5	–1.6
Vegetables	45.9	66.3	20.4	51.0	68.3	17.3	47.7	67.0	19.3
Fruits	2.7	9.3	6.6	4.1	12.9	8.8	3.2	10.5	7.3
Milk	38.7	54.8	16.1	55.6	61.4	5.8	44.7	57.0	12.3
Meat, fish and eggs	4.7	6.4	1.7	6.6	7.7	1.1	5.3	6.9	1.6
Share in total food expenditure (%)									
Cereals	48.7	31.9	–16.8	34.0	26.0	–8.0	42.7	29.6	–13.1
Pulses	5.7	5.4	–0.3	5.7	5.3	–0.4	5.7	5.4	–0.3
Edible oils	6.0	7.8	1.8	8.3	8.0	–0.3	6.9	7.9	1.0
Sugar	4.3	4.0	–0.3	4.2	3.5	–0.7	4.2	3.8	–0.4
Vegetables	7.6	11.8	4.2	8.8	11.8	3.0	8.1	11.8	3.7
Fruits	1.8	2.9	1.1	2.9	4.1	1.2	2.3	3.4	1.1
Milk	11.9	15.8	3.9	16.2	17.3	1.1	13.7	16.3	2.6
Meat, fish and eggs	5.3	8.0	2.7	7.0	8.1	1.1	6.0	8.1	2.1
Others	8.8	12.4	3.6	12.9	16.0	3.1	10.5	13.8	3.3
Percentage of total calorie intake by source									
Cereals	77.1	62.7	–14.4	67.2	55.2	–12.0	73.9	60.2	–13.7
Pulses	4.8	3.6	–1.2	5.9	4.1	–1.8	5.2	3.8	–1.4
Edible oils	3.9	6.5	2.6	7.6	8.4	0.8	5.1	7.1	2.0
Sugar	5.2	4.5	–0.7	6.4	4.9	–1.5	5.6	4.6	–1.0
Vegetables	3.3	11.5	8.2	3.7	13.1	9.4	3.4	12.0	8.6
Fruits	0.2	1.2	1.0	0.4	1.6	1.2	0.3	1.3	1.0
Milk	4.5	6.7	2.2	7.4	7.7	0.3	5.5	7.0	1.5
Meat, fish and eggs	0.6	0.9	0.3	0.9	1.1	0.2	0.7	1.0	0.3
Others	0.4	2.5	2.2	0.5	3.8	3.3	0.4	2.9	2.5
Percentage of total protein intake by source									
Cereals	75.4	65.2	–10.2	66.4	59.3	–7.1	72.4	63.3	–9.2
Pulses	11.1	9.6	–1.5	13.7	11.0	–2.8	12.0	10.0	–1.9

Sugar	0.1	0.1	0.0	0.1	0.1	0.0	0.1	0.1	–0.1
Vegetables	3.6	6.3	2.7	4.4	6.9	2.6	3.8	6.5	2.6
Fruits	0.1	0.5	0.4	0.1	0.7	0.6	0.1	0.6	0.5
Milk	5.8	10.4	4.6	9.6	11.6	2.0	7.0	10.8	3.8
Meat, fish and eggs	3.5	5.4	1.9	5.1	6.8	1.6	4.0	5.8	1.8
Others	0.4	2.6	2.2	0.6	3.7	3.1	0.5	3.0	2.5
Total expenditure on food and non-food commodities (%)									
Food	65.2	53 5	–11.7	59.4	44.4	–15.0	62.7	49.6	–13.1
Non-food	34.8	46.6	11.8	40.6	55.6	15.0	37.3	50.5	13.2
Calorie and protein intake per capita per day									
Calories (kcal)	2218	2245	27.0	1986	2182	196	2136	2223	87.0
Protein (g)	62.8	56.5	–6.3	56.2	54.4	–1.8	60.4	55.8	–4.6
Dynamics of poverty and undernourishment (head count ratio, %)									
Poor (BPL)	40.2	20.8	–19.4	38.6	31.1	–7.5	39.7	24.3	–15.4
Calorie-deficit	31.2	23.0	–8.2	26.4	13.6	–12.8	29.5	19.8	–9.7
Protein-deficit	28.9	33.7	4.8	37.2	38.3	1.1	31.8	35.3	3.5

BPL, below poverty line households

consumption has increased, from 39 kg to 55 kg in rural and from 56 kg to 61 kg in urban areas. The consumption of meat, fish and eggs has also shown an increasing trend, but their per capita annual consumption continues to remain low (6.4 kg in rural and 7.7 kg in urban households), because nearly two-thirds of the Indian population is vegetarian and derives more than 20% share of protein from milk and pulses.

During the period 1983–2004, the expenditure share of cereals in total food expenditure declined from 48.7% to 31.9% in rural and from 34.0% to 26.0% in urban areas. This share has been diverted to high-value food commodities. The food share in total expenditure has declined from 65% to 54% in rural and from 59% to 44% in urban areas. The average daily per capita intake has increased by 27 cal in rural and by 196 cal in urban households. The additional energy requirement is being met from non-cereals and non-crop commodities. The average per capita daily intake of protein has decreased by 6.3% in rural and 1.8% in urban areas during the period 1983–2004.

The increasing demand for livestock products (milk, meat, eggs and fish) will push up feed demand in the country. Dietary shifts towards high-value food commodities would have a profound impact on agricultural production, marketing, processing and retailing sectors. However, despite increasing demand for high-value commodities, the importance of cereals and pulses for attaining nutritional security in the country will continue, because food grains account for higher than a 75% share in total calorie and protein intake. The incidence of poor and calorie-deficit households has declined by 15.4% and 9.7%, respectively, during 1983–2004, however, about 23% rural and 14% urban households still remain calorie-deficient. In 2004, the share of the undernourished population was 34% in rural and 38% in urban areas. At the national level, the incidence of protein-deficient households has increased by 3.5% during the period 1983–2004. The decline in cereals consumption over time has not been compensated adequately by the increase in consumption of horticultural and livestock products.

The inequitable distribution of food among different segments of the population is one of the major factors responsible for undernourishment in India and has increased nutritional deficiency among both rural and urban households. Cereals continue to be the most important food for meeting nutritional requirements and are the cheapest source of energy and protein. The low levels of income prevent the households from substituting cereals with fruits and vegetables, milk, meat, fish, etc. The price of cereals plays an important role in providing food and nutritional security to households in India. The higher cereal prices might build buffer stocks of food grains with the government, but result in their reduced consumption, which is detrimental to household food security. Due importance should continue to be provided to the role played by cereals and pulses in achieving adequate nutritional and food security. The food consumption patterns have significant implications on future demand, research priority setting and resource allocations to achieve food and nutritional security in the country.

The additional energy requirement is met from non-cereals and non-crop commodities. The average per capita daily intake of protein has decreased by 6.3% in rural and by 1.8% in urban areas during 1983–2004.

To meet the food and nutritional requirements of the growing population, the nation will have to increase its current levels of food production with higher emphasis on better natural resources management, better postharvest management through technological breakthroughs and addressing climatic and environmental concerns.

Food demand elasticity

To estimate the income and price elasticities of demand for food commodities, several models are reported in the literature. The expenditure (income) and calorie elasticities based on the linear expenditure system (LEDS), transcendental logarithmic demand system (TLDS), normalized quadratic demand system (NQDS), food characteristic demand system (FCDS) and three-stage quadratic almost ideal demand system (3-stage QUAIDS) models were compared to obtain a realistic view of demand elasticities (Table 4.2). These estimates have shown that expenditure elasticities are lower for urban than for rural consumers. The magnitude of expenditure elasticities for cereals is much higher on using LEDS, TLDS, NQDS models compared to that obtained from FCDS and 3-stage QUAIDS models. It is strange to note that once the expenditure elasticities for rice and wheat are positive and significantly high in magnitude, the actual per capita cereal consumption does not increase with total expenditure!

A comparison of demand elasticities calculated by different models (Table 4.2) reveals that the value for calorie-income elasticity is lowest on using the FCDS model. Behrman and Deolalikar (1989) and Bouis and Haddad (1992) have presented empirical evidences for the Indian and Philippine populations, respectively, that calorie-income elasticity is not significantly different from zero across income-groups and regions. The poor households spend a high proportion of their income on food, and a large share of their total food expenditure is on a low-cost calorie staple to avoid going hungry. The rich households can afford to substitute a part of the low-cost calorie staple with high-cost calorie food without increasing calories. Thus, calorie-income elasticity would be highly inelastic, close to zero. Therefore, one can assume that the demand elasticities obtained from FCDS predict the consumer behaviour as observed in the data and may predict most reliable demand for food commodities. The studies which used FCDS-based elasticities could predict food demand in a highly credible range (see Kumar, 1998; Paroda and Kumar, 2000; Chand, 2007; Kumar *et al.*, 2007, 2009, 2010).

In the present study, FCDS was used for computing demand elasticities of various food commodities, i.e. rice, wheat, coarse grains and major commodity groups, such as pulses, edible oils, vegetables, fruits, milk, meat, fish and eggs and other food and non-food commodities across regions, rural/urban households and income groups. The national-level estimates of income and own price elasticities were computed as the weighted averages of the disaggregated elasticity (Table 4.3).

Table 4.3 revealed that demand elasticities vary widely across regions, rural/urban households and income groups due to changes in production environment, tastes

Table 4.2. A comparison of food and calorie income elasticities across different demand models, for rural and urban India (authors' calculations).

Food commodities	Households	Models LEDS	TLDS	NQDS	FCDS	3-Stage QUAIDS
Food-income elasticity						
Rice	Rural	0.45	0.71	0.57	0.03	0.02
	Urban	0.22	0.46	0.42	0.01	0.01
Wheat	Rural	0.44	0.63	0.55	0.07	0.03
	Urban	0.25	0.3	0.32	0.08	0.02
Coarse cereals	Rural	0.03	−0.55	−0.09	−0.12	−0.02
	Urban	−0.26	−1.62	−1.05	−0.17	0.00
Other foods	Rural	0.89	0.99	0.76	0.81	0.89
	Urban	0.84	0.98	0.88	0.67	0.77
Aggregate food-income elasticity						
All food commodities	Rural	0.72	0.8	0.65	0.29	0.35
	Urban	0.73	0.8	0.73	0.28	0.32
Calorie-income elasticity						
All food commodities	Rural	0.46	0.6	0.53	0.12	0.14
	Urban	0.42	0.51	0.49	0.12	0.14

Table 4.3. Expenditure and own price elasticities for food commodities in households by income groups in India (authors' calculations).

Food commodity	Expenditure elasticity				Own price elasticity			
	Poor households	Middle-income households	High-income households	All households	Poor households	Middle-income households	High-income households	All households
Rural households								
Rice	0.157	0.041	−0.017	0.049	−0.463	−0.291	−0.163	−0.289
Wheat	0.128	0.087	0.057	0.083	−0.531	−0.400	−0.251	−0.367
Coarse cereals	−0.178	−0.147	−0.091	−0.142	−0.353	−0.203	−0.102	−0.228
Pulses	0.499	0.285	0.111	0.248	−0.686	−0.507	−0.300	−0.448
Edible oils	0.657	0.389	0.174	0.333	−0.751	−0.576	−0.373	−0.509
Sugar	0.277	0.113	0.023	0.097	−0.580	−0.387	−0.222	−0.340
Vegetables	0.597	0.358	0.164	0.322	−0.729	−0.552	−0.359	−0.504
Fruits	0.716	0.508	0.286	0.408	−0.795	−0.660	−0.493	−0.583
Spices and beverages	1.150	0.972	0.720	0.880	−0.955	−0.929	−0.909	−0.924
Milk	0.799	0.558	0.288	0.423	−0.828	−0.690	−0.475	−0.577
Meat, fish and eggs	1.045	0.863	0.580	0.752	−0.911	−0.865	−0.786	−0.833
Urban households								
Rice	0.130	0.015	−0.029	0.008	−0.477	−0.330	−0.222	−0.293
Wheat	0.078	0.052	0.101	0.083	−0.485	−0.410	−0.342	−0.389
Coarse cereals	−0.163	−0.200	−0.109	−0.153	−0.415	−0.279	−0.167	−0.264
Pulses	0.501	0.260	0.090	0.176	−0.723	−0.557	−0.381	−0.462
Edible oils	0.572	0.310	0.123	0.208	−0.751	−0.586	−0.404	−0.479
Sugar	0.260	0.085	−0.010	0.040	−0.610	−0.440	−0.268	−0.346
Vegetables	0.525	0.296	0.127	0.211	−0.728	−0.587	−0.421	−0.496
Fruits	0.674	0.486	0.284	0.341	−0.814	−0.734	−0.610	−0.644
Spices and beverages	1.055	0.839	0.561	0.649	−0.956	−0.944	−0.913	−0.922
Milk	0.758	0.510	0.264	0.343	−0.840	−0.741	−0.577	−0.627
Meat, fish and eggs	0.946	0.726	0.469	0.578	−0.911	−0.872	−0.817	−0.840
All households								
Rice	0.146	0.028	−0.024	0.026	−0.469	−0.309	−0.200	−0.291
Wheat	0.104	0.071	0.082	0.083	−0.508	−0.404	−0.303	−0.379

Coarse cereals	−0.171	−0.176	−0.102	−0.148	−0.381	−0.244	−0.144	−0.248
Pulses	0.500	0.274	0.098	0.206	−0.699	−0.530	−0.349	−0.456
Edible oils	0.630	0.353	0.143	0.259	−0.751	−0.580	−0.392	−0.492
Sugar	0.271	0.100	0.004	0.064	−0.591	−0.411	−0.249	−0.343
Vegetables	0.573	0.330	0.141	0.256	−0.729	−0.568	−0.397	−0.499
Fruits	0.704	0.499	0.285	0.368	−0.801	−0.689	−0.562	−0.620
Spices and beverages	1.126	0.916	0.621	0.736	−0.955	−0.935	−0.912	−0.923
Milk	0.785	0.538	0.275	0.377	−0.832	−0.712	−0.533	−0.605
Meat, fish and eggs	1.009	0.800	0.515	0.651	−0.911	−0.868	−0.805	−0.837

and food preferences. The demand elasticities for staple food (rice, wheat, coarse cereals) have been found highly inelastic, close to zero, and even negative for coarse cereals. The magnitude of elasticity has shown a decline with rise in income across all income groups and is higher for rural than urban households. The expenditure elasticities have been found much higher for high-value food commodities, i.e. livestock and horticultural products. With growth in economy, the demand will increase faster for high-value food commodities than for the cereals.

Dietary Pattern Projections

For predicting dietary patterns to 2030, the average annual per capita consumption of different commodities during 2004–2005 was used as the base consumption. During the period 2010–2030, the annual consumption of rice will decline marginally and that of wheat will increase by 11.2–13.8% under different income scenarios (Table 4.4). The consumption of coarse cereals is predicted to decline by 7.1–8.4%, but that of total cereals will increase by 3.2–4.7% by the year 2030 over the year 2010. A slight shift in consumption from rice and a substantial shift from coarse cereals to wheat are predicted within the cereals.

Amongst high-value commodities, the consumption will increase by 20–26% of pulses, 26–35% of edible oils, 15–20% of sugar, 25–33% of vegetables, 29–38% of fruits, 34–46% of milk, 41–58 % of fish, 44–62% of meat, 43–60% of poultry and 37–50% of eggs by the year 2030 over the year 2010 under different income scenarios. Overall, a continuance of dietary diversification is predicted and will be substantial with the growth in economy.

The easy availability of wheat through the public distribution system under the food security programme of the Government of India is leading to an increase in wheat consumption even in the traditionally rice-eating states of the country. Despite a steep fall in the importance of coarse grains in the Indian diet, these constitute a large share of the cereals in the states of Karnataka, Rajasthan, Maharashtra, Gujarat, Himachal Pradesh and Madhya Pradesh. Coarse grains are also gaining importance as feed for the fast growing demand for livestock. Moreover, coarse grains are generally grown in an unfavourable rainfed environment, where other remunerative production choices are limited. The food basket in India will continue to diversify with increasing per capita consumption of milk, fruits, vegetables, meat, poultry products and fish. Consequently, the demand for horticultural, livestock, poultry and fishery products will rise considerably in the coming years.

Food Demand Projections for Human Consumption

The food demand for human consumption was computed by multiplying the projected per capita consumption by the projected population and is projected at two stages: (i) at home; and (ii) outside home. The projections for different food commodities at home and outside home under three income growth scenarios were computed by the year 2030 and are presented in Appendix Table 4.2 and Table 4.5.

By 2020, the food grains demand for human consumption is projected to be 238–241 Mt with a break-up of about 106 Mt of rice, 86–88 Mt of wheat, 27 Mt of coarse grains and 18 Mt of pulses. The food grains demand for human consumption to 2030 is projected to be in the range of 267–272 Mt, comprising 116 Mt of rice, 101–104 Mt of wheat, 28 Mt of coarse grains and 22–23 Mt of pulses. The demand projections for food grains across different GDP growth (GDPG) scenarios move in a narrow range because of highly inelastic demand elasticity for cereals.

Indirect Demand Projections for Food Grains

Increasing demand for livestock products (milk, meat, eggs) and fishery products can rapidly drive up the demand for feed grains.

Table 4.4. Projected annual per capita food consumption in India to 2030 (in kg; authors' calculations).

	Current GDPG				High GDPG			
Food commodity	2010	2020	2030	Change (%) 2010–2030	2010	2020	2030	Change (%) 2010–2030
Rice	78.7	78.8	78.6	–0.2	78.7	79.0	79.1	0.6
Wheat	62.1	65.0	69.0	11.2	62.1	65.7	70.6	13.8
Coarse cereals	20.8	20.6	19.1	–8.4	20.8	20.7	19.3	–7.1
Total cereals	161.6	164.4	166.6	3.2	161.6	165.4	169.1	4.7
Pulses	12.3	13.2	14.7	19.8	12.3	13.5	15.5	26.5
Food grains	173.8	177.6	181.3	4.3	173.8	178.9	184.6	6.2
Edible oils	9.5	10.4	12.0	26.3	9.5	10.7	12.8	35.1
Sugar	20.2	21.5	23.2	14.9	20.2	21.8	24.3	20.0
Vegetables	92.9	102.1	115.8	24.6	92.9	104.9	123.7	33.1
Fruits	43.0	47.5	55.3	28.7	43.0	49.0	59.4	38.3
Milk	91.8	104.0	122.7	33.7	91.8	107.7	133.6	45.6
Fish	5.0	5.6	7.0	40.6	5.0	5.9	7.8	58.2
Bovine meat	2.7	3.1	3.9	44.6	2.7	3.3	4.4	61.8
Poultry	1.6	1.8	2.3	42.6	1.6	1.9	2.5	59.5
Eggs	2.5	2.8	3.4	36.5	2.5	2.9	3.8	50.4
Others	6.5	7.2	8.5	30.9	6.5	7.4	9.1	40.7

GDPG, gross domestic product growth

Table 4.5. Projected household demand for food grains by commodities under different GDPG scenarios in India, 2010–2030 (in Mt; authors' calculations).

Food commodity	Year	GDP scenario		
		Current	Low growth	High growth
Rice	2010	93.67	93.67	93.67
	2020	105.99	105.69	106.29
	2030	115.57	114.8	116.39
Wheat	2010	73.89	73.89	73.89
	2020	87.44	86.58	88.32
	2030	101.53	99.25	103.92
Coarse grains	2010	24.8	24.8	24.8
	2020	27.69	27.59	27.79
	2030	28.08	27.77	28.46
Total cereals	2010	192.36	191.36	192.36
	2020	221.11	219.86	222.4
	2030	245.18	241.81	248.78
Pulses	2010	14.61	14.61	14.61
	2020	17.81	17.42	18.22
	2030	21.62	20.51	22.84
Food grains	2010	206.97	206.97	206.97
	2020	238.93	237.28	240.62
	2030	266.81	262.32	271.62

For projecting total demand for food grains, information on indirect demand for food grains as seed and feed, in industrial uses and wastage allowances is also needed in addition to the data on direct demand for human consumption at home and outside home. This section attempts to assess the requirement of food grains for seed, feed, industrial uses and as wastages by the year 2020 and 2030.

Seed

The requirement of seeds was estimated on the basis of projected area under a crop and the expected seed rate application. A look at the cropping pattern reveals very little change in area allocation across broad crop groups such as food grains, oilseeds and commercial crops (Kumar, 1998). It is only some specific crops within these groups that have predicted changes in their area allocation because of the changes in their relative profitability. However, the area elasticities with respect to the expected crop revenue and crop output price are highly inelastic and nearly zero for rice and wheat (Kumar and Rosegrant, 1997). Therefore, it was presumed that the projected crop area will not change and will continue to remain at the base-year 2004 level. The annual seed requirement was estimated to be 1.28 Mt for rice, 1.79 Mt for wheat, 0.55 Mt for coarse grains and 1.2 Mt for pulses. The total requirement of food grains as seed was estimated to be 4.4 Mt and was assumed to remain at the same level in the years to come.

Feed

The livestock and poultry sectors are the major consumers of feed grains and oilcakes. Some fish species under aquaculture environment are also fed fish meal, which contains rice bran, millets and oilcakes. In the present study, the demand for milk, meat, poultry, eggs and fish was estimated for the years 2010, 2020 and 2030 and was used to compute livestock output units (LOU) given in Table 4.6. The LOU, projected to be 26.3–27.4 million by the year 2020, will grow to 34.6–38.1 million by 2030.

Table 4.6. Demand projections for feed to 2030, India (authors' calculations).

GDP growth scenario	2010	2020	2030
Livestock output unit (million LOU)			
Current	20.45	26.31	34.60
Low	20.45	25.32	31.58
High	20.45	27.34	38.06
Total feed per unit of LOU	1.5	1.5	1.5
Total feed requirement (Mt)			
Current	30.67	39.46	51.90
Low	30.67	37.99	47.37
High	30.67	41.02	57.09
Demand for feed grains			
Rice			
Current	0.8	1.03	1.35
Low	0.8	0.99	1.23
High	0.8	1.07	1.48
Wheat			
Current	3.89	5.01	6.59
Low	3.89	4.82	6.02
High	3.89	5.21	7.25
Coarse cereals			
Current	9.38	12.07	15.88
Low	9.38	11.62	14.50
High	9.38	12.55	17.47
Pulses			
Current	1.56	2.01	2.65
Low	1.56	1.94	2.42
High	1.56	2.09	2.91
Total food grains (rice + wheat + coarse cereals + pulses)			
Current	15.64	20.12	26.47
Low	15.64	19.37	24.16
High	15.64	20.92	29.12

The feed conversion ratio, which expresses the quantity of feed used to produce one unit of livestock output, was 2.1 in the 1980s; it declined to 1.5 in the 1990s and continues to be around 1.5 even in recent years (Kumar and Mruthyunjaya, 1995). This decline was observed because of the increasing conversion efficiency due to adoption of improved livestock breeds under the recent breeding programmes in the country. The lower feeding ratio is also associated with the supply constraints as a result of increasing export of oilcakes and rising feed prices. Using feeding ratio of 1.5, the demand for feed was projected as shown in Table 4.6. The feed demand, which was 30.7 Mt in 2010, is projected to grow to 38–41 Mt by the year 2020 and 47–57 Mt by the year 2030. A still higher demand is expected for feed because of the fast shift in dietary pattern towards the livestock products with increase in income growth and urbanization.

The review of surveys conducted by the Institute of Agricultural Research Statistics (1963–1983) and of studies undertaken at National Dairy Research Institute (1973) and Indian Agricultural Research Institute (Kumar, 1998) has revealed that in the contents of feed, rice accounts for about 2.6%, wheat 12.7%, coarse grains 30.6%, pulses 5.1%, oilcakes 33%, cotton seed 11% and other concentrates (salt, *gur*, *methi*, oil, etc.) 5%. These estimates were used in deriving the requirement for feed of various food grains, i.e. rice, wheat, coarse grains and pulses (Table 4.6). The estimates have revealed that by the year 2020, the demand for food grains as feed will be in the range of 19.4–20.9 Mt, comprising 0.99–1.07 Mt of rice, 4.8–5.2 Mt of wheat, 11.6–12.6 Mt of coarse grains and 1.94–2.09 Mt of pulses. The demand for feed grains to 2030 is projected to be 24.2–29.1 Mt.

Industrial use

The National Commission on Agriculture had estimated the industrial use of food grains as 5 Mt for the year 2000, with shares of 40% each of rice and wheat, 13% of coarse grains and 7% of pulses. Using the base level for the year 2000, demand for food grains for industrial use was projected assuming annual growth rate of 2.0% for rice, 1.5% for wheat, 3.0% for coarse grains and pulses (Table 4.7). The demand of different grains for industrial use in year 2010 is estimated to be 2.4 Mt for rice, 2.2 Mt for wheat, 0.9 Mt for coarse grains and 0.45 Mt for pulses. The industrial use of food grains is projected to grow from 7.3 Mt in 2020 to 9.0 Mt by the year 2030 with an annual growth rate of 2.1%.

Table 4.7. Demand projections for food grains by crop for industrial use, India, 2010–2030 (Mt).

Crop	2010	2020	2030	Annual growth rate (%)
Rice	2.4	2.92	3.56	2.0
Wheat	2.23	2.58	3.00	1.5
Coarse grains	0.89	1.19	1.60	3.0
Pulses	0.45	0.61	0.82	3.0
Food grains	5.96	7.3	8.98	2.1

Wastages

The wastages of food grains have been derived as 1.1% in rice production, 3.0% in wheat production, 4.6% in coarse grains production and 2.2% in pulses, as assumed by the DES. The wastage allowances also included the grains not fit for human consumption and used as feed. To overcome the problem of double accounting, half of these allowances were accounted towards the feed. The wastage allowances for food grains have been estimated at 2.79 Mt in 2010 and will increase to 3.25 Mt in the year 2020 and 3.71 Mt by 2030 (Table 4.8).

Total Indirect Demand Projections

Table 4.9 presents the overall requirement of food grains for seed, feed and industrial uses along with wastage allowances. The overall requirement has been calculated to be 34.7–36.3 Mt by 2020 and 41.6–46.7 Mt by 2030. In the year 2020, the total indirect demand has been estimated to be 5.8–5.9 Mt for rice, 10.8–11.0 Mt for wheat, 14.3–15.3 Mt for coarse grains and 4.0–4.1 Mt for pulses. By the year 2030, the total indirect demand will rise to 6.7–7.0 Mt for rice, 12.5–13.8 Mt for wheat, 17.7–20.7 Mt for coarse grains and 4.7–5.2 Mt for pulses.

Demand Projections for Food Grains

The total demand for food grains, except for export, was arrived at by adding their direct demand (human food consumption at home and outside home) and indirect demand (seed, feed, industrial uses, and wastages) (see Appendix 4.2). The demand for each food grain has been projected under three GDP growth scenarios (current, low growth and high growth). The results at 10-year intervals are presented for the period 2010–2030 in Table 4.10. By the year 2020, the demand is projected to be about 112 Mt for rice, 97–99 Mt for wheat, 42–43 Mt for coarse grains, 22 Mt for pulses, 251–255 Mt for total cereals and 272–277 Mt for total food grains. By 2030, the demand for total food grains will grow to the level of 304–318 Mt comprising 121–123 Mt of rice, 112–118 Mt of wheat, 45–49 Mt of coarse grains and 25–28 Mt of pulses. During 2010–2020, the demand is projected to grow fastest for pulses with the annual growth rate of 1.7–2.2%, followed by wheat 1.6–1.8%, coarse grains 1.4–1.7% and slowest for rice, 1.2–1.3%. A deceleration in the growth rate of demand for all the food grains has been observed due to diversification of dietary pattern. Taking all the cereals and pulses together, under the assumption of constant income elasticity and current income growth, the annual growth in demand for food grains is projected to come down from 1.51% during 2010–2020 to 1.25% during 2020–2030. The demand for food grains will grow at a rate higher than the population growth rate. India will have to play a major role in maintaining its self-reliant status in production of cereals and pulses to meet the additional requirements. The demand for coarse grains will increase because of their growing demand for livestock sector and, consequently, for feed in India. The high growth in demand for pulses will put pressure on their supply and consequently on price.

Table 4.8. Wastages of food grains by crop, India, 2010–2030 (Mt).

Commodity	Wastage assumed (% of production)	2010	2020	2030
Rice	1.1	0.54	0.61	0.67
Wheat	3.0	1.23	1.45	1.69
Coarse cereals	4.6	0.82	0.95	1.06
Pulses	2.2	0.20	0.24	0.29
Food grains	–	2.79	3.25	3.71

Table 4.9. Demand projections for food grains for feed and seed (Mt), in industrial use and wastage allowances under different income growth scenarios by crop in India, 2010–2030.

		Income scenario		
Commodities	Year	Current	Low	High
Rice	2010	5.01	5.01	5.01
	2020	5.84	5.80	5.88
	2030	6.86	6.74	7.00
Wheat	2010	9.14	9.14	9.14
	2020	10.84	10.63	11.05
	2030	13.07	12.45	13.78
Coarse grains	2010	11.64	11.64	11.64
	2020	14.77	14.31	15.26
	2030	19.10	17.67	20.73
Pulses	2010	3.41	3.41	3.41
	2020	4.06	3.98	4.14
	2030	4.95	4.71	5.23
Food grains	2010	29.20	29.20	29.20
	2020	35.50	34.72	36.34
	2030	43.98	41.57	46.74

Table 4.10. Domestic demand for food grains by commodities in India.

		Domestic demand (Mt)			Demand growth (%)	
Commodities	GDP growth	2010	2020	2030	2010–2020	2020–2030
Rice	Current	98.7	111.8	122.4	1.26	0.91
	Low	98.7	111.5	121.5	1.23	0.87
	High	98.7	112.2	123.4	1.29	0.96
Wheat	Current	83	98.3	114.6	1.70	1.55
	Low	83	97.2	111.7	1.59	1.40
	High	83	99.4	117.7	1.81	1.71
Coarse grains	Current	36.4	42.5	47.2	1.54	1.06
	Low	36.4	41.9	45.4	1.41	0.81
	High	36.4	43.1	49.2	1.68	1.34
Total cereals	Current	218.1	252.6	284.2	1.48	1.19
	Low	218.1	250.6	278.7	1.40	1.07
	High	218.1	254.6	290.3	1.56	1.32
Pulses	Current	18.0	21.9	26.6	1.96	1.97
	Low	18.0	21.4	25.2	1.75	1.66
	High	18.0	22.4	28.1	2.18	2.3
Food grains	Current	236.2	274	310.8	1.51	1.25
	Low	236.2	272	303.9	1.42	1.11
	High	236.2	277	318.4	1.61	1.40

Demand Projections for High-value Commodities

The demand for edible oils, sugar and horticultural, livestock, poultry and fishery products was computed year-wise for the period 2005–2030 using 2004 as the base year, but is presented in this chapter for the years 2010, 2020 and 2030 under the three GDP growth scenarios stated above.

The disaggregated food demand at home and outside home, indirect demand and total demand have been projected and are given in Appendix 4.3. The demand projections to 2030 for high-value commodities at household level (for human consumption at home and outside home) are given in Table 4.11 and at national level (for human consumption at home, outside home and indirect demand) are given in Table 4.12 under different income scenarios.

Edible oils

The household demand for edible oils will grow faster than growth in population and food grains. It will grow at an annual compound growth rate of 1.9–2.5% during 2010–2020 and of 1.9–2.7% during 2020–2030. The requirement for edible oils for human consumption at household level is estimated to be 13.7–14.4 Mt by the year 2020 and 16.5–18.8 Mt by the year 2030. The total demand for edible oils at national level is projected to be 16.5–17.5 Mt by 2020 and 19.9–22.8 Mt by 2030. The demand for edible oils will continue to remain much higher

Table 4.11. Demand projections to 2030 for high-value commodities (Mt) at household level, India.

Commodities	GDP growth	2010	2020	2030	Annual growth (%) 2010–2020	Annual growth (%) 2020–2030
Edible oils	Current	11.28	14.05	17.60	2.22	2.28
	Low	11.28	13.67	16.49	1.93	1.90
	High	11.28	14.45	18.84	2.50	2.69
Sugar	Current	24.08	28.86	34.19	1.81	1.71
	Low	24.08	28.37	32.80	1.65	1.46
	High	24.08	29.37	35.69	2.01	1.97
Vegetables	Current	110.64	137.34	170.34	2.19	2.18
	Low	110.64	129.83	154.52	1.90	1.76
	High	110.64	141.11	181.95	2.46	2.57
Fruits	Current	51.19	63.93	81.39	2.25	2.44
	Low	51.19	62.06	75.87	1.94	2.03
	High	51.19	65.86	87.47	2.55	2.88
Milk	Current	109.27	139.81	180.55	2.50	2.59
	Low	109.27	135.03	166.34	2.14	2.11
	High	109.27	144.84	196.55	2.86	3.10
Fish	Current	5.90	7.59	10.26	2.55	3.06
	Low	5.90	7.26	9.18	2.00	2.38
	High	5.90	7.95	11.54	3.02	3.79
Meat	Current	3.22	4.23	5.75	2.79	3.10
	Low	3.22	4.05	5.16	2.34	2.45
	High	3.22	4.43	6.43	3.25	3.80
Poultry meat	Current	1.89	2.45	3.32	2.66	3.09
	Low	1.89	2.35	2.99	2.21	2.44
	High	1.89	2.56	3.72	3.12	3.76
Eggs	Current	2.98	3.80	5.03	2.46	2.84
	Low	2.98	3.66	4.58	2.07	2.28
	High	2.98	3.95	5.54	2.85	3.44

Table 4.12. Total demand projection to 2030 for high-value commodities (Mt) at national level, India (authors' calculations).

Commodities	GDP growth	2010	2020	2030	Annual growth 2010–2020	Annual growth 2020–2030
Edible oils	Current	13.63	16.97	21.26	2.22	2.28
	Low	13.63	16.51	19.92	1.93	1.90
	High	13.63	17.45	22.76	2.50	2.69
Sugar	Current	27.62	33.10	39.21	1.81	1.71
	Low	27.62	32.54	37.62	1.65	1.46
	High	27.62	33.69	40.94	2.01	1.97
Vegetables	Current	124.74	154.85	192.05	2.19	2.18
	Low	124.74	146.38	174.22	1.90	1.76
	High	124.74	159.09	205.14	2.46	2.57
Fruits	Current	64.78	80.90	102.99	2.25	2.44
	Low	64.78	78.54	96.00	1.94	2.03
	High	64.78	83.34	110.69	2.55	2.88
Milk	Current	111.91	143.19	184.91	2.50	2.59
	Low	111.91	138.29	170.36	2.14	2.11
	High	111.91	148.34	201.30	2.86	3.10
Fish	Current	6.40	8.23	11.12	2.55	3.06
	Low	6.40	7.87	9.95	2.00	2.38
	High	6.40	8.62	12.51	3.02	3.79
Meat	Current	3.25	4.28	5.81	2.79	3.10
	Low	3.25	4.10	5.22	2.34	2.45
	High	3.25	4.48	6.51	3.25	3.80
Poultry meat	Current	1.91	2.48	3.36	2.66	3.09
	Low	1.91	2.37	3.02	2.21	2.44
	High	1.91	2.59	3.76	3.12	3.76
Eggs	Current	3.45	4.40	5.82	2.46	2.84
	Low	3.45	4.24	5.31	2.07	2.28
	High	3.45	4.67	6.42	2.85	3.44

than their production in the country and will likely rely on their import in large quantities.

Sugar

The demand for sugar at household and national levels is estimated to be 28.4–29.4 Mt and 32.5–33.7 Mt, respectively, by the year 2020. At the national level, it will grow to 37.6–40.9 Mt by the year 2030 with the annual growth rate of 1.46–1.97% under different GDP growth scenarios.

Vegetables

Demand for vegetables in India under the scenarios of low and high income growths has been presented for the years 2020 and 2030. By 2020, the demand for vegetables has been projected to be 130–141 Mt at the household level and 146–159 Mt at the national level. This demand will grow to the level of 154–182 Mt at household level and 174–205 Mt at national level by 2030. The demand for vegetables is likely to grow at the annual rate of 1.9–2.5% during 2010–2020 and 1.8–2.6% during 2020–2030.

Fruits

The household demand for fruits is projected to be 51.2 Mt in the year 2010; it will rise to 62–66 Mt by 2020 and 76–87 Mt by 2030. Due to substantial postharvest losses and their industrial use for processed products, the total requirement of fruits at the national level is projected to be 78–83 Mt by 2020 and 96–111 Mt by the year 2030. The demand for fruits is likely to grow at the annual rate of 1.94–2.88% during 2010–2030.

Milk

Milk demand at the household level was 109 Mt in the year 2010; it will rise to 135–144 Mt by 2020 and to 166–197 Mt by 2030. The projected milk demand at the national level is assessed to be 138–148 Mt by 2020 and 170–201 Mt by 2030 with an annual growth rate of 2.1–3.1%.

Fish

The fish produced in different production environments comprises marine, freshwater capture and aquaculture. In the present study, fish demand, irrespective of species and environments, was assessed to the year 2030. Fish demand at the household level has been assessed to be 5.9 Mt in the base year 2010; it will rise to 7.26–7.95 Mt by 2020 and further to 9.18–11.54 Mt by 2030. The projected fish demand at the national level (which will include indirect demand also) is assessed to be 7.87–8.62 Mt by 2020 and 9.95–12.51 Mt by 2030. Under high income growth scenario, fish demand will rise with an annual growth rate of 3.0% during the period 2010–2020 and 3.8% during 2020–2030.

Meat

The meat demand, which includes the meat of goat, cattle and pig, is likely to increase at an annual growth rate of 2.3–3.2% during 2010–2020 and 2.5–3.8% during 2020–2030 at the national level. With growth in the economy, meat demand will increase substantially with high growth, as is evident from the results. In 2010, the meat demand at the national level is assessed as 3.25 Mt and it is estimated to be 4.1–4.5 Mt by 2020 and 5.2–6.5 Mt by 2030.

Poultry meat

The demand for poultry meat under different growth scenarios is assessed to be 1.9 Mt in 2010 and it is likely to increase to 2.4–2.6 Mt by 2020 and 3.0–3.8 Mt by 2030 with an annual growth rate of 2.2–3.1% during 2010–2020 and of 2.4–3.8% during 2020–2030. The poultry demand with higher income growth will accelerate its household consumption and total demand.

Eggs

The demand for eggs is projected to grow at an annual growth rate of 2.1–2.9% during 2010–2020 and 2.3–3.4% during 2020–2030. The household requirement was assessed to be 2.98 Mt in 2010 and it is likely to be 3.7–4.0 Mt by 2020 and 4.6–5.5 Mt by 2030. The total national demand for eggs is projected to be 4.2–4.6 Mt by 2020 and 5.3–6.4 Mt by 2030. The demand for eggs will grow much faster than the population growth and will accelerate pressure on the supply of coarse grains and oilcakes as poultry feed.

Supply Projections

With the biggest congregation of poor in the world, food security is a very sensitive issue in India. After nearly achieving self-reliance in staple food production, the Government of India has launched a number of programmes for production (supply side), distribution and consumption (demand side) of food across the country. Currently, around half of India's population is covered by one or the other scheme under which subsidized staple food is made available to the people.

India has made substantial progress in food grains production by adopting a new agricultural strategy. As a result, production of food grains has increased from 115.6 Mt in 1960–1961 to 241.4 Mt in 2010–2011. Horticulture has emerged as an indispensable part of agriculture, offering a wide range of choices to the farmers for crop diversification and much-needed nutritional security to the people.

To project the supply of food commodities, one needs reliable empirical knowledge about the degree of responsiveness of input demand and crop output supply to input–output prices and technological changes. The econometric application of production theory based on the duality relationship between production functions and variable profit/cost function represents a major step towards generating appropriate empirical estimates of input demand functions and agricultural commodity supply, which are crucial for the application of economic theory for agricultural development policy (Lau and Yotopolous, 1972; Binswanger, 1974; Sidhu, 1974; Yotopolous *et al.*, 1976). Further, the development of flexible functional forms by several authors (Chand and Kumar, 1986; Bewley *et al.*, 1987; Mundlak, 1988; Rosegrant and Kasryno, 1992; Kumar and Rosegrant, 1997; Kumar, 1998; Kumar and Mittal, 2003) permits the application of duality theory for a more disaggregated analysis of the production structure than has been possible by the traditional approaches.

Each supply response model has its specific merits and limitations. Ideally, the methodological framework should be based on a profit function or cost function, but this approach requires data on output quantities and prices, and also on input quantities and prices. Limitations regarding availability of data are often a major constraint to adoption of this approach to model supply at the national level. In the present study, the crop-related data were culled from the 'Comprehensive Scheme for the Study of Cost of Cultivation of Principal Crops' of the DES (2009). It provides time series-cum-cross-section data on yield, and use of inputs and their prices. This data set is useful in estimating the profit function or cost function to derive the factor demand and output supply elasticities. The factor demand and output supply elasticities for cereals, pulses, edible oilseeds, sugarcane, onion, potato, cotton and jute have been used to project the domestic supply of these commodities. For livestock (milk, meat), poultry (chicken meat, eggs) and horticultural commodities (vegetables and fruits) input–output data were not available; therefore, supply projections for these commodities have been based on the past growth trend in their production.

Domestic supply projections

Supply growth

Crop area, total factor productivity, supply elasticity and input–output price environments are the major sources of supply growth. Thus:

$$S = f(P, p, AERA, TFP) \tag{4.9}$$

where, S is supply, P is price of commodity, p is the vector of input price, $AERA$ is the acreage under the commodity and TFP is the total factor productivity of the commodity.

The supply growth equation for the commodity can be expressed as:

$$S_g = E^{P}{}_{s} P_g + \Sigma E^{pi}{}_{s} p_{ig} + AREA_g + \left(TFP_{gt} - TFP_{g0}\right) \tag{4.10}$$

where:

S_g = Supply growth for the commodity,
$E^{P}{}_{s}$ = Output supply elasticity with respect to the product price,
P_g = Output price growth,
$E^{pi}{}_{s}$ = Elasticity of factor demand for the ith input,
p_{ig} = Input price growth of the ith input,
$AREA_g$ = Area growth under crop,
TFP_{g0} = TFP growth in the base year and
TFP_{gt} = TFP growth in the projected year t.

The supply growth equations were used to predict the supply of various commodities under the following three scenarios:

S1 = Baseline assumptions as given in Appendix Table 4.4.
S2 = Baseline assumptions plus 50% acceleration in TFP growth by the projected year 2030.
S3 = Baseline assumptions plus 50% deceleration in TFP growth by the projected year 2030.

The average production during 2009–2010 (TE 2010) was used as the base year domestic supply. The domestic supplies of major commodities have been explored to 2030:

$$S_t = S_{t-1} * \left(1 + S_g\right) \qquad (4.11)$$

where S_t is the supply for a commodity in the year t and S_g is the predicted growth under various scenarios.

Supply response elasticities

The output supply elasticities for major crops were computed from the factor demand elasticities and are presented in Table 4.13. The output supply elasticities have shown the response of output prices and input prices on the supply of major crops of India. Among crops, supply elasticity with respect to its price was highest for coarse grains (0.53), followed by edible oils (0.51), rice (0.24), wheat (0.22), pulses (0.17) and sugarcane (0.12). The input price response on supply was highly inelastic, nearly zero. The crop price has shown a dominating response on the supply of commodities and, therefore, a positive price policy will enhance domestic supply of food commodities.

Domestic supply growth

The supply growth was projected for major crops by using supply response elasticities and baseline assumptions for input and output prices, crop acreage and TFP growth under the three TFP growth scenarios and results are given in Table 4.14.

Domestic supply projections for commodities

The supply for different commodities has been projected using TE 2010 as the base year production. The supply projections for different food commodities under different scenarios have been presented at 10-year intervals for the period 2010–2030. Food supply and demand gaps for food grains, edible oils and sugar are presented in Table 4.15 and for high-value commodities, i.e. vegetables, fruits, milk, meat, eggs and fish, are given in Table 4.16.

Rice

The domestic production of rice under the baseline scenario S1 is estimated to be 108.1 Mt by the year 2020 and 122.1 Mt by 2030. A look at the past trend reveals that India has been marginally surplus in rice production and has even been exporting rice in small volumes (2–4 Mt). Under the accelerating TFP growth scenario S2, the production of rice is expected to be 109.1 Mt by the year 2020 and 126.4 Mt by 2030. However, under the decelerating TFP growth scenario S3, rice supply is projected to be lower at 106.7 Mt in 2020 and 117.3 Mt in 2030. As per these projections, India is not likely to remain in rice surplus and may even become deficit in rice production to the extent of 3–5 Mt in the coming years.

Wheat

The domestic production of wheat under the baseline scenario S1 is estimated to be 104.2 Mt by 2020 and 128.8 Mt by 2030. Under the accelerating TFP growth scenario S2, the production of wheat is expected to be 104.9 Mt in 2020 and 132.5 Mt in 2030. Under the decelerating TFP growth scenario S3, supply of wheat is likely to decline to 103.1 Mt by 2020 and 124.6 Mt by 2030. A perusal at the supply–demand scenario reveals that wheat demand will continue to be met from domestic production and there may even be a marginal surplus of about 4.8–6.6 Mt by the year 2020, which is likely to grow to 9.9–17.9 Mt by 2030. It is observed that a shift in consumption from rice to wheat is taking place even in the traditionally rice-eating states of India. Therefore, the surplus wheat production is likely to substitute rice, leading to lower availability of surplus wheat, as predicted in the study.

Coarse cereals

The domestic production of coarse cereals is estimated to be about 50 Mt by the year 2020, which will grow to 63–65 Mt in 2030 under different growth scenarios. The supply–demand gap of coarse grains is projected to be 8 Mt by 2020, which may grow to a higher

Table 4.13. Supply response elasticities for different crops in India.

Crop	Output price (P)	Input price				
		w/P	b/P	m/P	r/P	i/P
Rice	0.2357	−0.0017	−0.0004	0.0004	0.0001	0.0017
Wheat	0.2164	0.0163	−0.0288	0.0095	−0.0095	0.0125
Coarse grains	0.5333	−0.1105	0.0952	0.0198	0.2791	0.0500
Pulses	0.1695	−0.0007	−0.0012	0.0020	−0.0013	0.0012
Edible oilseeds	0.5079	−0.0011	0.0021	0.0168	0.0062	−0.0240
Sugarcane	0.1216	0.0021	−0.0002	−0.0020	0.0045	−0.0044

b, cost of animal labour (Rs/h); i, cost of irrigation (Rs/ha); m, cost of machine labour (Rs/h); P, price of crop (Rs/100 kg); r, cost of fertilizer (NPK) (Rs/kg); w, wage (Rs/h)

Table 4.14. Supply growth for major food commodities under different TFP growth scenarios in India, 2010–2030.

Commodity	Baseline scenario (S1)	2010	2020	2030
S2: Baseline growth + 50% acceleration in TFP growth by 2030				
Rice	1.227	1.240	1.385	1.562
Wheat	2.146	2.157	2.278	2.426
Coarse grains	2.450	2.457	2.530	2.619
Pulses	2.479	2.483	2.516	2.539
Edible oilseeds	4.357	4.365	4.454	4.562
Sugarcane	1.890	1.891	1.901	1.915
S3: Baseline growth + 50% deceleration in TFP growth by 2030				
Rice	1.227	1.205	1.023	0.892
Wheat	2.146	2.128	1.975	1.866
Coarse grains	2.450	2.439	2.347	2.282
Pulses	2.479	2.475	2.433	2.399
Edible oilseeds	4.357	4.343	4.232	4.152
Sugarcane	1.890	1.888	1.874	1.865

level of 16–18 Mt by 2030. This projection of demand–supply balance of coarse grains has provided some valuable insights about the possible level of self-sufficiency in India in coarse grains production, particularly their availability for meeting the feed requirements of the fast-growing livestock sector in the country in the years to come.

Total cereals

In India, the domestic supply of total cereals, which is the summation of rice, wheat and coarse grains production, is projected to be 240–264 Mt by 2020, which will rise to 304–324 Mt by 2030 under different TFP growth scenarios. A look at the supply–demand balance for cereals reveals that their demand in future will be met by national production and there could even be a surplus of 7–12 Mt cereals by 2020 and of 20–40 Mt by 2030.

The contribution of TFP in cereals supply is predicted to be about 9 Mt by the year 2030 under scenario S2 of accelerating TFP growth. On the other hand, the supply of cereals will decline by 10 Mt under scenario S3 of decelerating TFP growth over the baseline scenario. To maintain cereals security, there is a need to strengthen efforts towards maintaining the TFP growth by enhancing respective TFP growth for rice, wheat and coarse cereals.

Table 4.15. Supply projections for major food grains (Mt) to 2030, edible oils and sugar, India (authors' calculations).

Commodities	Year	Supply scenario S1	S2	S3	Demand baseline	Demand–supply gap Minimum	Maximum
Rice	2010	95.69	95.69	95.69	98.7	−3.01	−3.01
	2020	108.10	109.09	106.73	111.8	−2.71	−5.07
	2030	122.12	126.35	117.29	122.4	3.95	−5.11
Wheat	2010	84.24	84.24	84.24	83	1.24	1.24
	2020	104.16	104.95	103.07	98.3	4.77	6.65
	2030	128.80	132.48	124.56	114.6	9.96	17.88
Coarse cereals	2010	39.55	39.55	39.55	36.4	3.15	3.15
	2020	50.38	50.61	50.06	42.5	7.56	8.11
	2030	64.18	65.27	62.90	47.2	15.70	18.07
Total cereals	2010	219.48	219.48	219.48	218.1	1.38	1.38
	2020	240.02	264.65	259.86	252.6	7.26	12.05
	2030	315.10	324.11	304.75	284.2	20.55	39.91
Pulses	2010	16.17	16.17	16.17	18	−1.83	−1.83
	2020	20.65	20.70	20.59	21.9	−1.20	−1.31
	2030	26.38	26.57	26.14	26.6	−0.03	−0.46
Food grains	2010	234.03	234.03	234.03	236.2	−2.17	−2.17
	2020	281.23	283.26	278.40	274.4	4.00	8.86
	2030	338.84	348.02	328.27	310.8	17.47	37.22
Edible oils	2010	8.15	8.15	8.15	13.63	−5.48	−5.48
	2020	12.49	12.56	12.39	16.97	−4.41	−4.58
	2030	19.13	19.52	18.68	21.26	−1.74	−2.58
Sugar	2010	27.70	27.70	27.70	27.62	0.08	0.08
	2020	33.41	33.43	33.37	33.1	0.27	0.33
	2030	40.28	40.39	40.16	39.21	0.95	1.18

Table 4.16. Demand, supply and demand–supply gap projections (Mt) to 2030 for high-value food commodities in India (authors' calculations).

Commodities		Production 2010	2020	2030	Availability 2010	2020	2030	Postharvest losses (%)
Vegetables	Supply	140.6	186.4	210.5	106.9	141.7	160.0	23.99
	Demand	124.7	154.8	192.0	124.7	154.8	192.0	
	Gap	15.9	31.6	18.5	−17.8	−13.1	−32.0	
Fruits	Supply	73.5	97.7	116.4	58.8	78.2	93.1	20.00
	Demand	64.8	80.9	103.0	64.8	80.9	103.0	
	Gap	8.7	16.8	13.4	−6.0	−2.7	−9.9	
Milk	Supply	116.5	156.6	188.7	110.6	148.7	179.2	5.03
	Demand	111.9	138.3	170.4	111.9	138.3	170.4	
	Gap	4.6	18.3	18.3	−1.3	10.4	8.8	
Poultry and bovine meat	Supply	4.4	6.6	8.4	4.2	6.3	8.0	4.98
	Demand	5.2	6.8	9.2	5.2	6.8	9.2	
	Gap	−0.7	−0.2	−0.7	−0.9	−0.5	−1.2	
Eggs	Supply	3.1	4.7	6.2	2.9	4.5	5.9	5.02
	Demand	3.5	4.4	5.8	3.4	4.4	5.8	
	Gap	−0.4	0.3	0.4	−0.5	0.1	0.1	
Fish	Supply	7.4	10.2	13.9	6.3	8.7	11.9	15.05
	Demand	6.4	8.2	11.1	6.4	8.2	11.1	
	Gap	1.0	2.0	2.8	−0.1	0.5	0.8	

Pulses

The domestic production of pulses is projected to be about 21 Mt in 2020 and 26 Mt in 2030, with marginal differences across different scenarios. The supply of pulses will fall short of the demand by about 2 Mt in the years to come and India will have to continue their imports to meet the domestic needs.

Food grains

In India, the domestic supply of total food grains, which is the summation of rice, wheat, coarse cereals and pulses, is projected to be about 281 Mt in the year 2020 under the baseline scenario S1 and will grow to 339 Mt by 2030. A higher supply is predicted with accelerated TFP growth scenario S2 and is estimated to be 283 Mt in 2020 and 348 Mt in 2030. A lower supply is estimated with decelerated TFP growth assumption under scenario S3 and is predicted to be about 278 Mt by 2020 and 328 Mt by 2030.

A look at the supply and demand balance of food grains in India reveals that their future domestic demand will be met with national production and there is likelihood of a marginal surplus of about 4–8 Mt in 2020 and trade surplus of 17–37 Mt is predicted by the year 2030.

Edible oils

The domestic production of edible oils is projected to be about 12 Mt by 2020 and 19 Mt by 2030, with only marginal differences across different TFP growth scenarios. By looking at the supply–demand scenarios of edible oils, it may be predicted that their domestic production would fall short of demand under all the scenarios. The deficit in edible oils supply is projected to be about 4–5 Mt by 2020, and it may reduce to about 2 Mt by 2030. Thus, India will continue to depend on imports of edible oils, even in the coming decades.

Sugar

The supply of sugar is projected to be about 33 Mt in 2020 and it is likely to increase to 40 Mt by 2030. The domestic supply of sugar will be able to meet the demand of sugar in India in the coming years and there could be a marginal surplus of about 1 Mt by 2030.

High-value commodities

The domestic supply projections to 2030 for high-value commodities, i.e. vegetables, fruits, milk, eggs, meat and fish, are presented in Table 4.16. Their availability, i.e. domestic supply, has been computed from production after adjusting postharvest losses.

Vegetables

The domestic supply of total vegetables is projected to be 141 Mt in 2020 and 160 Mt by 2030. The supply–demand gap in total vegetables reveals that there will be a substantial shortage of vegetables unless postharvest losses are minimized.

Fruits

The domestic supply of fruits is projected to be 78.2 Mt in 2020 and 93.1 Mt by 2030. Looking at the supply–demand gap, it appears that India will have a defecit of fruit 10 Mt by 2030.

Milk

The milk supply in the country is projected to be 149 Mt in 2020 and 179 Mt in 2030. The supply–demand gap in milk reveals that the country will be able to meet its national domestic demand with a trade surplus of 8.8 Mt by 2030.

Meat

The total meat production from cattle, buffalo, sheep, goat, pig and poultry at an all-India level increased from 1.85 Mt in 2000 to 4.2 Mt in 2010. Poultry meat has not only accounted for the highest contribution to total meat production but has also witnessed the highest acceleration since 2000. Looking at the past growth, the supply of total meat by

2020 is projected to be 6.3 Mt. The total meat supply will grow to 8.0 Mt by 2030. It appears that India will remain deficit in total meat production in the years to come.

Eggs

Domestic egg production is projected to be 4.5 Mt in 2020 and 5.9 Mt in 2030. It seems that India will be able to meet the domestic demand for eggs with a marginal surplus.

Fish

India is the second largest producer of fish in the world with a contribution of 5.54% to global production. The total fish production during 2010 is estimated at 8.03 Mt with a contribution of 5.07 Mt from the inland sector and 2.96 Mt from the marine sector. The value of output from the fisheries sector at current prices during 2010 was 4.9% of total output of agriculture and allied sectors. India's exports of marine product have, for the first time, exceeded US$2 billion. During 2010, the volume of fish and fish products exported was 0.753 Mt, registering the highest growth rate of 10% in volume of fish exports in recent years. The projected domestic supply of fish is about 6.3 Mt in 2010, 8.7 Mt in 2020 and 11.9 Mt in 2030. The supply–demand gap of fish is projected to be 0.4–0.7 Mt. It appears that the country will continue to remain self-reliant in fish supply and will also be able to undertake international trade at the present level.

Conclusions

Empirical studies on the dynamics of supply and demand of food crops are valuable for a country like India from the point of achieving food security, and often provide deep insights to policy planners regarding the existing state of affairs and future directions on food self-sufficiency. This study has estimated the factor demand and output supply elasticities for major food crops in India. The elasticities have provided insights on the responsiveness of output supply and factor demand to changes in product and factor prices. The estimates have been used to make supply projections of food crops to 2030. The projections have been made under different growth scenarios, crop area, total factor productivity and input–output prices and have essentially presented the changes in supply of major food commodities. An assessment of crop demand–supply balance under different scenarios provides valuable insights on the possible levels of self-sufficiency and trade potential for each of the selected crops in the coming years.

The study has observed that the demand for rice and wheat will be met with domestic production in the coming years, possibly with a marginal surplus/deficit under the scenarios of with or without TFP growth and acreage response. However, it is quite likely that pulses, edible oils and sugar would be short in supply of demand in the coming years and India will open for imports of these commodities. The policies that can help in maintaining the TFP growth in the long-run will be able to keep a balance between domestic production and demand for cereals, pulses, edible oils and sugar. This emphasizes the need for strengthening efforts at increasing production potential through public investments on irrigation, infrastructural development, agricultural research and efficient use of water and plant nutrients (Fan *et al.*, 1999; Chand *et al.*, 2011).

Note

[1] Each Indian state, based mostly on its economic development, declares its poverty line for rural and urban households. Therefore, PL for different states corresponding to various NSS rounds was used to classify the sample households into three income groups.

References

Behrman, J. and Deolalikar, A. (1989) Is variety the spice of life? Implications for calorie intake. *Review of Economics and Statistics* 31, 398.

Binswanger, H.P. (1974) A cost function approach to the measurement of factor demand elasticities of substitution. *American Journal of Agricultural Economics* 56(20), 377–386.

Bewley, R., Young, T. and Colman, D. (1987) A systems approach to modelling supply equation. *Journal of Agricultural Economics* 38, 151–166.

Bouis, H. and Haddad, L. (1992) Are the estimates of calorie income elasticities too high: A recalibration of the plausible range. *Journal of Development Economics* 53(June), 115–137.

Chand, R. (2007) Demand for foodgrains. *Economic and Political Weekly* (December).

Chand, R. and Kumar, P. (1986) Supply response functions for major crops of Punjab in green revolution period – A profit function approach. *Indian Journal of Economics* 67(1), 91–109.

Chand, R., Kumar, P. and Kumar, S. (2011) *Total Factor Productivity and Contribution of Research Investment to Agricultural Growth in India*. Policy Paper 25. National Centre for Agricultural Economics and Policy Research, New Delhi.

DES (Directorate of Economics and Statistics) (2009) *Agricultural Statistics at a Glance*. Department of Agriculture and Cooperation, Government of India, New Delhi.

Fan, S., Hazell, P.B.R. and Thorat, S. (1999) *Linkages between Government Spending, Growth, and Poverty in Rural India*. Research Report No. 110. International Food Policy Research Institute, Washington, DC.

Kumar, P. (1998) *Food Demand and Supply Projections for India*. Agricultural Economics Policy Paper 98-01. Indian Agricultural Research Institute, New Delhi.

Kumar, P. and Mittal, S. (2003) Productivity and supply of foodgrains in India. In: Mahendra Dev, S., Kannan, K.P. and Ramachandran, N. (eds) *Towards a Food Secure India: Issues & Policies*. Manohar Publishers and Distributers, New Delhi, pp. 33–58.

Kumar, P. and Mruthyunjaya (1995) Demand for fruits and vegetables in India. *Agricultural Economics Research Review* 8(2), 7–17.

Kumar, P. and Rosegrant, M.W. (1997) Dynamic supply response of rice and other major food crops in India. *Agricultural Economics Research Review* 10(1), 1–23.

Kumar, P., Mruthyunjaya and Birthal, P.S. (2007) Changing consumption pattern in South Asia. In: Joshi, P.K., Gulati, A. and Cummings Jr, R. (eds) *Agricultural Diversification and Smallholders in South Asia*. Academic Foundation, New Delhi, pp. 151–187.

Kumar, P., Joshi, P.K. and Birthal, P.S. (2009) Demand projections for foodgrains in India. *Agricultural Economics Research Review* 22(2), 237–243.

Kumar, P., Shinoj, P., Raju, S.S., Kumar, A., Rich, K.M. and Msangi, S. (2010) Factor demand, output supply elasticities and supply projections for major crops of India. *Agricultural Economics Research Review* 23(1), 1–14.

Lau, L.J. and Yotopolous, P.A. (1972) Profit, supply and factor demand functions. *American Journal of Agricultural Economics* 54(1), 11–18.

Mundlak, Y. (1988) Endogenous technology and the measurement of productivity. In: Cepalbo, S.M. and Antle, J.M. (eds) *Agricultural Productivity: Measurement and Explanation*. Resources for the Future, Washington, DC.

NSSO (National Sample Survey Organization) (various rounds) *Consumer Expenditure Survey*. Government of India, New Delhi.

Paroda, R.S. and Kumar, P. (2000) Food production and demand in South Asia. *Agricultural Economics Research Review* 13(1), 1–24.

Registrar General (1996) *Population Projections for India and States: 1996–2016*. Report of Technical Group on Population Projections, Government of India, New Delhi.

Rosegrant, M.W. and Kasryno, F. (1992) *Food Crop Production Growth in Indonesia: Supply Response with Choice of Technique*. International Food Policy Research Institute, Washington, DC.

Sidhu, S.S. (1974) Relative efficiency in wheat production in Indian Punjab. *American Economic Review* 64(4), 742–751.

Yotopolous, P.A., Lau, L.J. and Lin, W.-L. (1976) Micro-economics output supply and factor demand function in the agriculture of the province of Taiwan. *American Journal of Agricultural Economics* 58, 333–340.

Tables for Appendices

Table A4.1. Annual per capita consumption of food in India, 2004.

	Food (kg/capita/year)			% share in total	
Food commodity	At home (H)	Outside home (OH)	Total (H+OH)	At home	Outside home
Rice	76.71	2.38	79.09	96.99	3.01
Wheat	52.53	8.55	61.08	86.00	14.00
Coarse cereals	12.55	8.01	20.56	61.04	38.96
Pulses	9.17	2.60	11.77	77.91	22.09
Edible oils	6.59	2.35	8.94	73.71	26.29
Sugar	9.60	9.80	19.40	49.48	50.52
Vegetables	65.83	22.99	88.92	74.03	25.85
Fruits	10.30	29.80	40.10	25.69	74.31
Milk	58.02	27.17	85.19	68.11	31.89
Meat	1.48	1.11	2.59	57.14	42.86
Poultry	0.76	0.68	1.44	52.78	47.22
Eggs	0.77	1.15	1.92	40.10	59.90
Fish	2.68	2.02	4.70	57.02	42.98

Table A4.2. Demand for cereals and pulses (1000 t) in India (authors' computation).

Commodity and year	Household demand			Indirect demand					Total domestic demand
	At home	Home away	Total	Seed (S)	Feed (F)	Wastages (W)	Industrial uses (IU)	Total SFWIU	
Rice									
Scenario 1									
2004 (Base year)	82,892	2,575	85,466	1,280	674	493	2,127	4,574	90,040
2010	90,713	2,960	93,673	1,280	797	540	2,396	5,013	98,686
2020	102,419	3,566	105,985	1,280	1,026	612	2,920	5,838	111,823
2030	111,331	4,242	115,573	1,280	1,349	670	3,560	6,859	122,432
Scenario 2									
2020	102,197	3,490	105,687	1,280	988	610	2,920	5,798	111,485
2030	110,768	4,028	114,796	1,280	1,232	665	3,560	6,737	121,533
Scenario 3									
2020	102,646	3,644	106,290	1,280	1,066	614	2,920	5,880	112,170
2030	111,921	4,472	116,393	1,280	1,484	675	3,560	6,999	123,392
Wheat									
Scenario 1									
2004 (Base year)	55,757	9,083	64,840	1,790	3,291	1,079	2,035	8,196	73,035
2010	63,138	10,750	73,888	1,790	3,895	1,227	2,225	9,137	83,025
2020	74,096	13,346	87,442	1,790	5,011	1,452	2,583	10,836	98,279
2030	85,065	16,465	101,530	1,790	6,591	1,694	2,997	13,072	114,602
Scenario 2									
2020	73,556	13,029	86,584	1,790	4,824	1,437	2,583	10,634	97,218
2030	83,687	15,559	99,246	1,790	6,016	1,651	2,997	12,454	111,700
Scenario 3									
2020	74,642	13,675	88,317	1,790	5,209	1,468	2,583	11,050	99,367
2030	86,478	17,446	103,924	1,790	7,251	1,739	2,997	13,778	117,702
Coarse cereals									
Scenario 1									
2004 (Base year)	13,751	8,777	22,528	550	7,930	730	744	9,954	32,482

Continued

Table A4.2. Continued.

Commodity and year	Household demand: At home	Household demand: Home away	Household demand: Total	Indirect demand: Seed (S)	Indirect demand: Feed (F)	Indirect demand: Wastages (W)	Indirect demand: Industrial uses (IU)	Indirect demand: Total SFWIU	Total domestic demand
2010	14,626	10,171	24,797	550	9,384	819	888	11,642	36,439
2020	15,415	12,270	27,685	550	12,074	955	1,193	14,772	42,458
2030	14,411	13,671	28,082	550	15,882	1,061	1,604	19,096	47,178
Scenario 2									
2020	15,607	11,981	27,588	550	11,623	942	1,193	14,309	41,897
2030	14,813	12,952	27,765	550	14,496	1,022	1,604	17,671	45,436
Scenario 3									
2020	15,226	12,568	27,794	550	12,551	968	1,193	15,262	43,056
2030	14,022	14,439	28,461	550	17,471	1,106	1,604	20,731	49,192
Pulses									
Scenario 1									
2004 (Base year)	9,904	2,804	12,708	1,200	1,322	172	378	3,072	15,780
2010	11,328	3,281	14,609	1,200	1,564	196	452	3,412	18,021
2020	13,768	4,045	17,814	1,200	2,012	238	607	4,058	21,871
2030	16,643	4,980	21,623	1,200	2,647	289	816	4,952	26,575
Scenario 2									
2020	13,478	3,945	17,423	1,200	1,937	233	607	3,977	21,400
2030	15,824	4,689	20,513	1,200	2,416	274	816	4,706	25,219
Scenario 3									
2020	14,070	4,150	18,220	1,200	2,092	243	607	4,142	22,362
2030	17,541	5,295	22,837	1,200	2,912	305	816	5,233	28,070

Scenario 1, current GDP growth; Scenario 2, low GDP growth; Scenario 3, high GDP growth

Table A4.3. Demand for high-value commodities (1000 t) in India (authors' computations).

Commodity and year	Household demand			Indirect demand	Total demand
	At home	Home away	Total		
Edible oils					
Scenario 1					
2004 (Base year)	7,155	2,548	9,703	2,016	11,719
2010	8,283	3,001	11,284	2,345	13,629
2020	10,313	3,737	14,050	2,919	16,969
2030	12,923	4,682	17,604	3,658	21,262
Scenario 2					
2020	10,030	3,637	13,666	2,839	16,506
2030	12,104	4,386	16,491	3,426	19,917
Scenario 3					
2020	104,626	36,483	141,109	17,984	159,093
2030	135,516	46,435	181,951	23,190	205,140
Sugar					
Scenario 1					
2004 (Base year)	10,380	10602	20,982	3,083	24,065
2010	11,641	12,439	24,079	3,538	27,618
2020	13,523	15,340	28,863	4,241	33,104
2030	15,353	18,835	34,188	5,024	39,212
Scenario 2					
2020	13,416	14,955	28,370	4,169	32,539
2030	15,071	17,727	32,798	4,819	37,618
Scenario 3					
2020	13,634	15,739	29,372	4,316	33,688
2030	15,653	20,040	35,692	5,245	40,937
Vegetables					
Scenario 1					
2004 (Base year)	70,808	24,729	95,537	12,176	107,713
2010	81,758	28,879	110,637	14,101	124,738
2020	101,747	35,595	137,342	17,504	154,846
2030	126,594	43,742	170,336	21,709	192,045
Scenario 2					
2020	98,998	30,836	129,833	16,547	146,381
2030	118,623	35,900	154,523	19,694	174,217

Continued

Table A4.3. Continued.

Commodity and year	Household demand			Indirect demand	Total demand
	At home	Home away	Total		
Scenario 3					
2020	104,626	36,483	141,109	17,984	159,093
2030	135,516	46,435	181,951	23,190	205,141
Fruits					
Scenario 1					
2004 (Base year)	11,190	32,391	43,582	11,569	55,151
2010	13,083	38,108	51,190	13,589	64,779
2020	16,606	47,319	63,925	16,970	80,895
2030	21,564	59,825	81,389	21,606	102,995
Scenario 2					
2020	16,042	46,023	62,065	16,476	78,541
2030	19,885	55,980	75,865	20,139	96,005
Scenario 3					
2020	17,197	48,664	65,861	17,484	83,345
2030	23,443	64,026	87,470	23,220	110,690
Milk					
Scenario 1					
2004 (Base year)	62,885	29,451	92,336	2,232	94,568
2010	74,423	34,850	109,273	2,641	111,915
2020	96,293	43,519	139,812	3,379	143,191
2030	126,046	54,500	180,547	4,364	184,911
Scenario 2					
2020	92,691	42,338	135,029	3,264	138,292
2030	115,279	51,064	166,343	4,021	170,364
Scenario 3					
2020	100,092	44,743	144,835	3,501	148,336
2030	138,300	58,250	196,550	4,751	201,300
Fish					
Scenario 1					
2004 (Base year)	2,864.5	2,146.3	5,010.8	421.1	5,431.9
2010	3,430.1	2,472.6	5,902.7	496.1	6,398.8
2020	4,627.2	2,964.8	7,592	638.1	8,230.1
2030	6,629.4	3,628.6	10,258	862.1	11,120.1

Scenario 2					
2020	4,362.4	2,894.2	7,256.5	609.9	7,866.4
2030	5,756.2	3 422.2	9,178.4	771.4	9,949.8
Scenario 3					
2020	4,912.7	3,037.9	7,950.6	668.2	8,618.8
2030	7,683.8	3,853.2	11,537	969.6	12,506.6
Bovine meat					
Scenario 1					
2004 (Base year)	1,544.8	1,157.5	2,702.3	31.3	2,733.6
2010	1,865.8	1,350.5	3,216.3	37.2	3,253.5
2020	2,563.1	1,671.4	4,234.5	49	4,283.5
2030	3,670.7	2,075.8	5,746.5	66.5	5,813
Scenario 2					
2020	2,420.2	1,631.3	4,051.6	46.9	4,098.4
2030	3,204	1,958.6	5,162.6	59.7	5,222.3
Scenario 3					
2020	2,716.5	1,712.9	4,429.4	51.2	4,480.7
2030	4,228	2,202.9	6,430.9	74.4	6,505.3
Poultry meat					
Scenario 1					
2004 (Base year)	847.3	743.9	1,591.2	18.4	1,609.6
2010	1,020.4	866	1,886.4	21.8	1,908.2
2020	1,388.9	1,063.2	2,452.1	28.4	2,480.4
2030	1,991.2	1,332.2	3,323.4	38.4	3,361.8
Scenario 2					
2020	1,311.4	1,035.3	2,346.6	27.1	2,373.8
2030	1,737.3	1,249	2,986.3	34.5	3,020.8
Scenario 3					
2020	1,472.1	1,092.2	2,564.2	29.7	2,593.9
2030	2,294.4	1,423	3,717.4	43	3,760.4
Eggs					
Scenario 1					
2004 (Base year)	1,016.9	1,515.6	2,532.5	400.5	2,933
2010	1,219.6	1,761.3	2,980.8	471.4	3,452.2
2020	1,645.1	2,154	3,799.1	600.8	4,399.9
2030	2,336.4	2,691.1	5,027.5	795.1	5,822.6

Continued

Table A4.3. Continued.

Commodity and year	Household demand			Indirect demand	Total demand
	At home	Home away	Total		
Scenario 2					
2020	1,556.8	2,100.2	3,657.1	578.4	4,235.4
2030	2,050.2	2,531.5	4,581.6	724.6	5,306.2
Scenario 3					
2020	1,739.6	2,209.6	3,949.2	624.6	4,573.8
2030	2,675.3	2,864.9	5,540.2	876.2	6,416.4

Scenario 1, current GDP growth; Scenario 2, low GDP growth; Scenario 3, high GDP growth

Table A4.4. Annual growth of input–output prices, area and TFP for crops in India, 1981–2006 (authors' computations).

Commodity	Output price growth (P)	Input price growth					Area growth (2000–2010)	TFP growth
		w/P	b/P	m/P	r/P	i/P		
Rice	6.78	4.18	5.19	–0.64	–1.60	–0.91	–0.36	0.67
Wheat	7.67	3.29	4.30	–1.53	–2.49	–1.80	0.57	0.56
Coarse grains	6.87	4.09	5.10	–0.73	–1.69	–1.00	–0.71	0.34
Pulses	8.90	2.06	3.07	–2.76	–3.72	–3.03	0.98	–0.12
Edible oilseeds	6.73	4.23	5.24	–0.59	–1.55	–0.87	0.93	0.41
Sugarcane	9.49	1.47	2.48	–3.35	–4.31	–3.63	0.73	0.05

b, cost of animal labour (Rs/h); i, cost of irrigation (Rs/ha); m, cost of machine labour (Rs/h); P, price of crop (Rs/100 kg); r, cost of fertilizer (NPK) (Rs/kg); w, wage (Rs/h)

5 Indian Economic Growth and Trade Agreements: What Matters for India and for Global Markets?

Geert Woltjer* **and Martine Rutten**
Public Issues Division, LEI Wageningen UR, The Hague, The Netherlands

Introduction

India is one of the fastest growing economies in the world. Despite global downturn, India has recorded an average growth rate of 7.6% per year over the past decade (IMF, 2012), with growth expected to rise again to 6.3% in 2014. This growth rate is well above the 2.1% growth estimated for the advanced economies (including the USA, the Euro area and Japan) and the 3.8% growth for the world average (IMF, 2013). Given that India has a relatively large population of over 1.2 billion, which is estimated to grow to around 1.6 billion by 2050 (UN, 2012), it is very likely that the development path followed by India influences the world economy. India's trade with the world is estimated at €815 billion in 2012, which represents a fairly small share of less than 2% in world trade (WTO, 2012; European Commission, 2013). Although India is currently little integrated with the world economy, merchandise trade is growing fast at a rate of 17% per year since 2005, which is more than twice the global growth rate (WTO, 2012). If it opens itself up more to the world market, India may also become more dependent on and vulnerable to developments in the world.

One major area of concern is the influence of recent global food crises on domestic food price inflation in India, with damaging impacts, on poor households especially, on their net food purchasing. Although India is currently not a main food importer, it suffered from domestic food price inflation with the country recording the second largest increase in wheat prices (after Sudan) and the third largest increase in rice prices (after Malawi and Rwanda) in the world in 2012 (World Bank, 2012). This is a major concern, given that households in India on average spend close to 40% of their total expenses on food. A large part is caused by the frequent occurrence of droughts, reducing rural households' income and raising poverty levels considerably. Other causal factors are said to have been cost increases (also from rising oil prices) and speculative activity, the impacts of which may intensify in an increasingly globalizing world (Chandrasekhar, 2012).

This chapter investigates the impacts of alternative growth and trade policies on the Indian economy and the rest of the world, notably the European Union (EU), using a global trade model. The chapter starts with a description of the model used and adjustments made in the model to make it suitable for analyses on India in a global context.

* Corresponding author, e-mail: Geert.Woltjer@wur.nl

 International Trade and Food Security (eds F. Brouwer and P.K. Joshi)

The next section focuses on alternative growth and trade policies that India could pursue, specifically that of a high growth agenda (increasing annual growth of India from 6% to 8%, and finally to 10%) and that of a Free Trade Agreement with the EU. The latter will be compared with a multilateral trade agreement in the context of the World Trade Organization (WTO). The behaviour of the model over time in the baseline, used as the benchmark or reference scenario to which all other scenarios have been compared, is also described. The fourth section presents the results of the scenarios. The final section presents conclusions.

Model

For empirical analyses in this chapter, we have used a global Computable General Equilibrium (CGE) model called MAGNET (Modular Applied GeNeral Equilibrium Tool). The MAGNET model has been widely used as a tool for global trade analyses (see, for example, Francois *et al.*, 2005; Rutten *et al.*, 2013). The MAGNET model is based on the Global Trade Analysis Project (GTAP) model but can be extended in various directions in a modular fashion, depending on the policy questions at hand. The GTAP core model is described in the following section, after which the key features introduced in the MAGNET model are described to make it suitable for analyses on India in a global context.

The GTAP core and data

The MAGNET's GTAP core accounts for the behaviour of households, firms and the government in the global economy and how they interact in markets (Hertel, 1997).

Households' behaviour is captured via a 'representative regional household', which in order to maximize its utility, collects all income that is generated in the economy and allocates it over the private household and government expenditures on commodities, and savings. Income comes from payments by firms to the regional household for the use of endowments of skilled and unskilled labour, land, capital and natural resources. The regional household also receives income from (net) taxes paid by the private household (on private consumption and income), firms (taxes on intermediate inputs and production) and the government (on its expenditures). Firms, to maximize profits, produce commodities by employing the aforementioned endowments and intermediate inputs from other firms using constant returns to scale production technology so as to sell them to private households, the government and other producers. The domestically produced goods can either be sold in the domestic market or to other regions in the world. Similarly, the demand of domestic intermediate, private household and government for goods can be satisfied by domestic production or by imports from other regions in the world. These come with their own import and export taxes. Sourcing of imports happens at the border, after which – on the basis of the resulting composite import price – the optimal mix of import and domestic goods is derived. The demand for and supply of commodities and endowments meet in the markets, which are perfectly competitive and clear via price adjustments. Natural resources are assumed to adjust sluggishly between sectors. The assumptions regarding labour, land, capital markets and investments have been discussed below as they are different from standard GTAP. With all markets in equilibrium, firms earning zero profits and households being on their budget constraint, global savings must equal global investments. In GTAP, global savings determine global investments, i.e. the macro closure is savings-driven and essentially neoclassical in nature. Since the CGE model can only determine relative prices, the gross domestic product (GDP) deflator is set as the numéraire of the model, against which all other prices are benchmarked. Changes in prices resulting from the model simulations thus constitute the real price changes.

The model has used the most recent GTAP database version 8 (final release), which contains value data for 2007. The 129 countries and/or regions and 57 sectors of the GTAP database have been aggregated into more manageable categories and close to those distinguished in the national CGE database for India (Woltjer, 2013), namely 22 regions and 44 sectors (Table 5.1).

The sectoral division distinguished the 11 primary agricultural commodities (1–11) available in GTAP at the highest level of detail.

Table 5.1. Commodity aggregation (Woltjer, 2013).

1	Rice
2	Wheat
3	Other cereal grains
4	Oilseeds
5	Sugarcane, sugarbeet
6	Vegetables, fruit, nuts
7	Plant-based fibres
8	Other crops
9	Cattle, sheep, goats, horses, wool
10	Pigs, poultry
11	Raw milk
12	Dairy products
13	Sugar
14	Molasses as by-product of sugar
15	Crude vegetable oil
16	Oil cake as by-product of vegetable oil
17	Vegetable oils products
18	Meat of ruminants
19	Other meat
20	Animal feed
21	Other foods
22	Beverages and tobacco
23	Fishing
24	Forestry
25	Coal
26	Crude oil
27	Other minerals
28	Petroleum, coal products
29	Biodiesel
30	Ethanol
31	Dried distillers grains with solubles
32	Natural gas mining
33	Gas manufacture and distribution
34	Textiles
35	Chemical industry
36	Fertilizer
37	Labour-intensive manufacturing
38	Capital-intensive manufacturing
39	Construction
40	Trading sector
41	Transport services
42	Electricity
43	Water supply
44	Other services

Furthermore, we distinguished seven processed food categories (12–22), which had strong links with the aforementioned primary sectors, and aggregated the remaining sectors into various manufacturing (23–39) and services (40–44) categories. Note that textiles was separated as this is an important export product of India. Moreover, for the analysis of biofuel policies and agricultural intensification, a number of sectors were added. These included biodiesel and biogasoline (i.e. ethanol), dried distillers grains with solubles (DDGS) as a by-product of ethanol production from maize, molasses as a by-product of sugar production and an important input for biogasoline production in India, a crude vegetable oil sector that splits oilseeds into vegetable oils and oilcake, and a refined vegetable oil sector that further refines the vegetable oil to sophisticated products for consumers. Finally, we distinguished feed from other foods, and fertilizer from the chemical sector. The regional division distinguished India and the remaining BRIC countries (Brazil, Russia, India and China). Due to limited data reliability, its most important neighbouring countries were taken together in a Rest of South Asia region. EU countries were also put together in an EU aggregate to facilitate the simulation of trade policies. The model retained the standard GTAP specification of five factors of production, including skilled and unskilled labour, capital, land and natural resources.

MAGNET model and data adjustments

For this study, MAGNET, compared to GTAP, incorporated five modules that are fairly standard in MAGNET applications (Banse *et al.*, 2008; Stehfest *et al.*, 2013). These included:

- A sophisticated production structure, accounting for the inherent difference in the ease of substitution between land and non-land factors of production, and within the latter, between capital and energy, versus land, labour and national resources.

- A sophisticated consumption structure, allowing for a better depiction of changes in diets observed over time, for India calibrated on income and price elasticities from Chapter 4 of this book.
- Segmented labour and capital markets, allowing for differences in factor remunerations between agricultural and non-agricultural sectors.
- Modelling of land market, accounting for land-use supply changes in response to price changes and an allocation of land over sectors, taking into account the inherent difference in the ease of substitution between different types of land.
- A biofuel directive module, which allows for the opportunity to set blending targets for the petroleum industry.

Next to these standard additions, two modules particularly relevant for analyses on India in a global context have been purposely developed, including dynamic international capital flows and investments and improved tariff data and the incorporation of tariff shocks. These are discussed in more detail below.

Dynamic international capital flows and investments

The GTAP model does not explicitly capture international capital flows and there is no link between investments and capital stocks over time. These are important for a strongly growing and increasingly open economy such as India that is likely to increase its savings and seek investment possibilities abroad. We have introduced international capital flows in the MAGNET model, using Ianchovichina and McDougall's (2001) international system of capital flow accounting. In contrast to their approach, we have derived international capital income flows from the international wealth positions instead of doing vice versa. The dynamic international investment module captures a two-stage decision process:

Stage 1: Invest available domestic funds domestically or internationally.
Stage 2: Distribute international investment funds over different regions.

In the first stage, the current distribution of investment streams is taken as the point of departure, with the relative profitability of domestic versus international funds gradually redirecting the investment streams. It assumes a fixed share of current domestic and international wealth reallocation as well as new savings. In the second stage, an international trust allocates the investment streams over different regions on the basis of past and current relative profitability (including risk premiums). The sum of domestic and international capital streams determines the funds available for investment and, therefore, the change in capital stock and the production of the capital goods sector in each region.

Improved tariff data and incorporation of tariff shocks

The GTAP database does not incorporate the latest tariff data, including bound and applied tariff data, which are important to analyse the impacts on India and the rest of the world of multilateral and/or bilateral trade liberalization. We have incorporated six-digit import tariff data of the Tariff Analytical and Simulation Tool for Economists (TASTE) program at the lowest aggregation level (Horridge and Labordo, 2008) and incorporated a novel procedure to calculate the applied tariffs of given bound tariffs rates, the latter often being reduced in negotiations. The resulting tariffs and tariff shocks will be discussed in the section on free-trade scenarios.

Scenarios

The scenarios help policy makers, researchers and other stakeholders to envision what the future may look like and guide the formulation of policies that are contingent on future expectations. They are by no means accurate reflections or predictions of the future, but present storylines. In this chapter, we have analysed the impacts of a set of alternative growth and trade liberalization scenarios on the Indian economy and the rest of the world. The first set of growth scenarios includes a 6%, 8% or 10%

annual growth path for the Indian economy, 8% being the growth realized by India in the baseline scenario. The second set of trade liberalization scenarios includes a Free Trade Agreement (FTA) with the EU or a (multilateral) WTO agreement. The results of each of these sets of scenarios have been presented relative to the baseline scenario which reflects the common expectations on how the (global) economy will develop with no new policies being implemented. Since the base year of MAGNET is 2007, the baseline scenario has been run for the period 2007–2010, and then in consecutive 5-year periods forward to 2030. Below follows a discussion on the most important assumptions underlying each of the scenarios.

The reference scenario: the baseline

The reference scenario is a 'business-as-usual' scenario and reflects a future in which major socio-economic drivers will follow the current trends. It assumes that there will be no major policy changes (e.g. WTO agreement, biofuels, etc.). Furthermore, yields will keep on increasing at the same pace as in the past and climate change is assumed not to have a significant impact on agricultural productivity and economic growth.

The GDP and population growth projections have been taken from USDA-ERS (2012), which assumes a return toward long-run steady growth after the global recession and financial crisis, and decreasing population growth across the world. The labour supply is assumed to follow the growth path of the population, and the natural resources will grow with 25% of GDP growth. For India, the labour and GDP growths are based on the assumptions in the national CGE model of India. Specifically, the labour supply is assumed to grow, with 8.2% for the period 2010–2015 and 6.5% for the period 2015–2020, after which it will decrease to 4.7% for 2020–2025 and 3.9% for 2025–2030. The GDP in India is assumed to grow by 8% per year over the whole period. Capital and land supply are the endogenous outcomes for the model and therefore do not need to be projected forward. Land productivity (i.e. yield) projections have been derived from IMAGE (Integrated Model to Assess the Global Environment) model and are based upon FAO projections up to 2030 (Bruinsma, 2003; MNP, 2006). The technological progress is assumed to be labour-saving and faster in manufacturing than in agriculture, and faster in agriculture than in services. These assumptions are in contrast with the standard assumptions in most MAGNET papers till now, but are consistent with the more pessimistic views about the future of agricultural productivity as represented by predictions of stable or even rising real agricultural prices in the future.

Growth scenarios for India

The alternative growth scenarios for India are meant to illustrate how they affect India and the rest of the world. The annual Indian growth has been varied at 6%, 8% (as in the baseline) and 10%, whilst keeping the growth of the rest of the world the same. The higher Indian growth is assumed to stem from technological progress, distributed over sectors and inputs in the same way as in the baseline scenario. They can thus be interpreted as alternative baselines, in which India is doing better and better as the growth percentage increases compared to the rest of the world.

Free trade scenarios: a multilateral WTO agreement and a bilateral free trade agreement with the EU

The trade policy scenarios reveal the respective impacts of bilateral and multilateral trade agreements on India and the rest of the world. Here, we have investigated the impacts of a free trade agreement between India and the EU and we have contrasted it with a multilateral agreement under the WTO. We have focused on tariff liberalization in agriculture and industry, excluding services. The impacts of non-tariff measures and trade liberalization in services have been taken up in other studies (e.g. Ecorys, 2009). Our study, instead, provides more details in agriculture.

Description of tariff barriers

The tariffs used for this analysis have been taken from the MacMap database developed by the International Trade Centre (ITC) at the six-digit level of the Harmonized System (HS).[1] The HS comprises approximately 5300 article or product descriptions. There are often significant differences between bound tariffs, i.e. the maximum allowable Most Favourite Nation tariffs that WTO members have agreed upon, and applied tariffs, that traders actually face. This is also true for India vis-à-vis the rest of the world and the EU in particular. The existence of such a gap (known as the 'binding overhang') provides the Indian government with significant policy space with respect to trade and agricultural prices. It allows India to raise and lower tariffs in response to world price changes and changing conditions in the domestic economy. Being an emerging or developing economy, India qualifies for the EU's generalized system of preferences (GSP), which offers duty-free access for imports from developing countries (non-sensitive products), or reductions in the otherwise applicable standard tariffs (sensitive agricultural products). When certain products grow in market share, the GSP preferences are abolished as soon as a certain threshold is reached (12.5% or 15%, depending on the product category). This is the case for Indian textiles (Chapter 50–00 of the HS). According to the latest statistics of the EU, 47.7% of the Indian exports are imported under the GSP.[2] In practice, preferential market access into the EU for India, however, is limited due to the exclusion of many, often 'sensitive', agricultural products and the often limited tariff reductions offered.

Figure 5.1 and Table 5.2 show the applied tariffs levied by India and the EU, respectively, for commodity groupings and at the commodity detail available in the model.

Figure 5.1 shows that the average tariffs levied by India are much higher than those levied by the EU and that, for both India and the EU, the tariffs are higher on primary agriculture than on industry. No tariffs on services are available in the database. Especially for processed agricultural commodities, the Indian tariffs are much higher on the products imported from the EU than the tariffs on commodities imported from the rest of the world. The EU levies higher taxes on imports from India than on those from the rest of the world for primary agricultural commodities, although these tariffs remain far below 10%, while Indian tariffs are around 35%.

Table 5.2 shows the applied tariffs at a more detailed level, i.e. the commodity aggregation as used in the simulations, excluding the commodities with zero tariffs, and reveals that they differ greatly. It shows that the high tariffs levied by India on processed agricultural goods from the EU (Fig. 5.1) especially

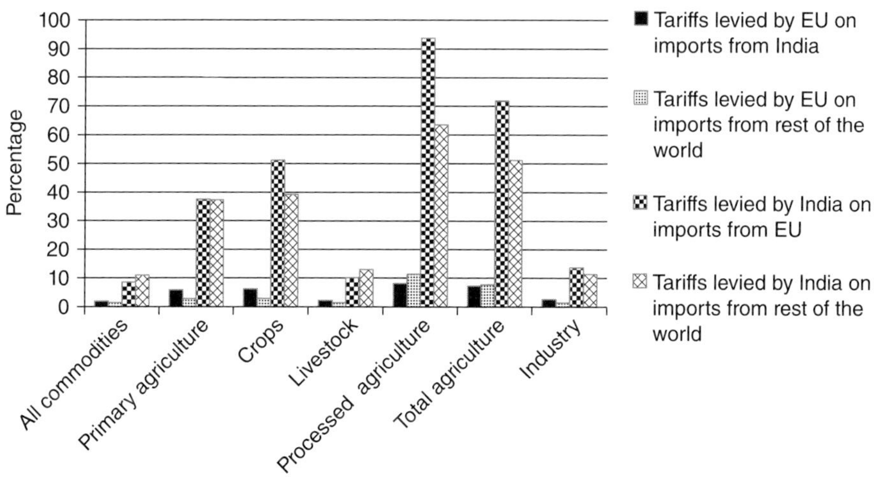

Fig. 5.1. Applied import tariffs by the EU and India in 2007.

Table 5.2. Detailed applied import tariffs by the EU and India in 2007.

	Tariffs levied by EU on imports from:		Tariffs levied by India on imports from:	
Commodity	India	Rest of world	EU	Rest of world
Rice	13.8	18.6	58.2	41.9
Wheat	5.4	5.0	99.5	99.3
Other cereal grains	10.2	3.1	0.0	21.0
Oilseeds	0.0	0.0	29.0	46.8
Vegetables, fruits, nuts	3.0	4.6	47.7	33.8
Plant-based fibres	0.0	0.0	13.3	9.8
Other crops	1.5	0.9	25.2	53.5
Cattle, sheep, goats, horses, wool	0.1	1.2	18.1	14.8
Pigs, poultry	3.4	1.7	5.3	6.6
Dairy products	3.1	23.9	32.2	30.1
Sugar	49.8	71.6	69.7	89.5
Vegetable oils products	1.1	4.6	38.6	57.7
Meat	10.3	29.5	29.9	14.9
Other foods	6.2	5.8	38.5	35.1
Beverages and tobacco	14.9	5.6	146.0	63.9
Fishing	3.5	2.8	29.3	12.3
Forestry	1.5	0.1	9.2	6.6
Crude oil	0.0	0.0	0.2	9.9
Petroleum, coal products	0.0	0.3	14.8	13.9
Natural gas mining	0.0	0.0	0.0	10.0
Coal	0.0	0.0	8.3	32.0
Chemical industry	0.5	1.7	15.7	14.4
Labour-intensive manufacturing	2.0	2.8	14.2	12.6
Capital-intensive manufacturing	2.6	2.0	13.0	11.6
Other services	0.0	0.0	0.0	0.0
Textiles	7.4	3.7	15.8	15.5
Other minerals	0.0	0.0	14.9	6.7

stemmed from high tariffs on beverages and tobacco products (applied tariff rate of 146% for the EU relative to 64% for the rest of the world).

A Free Trade Agreement between India and the EU

In a Free Trade Agreement (FTA) all tariffs between the EU and India are abolished, except for tariffs on sensitive products, i.e. products that are particularly susceptible to competition from imports from other country suppliers. Sensitive products for India are dairy products, animal products, sugar, spirits and wines and honey, while sensitive products for the EU are sugar, rice, cattle, beef and non-ruminant animal products. The tariff reductions are summarized in Fig. 5.2.

The average import tariff rate by the EU for commodities from India has been reduced from 2.0% to 0.2%, and the average import tariff rate by India for commodities from the EU has been reduced from 8.6% to 0.7%. The Indian import tariff reductions for crops and processed agricultural commodities are very substantial. Indian import tariffs for industry have been reduced from 13.7% to 0.3%. The EU tariff reductions are minor compared to these shocks, so the reduction in import tariffs is happening mainly on the Indian side. This can be explained by much higher initial rates of protection by India compared to the EU.

If we look into more details for primary agricultural commodities (Fig. 5.3), we see that reductions in Indian import tariffs are especially large for vegetables, fruits, nuts (45%), and to a lesser extent for oilseeds, plant-based fibres and ruminant animal products. For industry, the differences in tariff changes are much less significant; for India tariffs on industrial products have been reduced by about 15%, and for the EU the largest reductions are in tariffs on textiles (7%) and on other manufacturings (2%).

A WTO agreement

The WTO scenario follows the Falconer proposals and is summarized in Table 5.3. As shown, current bound tariffs are divided in bands of tariffs depending on the size of the tariff. Since rich countries have, on average, lower tariffs and more possibilities to reduce tariffs, the tariff cuts are higher and the tariff bands are smaller. Some commodities are exempted from tariff cuts because they fall under the sensitive category. Sensitive product exemptions are implemented for the EU (sugar, cattle, other agricultural products), the North American Free Trade Agreement (NAFTA) region (sugar, dairy) and India (rice and sugar). If a product is identified to be sensitive, only one-third of the tariff cuts, as defined in the Falconer proposal, are applied.

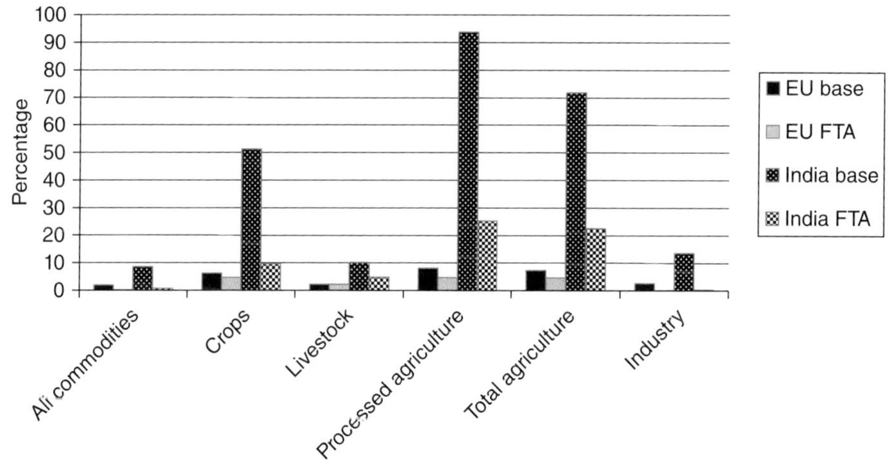

Fig. 5.2. Bilateral import tariffs in an EU–India free trade agreement.

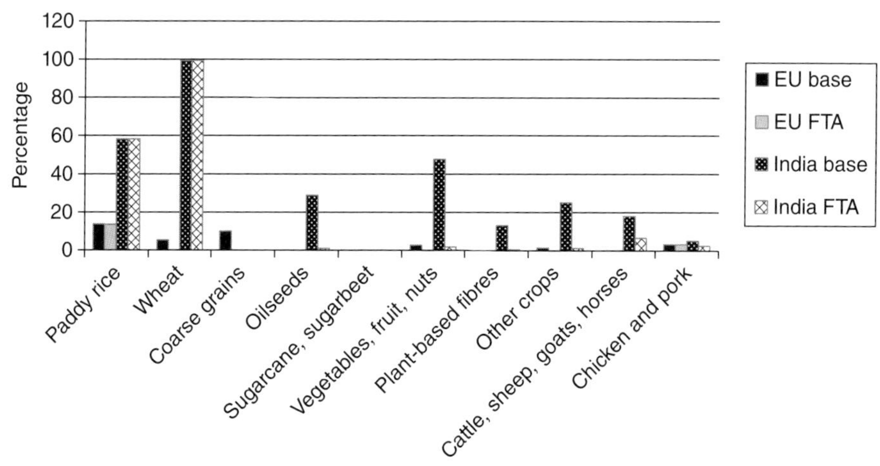

Fig. 5.3. Bilateral import tariffs on primary agricultural commodities in an EU–India free trade agreement.

The projected change in applied tariffs resulting from the reductions in bound tariff rates in 2015 and 2020 are shown in Figs 5.4 and 5.5, respectively, and in Table 5.4.

Considering the tariff changes by the EU for imports from India, we see that in general the WTO tariff reduction is smaller than the FTA tariff reduction, except for livestock products (Fig. 5.4). This is because in the FTA, the tariffs for non-sensitive commodities are reduced by 100%, while for the non-sensitive products in the WTO agreement, the bound tariffs are reduced by much less than 100%, where the effective applied rate reduction may even be smaller if the applied tariffs are below the bound tariffs (i.e. in the presence of a binding overhang).

Table 5.3. A multilateral reduction in tariffs under a WTO agreement.

Developed countries (including EU)		Developing countries (including India)	
Band	Tariff cut	Band	Tariff cut
<20	50	<30	33
20–50	57	30–80	38
50–75	64	80–130	43
>75	70	>130	47
Average	54	Average	36

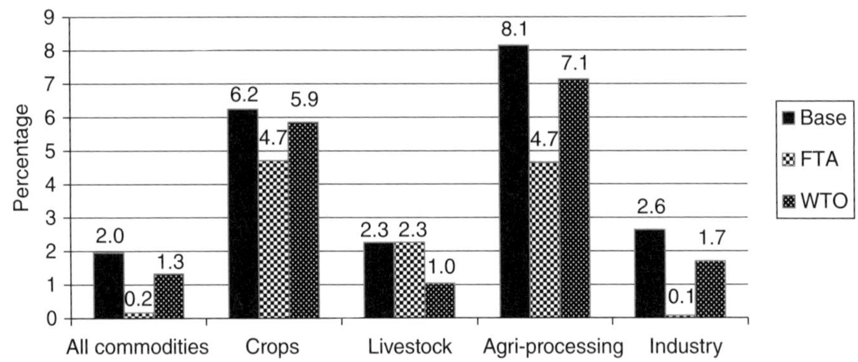

Fig. 5.4. EU applied import tariffs for commodities from India, 2015.

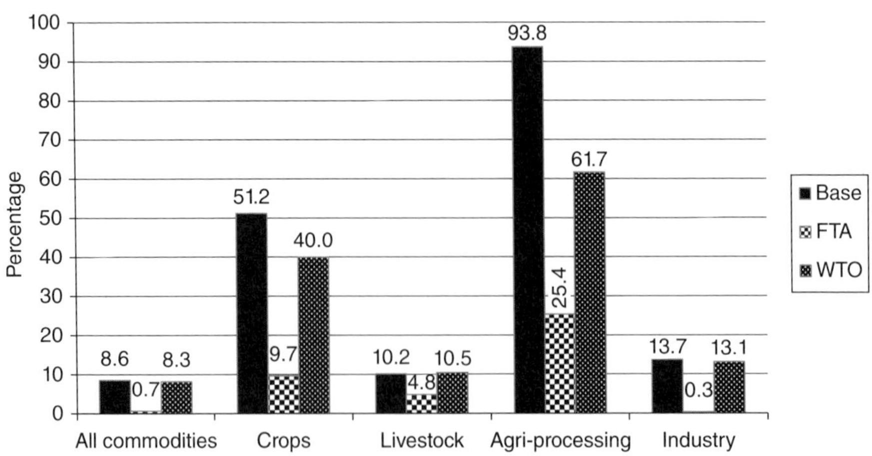

Fig. 5.5. Indian applied import tariffs for commodities from the EU, 2015.

With respect to Indian import tariffs for EU products, the reduction is larger for all commodities under a FTA compared to a WTO agreement, and also compared to the EU's import tariff reductions under a FTA (Fig. 5.5). The WTO agreement implies much smaller applied tariff reductions for India, also compared to those of the EU, as India has a much higher binding overhang; a large reduction of bound tariffs has only a small impact on the applied tariffs of India. Furthermore, for developing countries such as India, all bound tariff reductions are below 50%, while they are all above 50% for the developed countries. At a more detailed level it can be seen that as a consequence of changes in weights, the average tariff rate for livestock actually becomes higher with a WTO agreement than in the baseline scenario; this is the consequence of a faster increase in imports of commodities with a higher tariff rate compared to those with a lower tariff rate.

The general pattern is the same for detailed agricultural commodities (Table 5.4); in general, the WTO agreement gives a much smaller reduction in tariffs for Indian imports from the EU than the FTA gives. Only commodities that can be treated as sensitive under a FTA, but not under a WTO agreement like wheat and dairy products, keep higher tariffs under a FTA.

Table 5.4. Indian applied import tariffs for agricultural imports (including processing) from the EU under different scenarios.

Commodity	Base scenario	FTA	WTO agreement
Rice	58	58	50
Wheat	99	99	58
Oilseeds	29	1	29
Vegetables, fruit, nuts	48	2	31
Plant-based fibres	13	1	13
Other crops	25	1	23
Sheep, goat, cattle	18	7	18
Chicken and pork	5	3	5
Dairy products	32	32	28
Sugar	70	70	70
Vegetable oil	51	19	46
Meat	30	10	29
Other foods and feed	39	4	32
Beverages and tobacco	146	146	81
Fish	29	28	29

Results and Discussion

In order to understand the impacts of alternative growth and trade policy scenarios for India and the rest of the world, it is crucial to understand the developments in the baseline scenario. Specifically, this will make clear what drives the economic growth in India in the first place and how the economy will change over time and in a global context. The outcomes under the baseline scenario are first discussed, followed by the results of the alternative growth and trade policy scenarios.

The baseline scenario: what drives Indian growth and how does it affect the economy?

Table 5.5 summarizes the patterns emerging in the baseline scenario in 2030. It shows that India will be relatively fast growing compared to the rest of the world, mostly due to higher technological progress in the manufacturing sector (both food processing and other industry). As a result, in terms of Indian production and exports, industry, including agri-processing, will become more important at the cost of primary agricultural commodities. The latter will also be constrained by land and water availability, and will display rising unit costs and prices over time. Part of the explanation also lies in the changing Indian demand patterns, which will shift away from food towards manufactured goods and services (not

Table 5.5. Baseline trends: percentage change in 2030 relative to 2010 (MAGNET, Woltjer *et al.*, 2014).

Particular	India	EU	Rest of the world
GDP per capita	272	42	71
Land price	361	6	155
Fertilizer input per hectare	320	15	101
Production per hectare	79	15	53

shown). The agricultural sector, however, will continue to grow, and given preference shifts and developments in comparative advantages, the sheep, goats and chicken, wheat, milk and cotton sectors benefit especially (not shown). With respect to trade relations, in terms of agricultural and overall trade, the EU would become relatively less important for India, with the exception of agricultural imports, where the EU's importance as a source region will rise slightly. More detailed figures show that even in 2030 there will be potential for growth in exports from India to the EU for vegetables, fruits, plant-based fibres, other crops and animal products (not shown). Increased demands for land and so rising land prices would lead to an intensification in land-use, which will ensure that it would be necessary for India to grow fast without becoming too dependent on the world market for food. The consequences of the development of India for incomes would depend to a large extent on the possibility to increase mobility of labour from rural to urban areas.

The changes in the Indian economy observed over time in the baseline scenario have generally followed the structural changes undergone by the industrialized world and are in line with the findings from other studies (Table 5.6; Alexandratos and Bruinsma, 2012; Binswanger-Mkhize, 2012; Binswanger-Mkhize *et al.*, 2012).

The implications of Indian growth

Higher growth in India implies a faster transition towards a developed society, thus strengthening the patterns observed in the baseline scenario. The cumulative pattern of technological change would not only lead to a much higher GDP per capita (a 6% annual growth leads to a 31% lower GDP per capita in 2030 compared to the baseline, whereas a 10% annual growth leads to a

Table 5.6. Baseline sectoral trends in different indicators for India, EU and rest of the world: % change in 2030 relative to 2010 (MAGNET, Woltjer *et al.*, 2014).

	Sectoral change 2010–2030 (%)				
Place	Total	Primary agricultural products	Agri-processing	Other industry	Services
Total factor productivity					
India	137	79	272	300	67
EU	21	21	30	41	13
Rest of the world	39	50	65	74	22
Price					
India	−14	31	−11	−22	−8
EU	4	−3	−3	0	8
Rest of the world	0	10	−4	−4	3
Production volume					
India	352	79	86	444	350
EU	41	16	17	40	44
Rest of the world	111	66	60	136	97
Indian exports to					
EU	359	−58	74	520	129
Rest of the world	502	−24	120	629	207
Indian imports from					
EU	186	406	48	150	245
Rest of the world	228	363	124	215	314
Percentage share of sectoral income in GDP (absolute)					
2010	100.0	14.5	1.6	27.4	56.5
2030	100.0	10.0	0.5	25.9	63.5

44% higher GDP per capita), but may also change the relative competitive positions of different sectors, benefiting industry and agri-processing sectors relatively more. Combined with consumption patterns changing away from food towards industrial commodities and services (and within food towards more processed food), the net trade position of sectors would change quite considerably (Fig. 5.6). With a rise in annual growth in India from 6% to 10%, the industry would change from a net importing position to a net exporting position, whereas the services sector would change from a net exporting position to a net importing position. Similarly, livestock and crops' net importing position would increase, whereas the relative net exporting position of agro-processing would fall.

The impact of Indian growth on production in the rest of the world would be relatively small. The largest impact would be felt in industry, with industrial production being 2.3% lower in the EU and 1.7% lower in the rest of the world in 2030 if India grows by 10% per year compared to 6% per year. The EU, as Africa, will benefit relatively more from rising agricultural imports by India than other regions.

The rise in imports by India may not be enough to satisfy the rising Indian demands for food so that the pressure on land would increase, resulting in fast rising land, crop and livestock prices and leading to an intensification in the use of land (Fig. 5.7).

The crop prices in India are shown to increase by 50% more compared to the rest of the world over the period 2010–2030 in the case of 10% annual growth in India. This may necessitate a loosening of import restrictions, and perhaps may lead to a stronger reaction of imports on changes in relative domestic prices in the future.

The intensification in the use of land is particularly visible in the use of fertilizer and capital, the former rising by 70% and the latter by 33%. The labour use may be reduced as a consequence of higher wages and labour-saving productivity may increase. Agricultural investments to enable a higher land productivity and, more generally, technological progress in agriculture would be another way to mitigate the rising land and food prices and may soften the impacts that a faster Indian growth has on resource scarcity. This finding is corroborated by the observation of Binswanger-Mkhize *et al.* (2012), that if imports are constrained to levels only

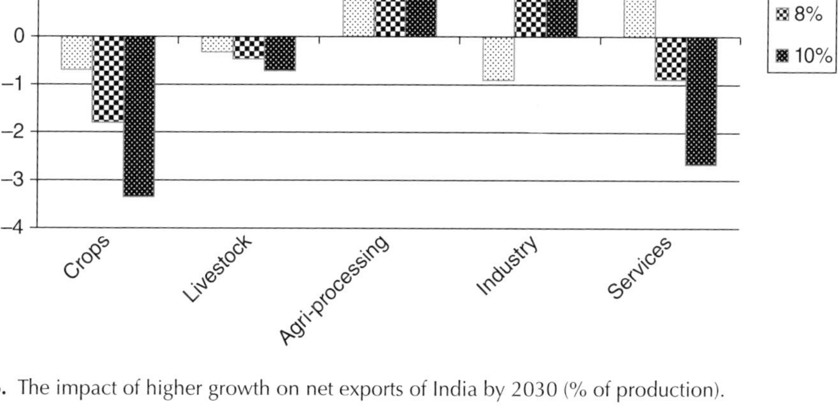

Fig. 5.6. The impact of higher growth on net exports of India by 2030 (% of production).

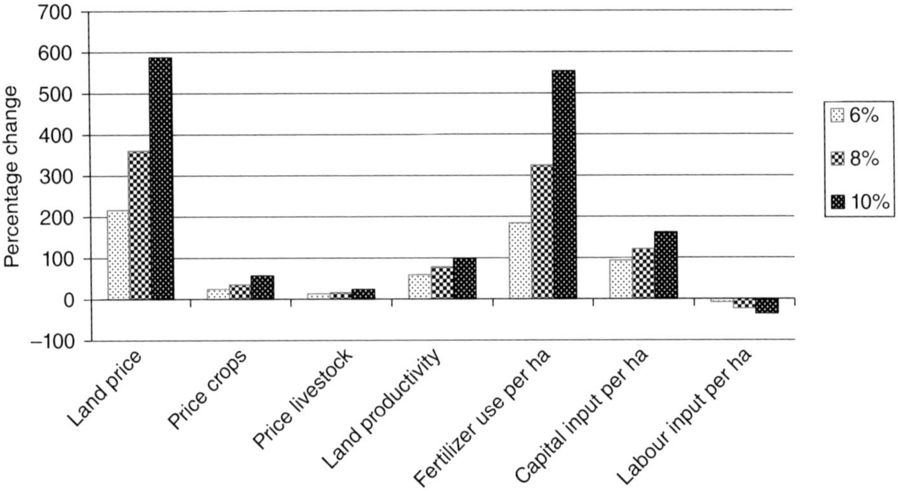

Fig. 5.7. The impact of higher growth on land scarcity and intensification (% change over 2010–2030).

slightly higher than the present levels, an agricultural growth rate of 4% or more is needed to support GDP growth rates in excess of 8%. Climate change, not modelled in this study, would further necessitate such growth.

The implications of multilateral and bilateral Free Trade Agreements for India and the rest of the world

This section presents the results of the FTA and WTO scenarios. Since the implementation of the FTA and WTO agreement is in the period 2010–2015, we will present results for 2015, showing the impact of these agreements in difference from the baseline outcome in 2015. The longer term impacts beyond 2015 are included while discussing the welfare impacts.

For both the EU and India GDP, the impacts of a FTA are relatively small – an increase in real GDP of US$5 billion (0.23% of GDP) for India and of US$588 million (0.003% of GDP) for the EU. The GDP effect of a WTO agreement is a little bit smaller than of a FTA for India (US$4 billion, being 0.17% of GDP), but a little bit more so for the EU (US$9 billion, being 0.05% of GDP). According to Ecorys (2009), most of the gains for India will stem from non-tariff barrier reductions, which are not included in our model. Non-tariff barriers will not be very important for the primary agricultural commodities, because these are relatively homogeneous, and much more important for processed foods like dairy products.

Table 5.7 shows that total imports by India of products from the EU would change only slightly in the presence of a WTO agreement (2.7%), and would increase considerably (by 52%) with a FTA in 2015. Especially, the imports of processed agricultural products could increase greatly (a close to eight-fold increase), while the import of industrial commodities would almost double. In relative terms, crop imports will also show a large (three-fold) increase in the presence of a FTA. In general, the observed patterns are in line with what we would expect from the size of the tariff shocks discussed before, with WTO tariff reductions by India being relatively minor, and FTA tariff reductions being considerable and especially for crops, agri-processing and industry. Table 5.7 also shows that a large part of the increase in Indian imports from the EU will not influence the total Indian imports that much. About half of the changes in imports from the EU would be trade-diversion, where India imports more from the EU at the cost of other

Table 5.7. Volume of Indian exports and imports to and from the EU and the rest of the world (% difference with baseline in 2015).

Exports/Imports		EU FTA	EU WTO	Rest of world FTA	Rest of world WTO
Exports	All commodities	15.7	2.4	1.0	0.6
	Crops	6.1	1.8	–1.8	–0.2
	Livestock	–0.3	2.8	–0.4	3.3
	Agri-processing	12.7	7.6	–0.5	3.2
	Industry	23.5	3.7	1.6	0.8
	Services	–1.0	–0.9	–1.3	–0.3
Imports	All commodities	51.6	2.7	–6.4	0.6
	Crops	202.4	54.7	–0.7	16.3
	Livestock	34.4	5.8	–2.8	–5.6
	Agri-processing	686.2	38.0	–13.6	0.6
	Industry	82.3	3.5	–7.5	0.5
	Services	0.4	0.9	0.9	–0.1

countries (rest of Europe 10%, rest of Asia 9%, America and Oceania 6% and Africa 3%). As a consequence of cheaper imports, the average market price of Indian commodities would decrease and this will give the opportunity to export more, also to non-Indian regions. This pro-competitive effect would be strongest for industry, where India already has a strong comparative advantage. These findings are corroborated by Ecorys (2009). As India would be importing less from other regions in case of a FTA with the EU, and exporting more, these regions will try to sell their (industrial) commodities to the EU. Ecorys (2009) has found very limited third-country effects on the neighbouring countries, mainly due to limited export volumes and the fact that some of India's neighbouring countries already enjoy preferential treatment through GSP+ and Everything But Arms (EBA). The remaining negative impact would mostly be due to losses in market shares in textiles, important for those countries and largely overlapping with India's export interests.

The most sensitive part of a trade agreement is its implications for production of important sectors in the economy. Table 5.8, which is sorted on impact of a FTA on India in 2015 and has only included commodities where the production effect is more than 0.1%, shows that, in the case of a FTA with the EU, vegetable oil and oilseeds is the main sector that would lose in India, while textile and plant-based fibres is the main sector that would gain, at the cost of EU textile production. Manufacturing would gain slightly, while meats would lose out slightly. These outcomes are in line with the Ecorys (2009) study, which has also found that the greatest gains would be in clothing and leather products with around 20% increases in output in the short-run as these sectors benefit from improved market access into the EU. Ecorys has also found other industry expanding (light and heavy), but by less, and the primary agricultural sectors contracting (by up to 0.5% in the short-run). The surge in machinery and equipment production and export would support the rise in investment in India and the subsequent dynamic gains. The cheaper imports are also found to fuel the expansion of the domestic industry. For the EU, the vegetable oil sector would be the main winner, while the textile industry may be a loser.

For a WTO agreement, the positive effects for textile industry in India are less than for a FTA. This is both because the tariff cuts are smaller and also competing producers obtain better tariff rates. India may lose on wheat production, not being able to keep wheat as a sensitive product, while India would focus more on chicken meat and less on sheep and goat meat. For the EU a WTO agreement would mainly reduce the animal cattle production, and to a lesser extent, refined vegetable oil production.

Table 5.8. Percentage change in volume of production as a consequence of trade agreements (% difference with baseline in 2015).

	FTA			WTO		
Sector	India	EU	Rest of world	India	EU	Rest of world
Textile	5.4	−1.7	−0.2	1.8	−2.9	0.5
Plant-based fibres	3.6	0.1	0.1	2.7	2.4	−0.1
Biodiesel	2.6	0.0	0.1	−4.6	−1.4	1.8
Fertilizer	0.9	−0.3	0.0	−2.0	−0.1	−0.2
Labour-intensive manufacturing	0.7	−0.1	0.0	0.1	−0.4	0.2
Construction	0.5	−0.1	0.0	0.1	0.2	0.1
Capital-intensive manufacturing	0.4	0.1	−0.1	0.1	−0.1	0.1
Other foods	0.2	−0.1	0.0	0.4	0.3	−0.1
Wheat	0.2	−0.1	0.0	−5.4	0.7	0.1
Animal feed	0.1	−0.1	0.0	1.9	−3.4	0.5
Beverages/tobacco	0.0	0.0	0.0	−0.4	0.2	0.0
Chicken and pork	0.0	−0.1	0.0	0.5	−0.6	0.0
Gas distribution	0.0	−0.1	0.0	−0.5	0.2	0.0
Rice	0.0	−0.2	0.0	−1.1	1.9	−0.8
Coal	0.0	0.0	0.0	−0.6	0.2	0.1
Milk	0.0	0.0	0.0	0.4	−0.6	0.2
Sugarbeet/cane	0.0	−0.1	0.0	1.0	−0.2	−0.1
Sugar	−0.1	−0.1	0.0	1.4	−0.3	−0.6
Other meat	−0.3	−0.1	0.0	4.3	−1.2	0.2
Minerals	−0.5	1.4	−0.1	0.0	0.1	0.0
Cattle/sheep/goat meat	−0.5	−0.1	0.0	−2.6	−19.9	2.8
Chemicals	−0.6	0.2	0.0	−0.2	0.1	−0.2
Crude vegetable oil	−1.1	−0.1	−0.2	−0.2	−0.6	0.1
Oilseeds	−1.6	0.6	−0.1	0.7	1.3	−0.9
Refined vegetable oils	−6.0	5.7	−0.3	0.8	−3.7	0.1

The contraction of several primary agricultural sectors in India may imply farmers losing production and income, which has implications for food security, especially for the marginalized farmers. In this context, government support may be required. In general, however, the impacts are relatively small, although they may be higher for some specialized commodities. Note that non-tariff barrier reductions, which may be quite considerable, have not been considered, nor have longer term dynamic effects been included at this stage.

Figure 5.8 gives a rough idea on the influence of the trade agreements on food security over time. The rise in unskilled labour wages in agriculture (as an indicator of income for the poor in rural areas) and non-agriculture (as an indicator of income for the poor in urban areas) sectors has been compared with the price increase in food crops (cereals, vegetables and fruits). The graph shows that both the agreements are beneficial to the rural as well as urban poor, because the wages for unskilled labour increase more or decrease less than prices of essential food commodities (although we must be aware that textile prices rise more than wages under a WTO agreement), where the real wages for unskilled labour tend to be able to buy about 1% more food than without the agreements. Because of the dynamics in the labour market, the benefits are larger in the long-term than in the short-term.

Conclusions

A faster growth of India implies a faster transition towards a developed society. The consumption patterns have depicted a change away from food towards industrial commodities and services, as does production. Because technological development is

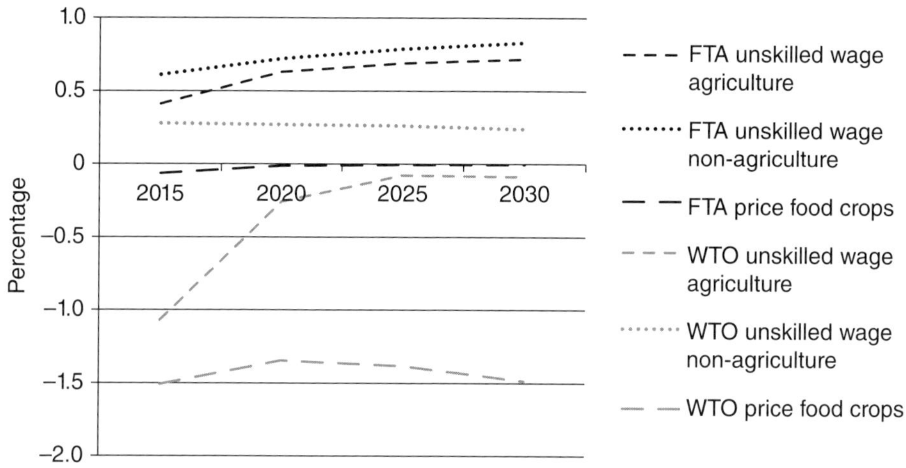

Fig. 5.8. Food security impacts: percentage difference in unskilled labour wages and price of food crops compared to baseline scenario.

faster in industry than in services, there is a tendency of the services share in national income to increase. Related to the fast technological change in industry, India would become a more important net exporter of industrial commodities and a net importer of services by 2020. With respect to agriculture, the restrictions on land and water availability, in combination with a rise in demand for foods, imply that the net imports of crops would increase much more in the times to come. The rise in imports is, however, not enough to satisfy the rising Indian demands for food, so the pressure on land would increase, resulting in fast rising land and crop (and livestock) prices, much more than in the rest of the world. Rising land prices will lead to intensification in agriculture.

The exercise with the trade agreements has shown that trade agreements in general are good for food security, with the improvements increasing over time. A FTA with the EU will not influence food prices much, but will increase income of the poor as the wages for unskilled labour would rise. A WTO agreement has depicted more influence on agricultural prices because trade with the EU is only a small part of Indian agricultural trade, and this price reduction is larger than the reduction in unskilled labour wages in agriculture. The reduction in unskilled labour wages in agriculture will fade away over time, while the price reduction remains. A WTO agreement will also reduce the pressure on land in India by a small amount.

The results of the trade scenarios may underestimate the benefits of trade agreements, because non-tariff measures, measures in services and measures on foreign direct investment are not considered. The simulations have shown that pressures on land may become very high and that measures to reduce this pressure through investments in agriculture and further opening up trade are essential ingredients to reduce poverty and to prevent unnatural high food prices. With climate change, the effects of which are not considered in this study, this may become even more important.

Notes

[1] http://www.macmap.org

[2] http://trade.ec.europa.eu/doclib/docs/2010/march/tradoc_145945.pdf

References

Alexandratos, N. and Bruinsma, J. (2012) *World Agriculture Towards 2030/2050: The 2012 Revision*. ESA Working paper No. 12-03. Food and Agriculture Organization, Rome.

Banse, M., van Meijl, H., Tabeau, A. and Woltjer, G. (2008) Will EU biofuel policies affect global agricultural markets? *European Review of Agricultural Economics* 35, 117–141.

Binswanger-Mkhize, H. (2012) India 1960–2010: Structural change: The rural non-farm sector and prospects for agriculture. Paper presented at *Stanford Symposium Series on Global Food Policy and Food Security in the 21st Century*, 10 May.

Binswanger-Mkhize, H., Mukherjee, N. and Parikh, K. (2012) India 2039 – Transforming agriculture: Productivity, markets, and institutions. *Global Journal of Emerging Market Economies* 4, 197–226.

Bruinsma, J. (2003) *World Agriculture: Towards 2015/30, an FAO Perspective*. Earthscan, London.

Chandrasekhar, P. (2012) The food crisis and India. *The Hindu*, 3 September.

Ecorys (2009) *Trade Sustainability Impact Assessment for the FTA between the EU and the Republic of India.* TRADE07/C1/C01 – Lot 1 Final Report. Available at: https://trade.ec.europa.eu/doclib/docs/2009/june/tradoc_143372.pdf (accessed 12 May 2016).

European Commission (2013) *India EU Bilateral Trade and Trade with the World.* Available at: http://ec.europa.eu/trade/policy/countries-and-regions/countries/india (accessed 25 July 2013).

Francois, J., van Meijl, H. and van Tongeren, F. (2005) Trade liberalization in the Doha Development Round. *Economic Policy* 20, 349–391.

Hertel, T.W. (ed.) (1997) *Global Trade Analysis: Modeling and Applications.* Cambridge University Press, Cambridge, UK.

Horridge, J.M. and Laborde, D. (2008) TASTE: A Program to Adapt Detailed Trade and Tariff Data to GTAP-related Purposes. GTAP resource 2666. Available at: https://www.gtap.agecon.purdue.edu/resources/res_display.asp?RecordID=2666 (accessed 25 July 2013).

Ianchovichina, E. and McDougall, R. (2001) Theoretical Structure of Dynamic GTAP. GTAP technical paper 17. Available at: https://www.gtap.agecon.purdue.edu/resources/download/1582.pdf (accessed 25 July 2013).

IMF (International Monetary Fund) (2012) *World Economic Outlook Database*. IMF, Washington, DC. Available at: http://www.imf.org/external/pubs/ft/weo/2012/02/weodata/index.aspx (accessed 25 July 2013).

IMF (International Monetary Fund) (2013) *World Economic Outlook Update*. IMF, Washington, DC.

MNP (Netherlands Environmental Assessment Agency) (2006) *Integrated Modelling of Global Environmental Change: An Overview of IMAGE 2.4.* Bilthoven, The Netherlands.

Rutten, M., Shutes, L. and Meijerink, G. (2013) Sit down at the ballgame: How trade barriers make the world less food secure. *Food Policy* 38, 1–10.

Stehfest, E., Van den Berg, B., Woltjer, G., Msangi, S. and Westhoek, H. (2013) Options to reduce the environmental effects of livestock production – Comparison of two economic models. *Agricultural Systems* 114, 38–53.

UN (United Nations) (2012) World Population Prospects: The 2012 Revision. Online database. Available at: http://esa.un.org/unpd/wpp/unpp/panel_population.htm (accessed 25 July 2013).

USDA-ERS (United States Department of Agriculture – Economic Research Service) (2012) International Macroeconomic Data Set. Available at: http://www.ers.usda.gov/data-products/international-macroeconomic-data-set.aspx#BaselineMacroTables (accessed 25 July 2013).

Woltjer, G. (2013) *Report on Global CGE model, TAPSIM Deliverable 6.1*. LEI/WUR, The Hague, The Netherlands, 85 pp.

Woltjer, G., Kuiper, M., Kavallari, A., Meijl, H. van, Powell, J.P., Rutten, M.M., Shutes, L.J. and Tabeau, A.A. (2014) *The MAGNET Model, Module Description.* LEI Report 14-057. LEI–Wageningen UR, The Hague, The Netherlands.

World Bank (2012) *Food Price Watch August 2012.* Poverty Reduction and Economic Management Network, The World Bank, Washington, DC. Available at: http://siteresources.worldbank.org/EXTPOVERTY/Resources/336991-1311966520397/Food-Price-Watch-August-2012.pdf (accessed 25 July 2013).

WTO (2012) WTO Statistic Database: Trade Profile for India. Available at: http://stat.wto.org/CountryProfile/WSDBcountryPFExport.aspx?Language=E&Country=IN (accessed 25 July 2013).

6 India: Economic Growth and Income Distribution in Rural and Urban Areas

G. Mythili*
Indira Gandhi Institute of Development Research, Mumbai, India

Introduction

India has registered an impressive gross domestic product (GDP) growth of 7–8% after launching economic liberalization in 1991. Reforms were initiated in the public sector, financial sector and investment and trade regimes and this helped in better integration with the global economy. However, its crucial sector, agriculture, which is the source of livelihood for nearly two-thirds of the population, has lost its momentum. The sector is lagging behind at less than 3% growth and its share in GDP is falling sharply over the years. Moreover, the economy has witnessed a significant structural transformation in both agricultural production and consumption in the past decade or so.

The performance of the economy over the years has been marked by higher rates of savings, investments, and improvements in many other macroeconomic indicators. The investment to GDP ratio went up to 34% in 2006 coupled with increases in domestic and national savings. A notable growth in trade was also witnessed. As regards exports of goods and services, their ratio over GDP increased from a mere 6% of GDP in the early-1980s to 23% in 2006, while the share of imports of goods and services in GDP rose from 8.7% in 1981 to around 26% in 2006. It is also to be noted that due to trade reforms, tariffs were drastically reduced on consumer goods as compared to intermediate and capital goods.

Over the past two decades, there has generally been a reduction in poverty in both rural and urban areas, particularly visible during the post-reform period. If the growth continues for the next decade or so, what would be its impact on poverty and income inequality? This is the question we have addressed in this chapter.

Growth, Poverty and Inequality: A Review

Several studies have analysed the poverty trend in India and its association with other economic variables using the data of the National Sample Survey Organization (NSSO). During the post-reform period, with increasing GDP growth, the head-count ratio has seen a significant decline based on the analysis for three periods, 1983, 1993/94 and 2004/05 (Mahendra Dev and Ravi, 2007). On studying the period 1993/94 to 2004/05, it was revealed that the reduction was more significant between 1993/94 and

* Corresponding author, e-mail: mythili@igidr.ac.in

1999/2000 than between 2000/01 and 2004/05, which could be attributed to the increase in GDP growth. However, as per the statistics provided in this study, inequality based on Gini coefficients has increased during the post-reform period in both rural and urban areas, and with a higher rate in urban areas. The study has revealed that even though the growth of agriculture has decreased in the second sub-period of post-reform, factors such as lower inflation and higher employment growth in the non-farm sector were cited as the reasons for a higher rate of decline in rural poverty in this period. The relative food-grain prices, which recorded a significant decline in this sub-period, were also responsible for such a decline in poverty. There is overall consensus among the studies that the inequality significantly increased between 1993 and 2005 (Table 6.1). However, no visible changes in inequality were seen between the two sub-periods, viz. 1993–2000 and 2000–2005, and no significant difference was seen in the rate of decline of poverty between pre-reform and post-reform periods in the rural areas. However, in the urban areas, the rate of decline has become slower in the post-reform period. On decomposing the changes in poverty, it is revealed that the increase in inequality has slowed down the rate of poverty reduction in the post-reform period.

Over the years, the Government of India has come up with many poverty alleviation programmes and the reduction in poverty could be partly attributed to such initiatives by the government.

The analysis of growth in wages in rural and urban areas for the three periods, 1983/84, 1993/94 and 2004/05 (Sarkar and Mehta, 2010), has revealed that the wages of the regular workers category have increased faster in the rural than urban areas between 1983 and 2004. The casual workers' wages also followed a similar pattern, but at a rate slower than the regular workers' wages (Table 6.2). During the post-reform period, the casual workers' wages have shown a faster increase in the rural areas.

Even though the regular workers' real wage (base 1993/94) in the non-agriculture sector is double that of the wage in the agriculture sector, the post-reform period has seen a higher growth rate of wage in the agriculture as compared to the non-agriculture sector. A decline in the attached labour practice has been cited as the main reason for the faster growth in agricultural wages (Sarkar and Mehta, 2010).

It was also documented that in the urban areas the share of consumption expenditure in the bottom 30% segment (based

Table 6.1. Poverty ratios (head-count ratio) and Gini coefficients in India (based on surveys of 30-day uniform reference period) (Mahendra Dev and Ravi, 2007).

	Poverty head-count ratio		
Population type	1983	1993/94	2004/05
Rural areas			
Poor	45.76	37.26	29.18
Very poor	25.52	15.38	9.64
Urban areas			
Poor	42.27	32.58	26.02
Very poor	22.45	16.00	12.00
All India			
Poor	44.93	36.02	28.27
Very poor	24.79	15.54	10.32
Gini coefficient			
Rural India	0.3079	0.2855	0.3045
Urban India	0.3406	0.3431	0.3751

'Poor' are those who are below the poverty line and 'very poor' are those who are below 75% of the poverty line. Hence 'very poor' is only a subset of 'poor'

Table 6.2. Growth rate (percentage) in daily wages of regular and casual workers in rural and urban India, 1983–2004 (Sarkar and Mehta, 2010).

Area	1983/93	1993/2004	1983/2004
Regular workers			
Rural	3.43	2.89	3.15
Urban	2.57	2.50	2.53
Casual workers			
Rural	2.03	2.98	2.52
Urban	2.15	1.67	1.90

on MPCE) of the total population has been declining and that of the top 30% of the population has been increasing as per the data of various NSSO rounds covering the period 1973–2000. However, in rural areas, both the bottom 30% and top 30% segments have shown an upward trend (Chand, 2007).

As per NSSO 2004/05 data, a typical median urban household earns more than two times the income of a median rural household. Using the data of consumption surveys from 1951 to 1991, Ravallion and Datt (1996) derived a new series of poverty measures for urban and rural India. The study examined the impact of economic growth and sectoral composition on urban–rural poverty. It re-emphasized the significant role of rural economic growth in overall poverty reduction. It was also found both rural and urban poor benefited from rural economic growth, whereas urban economic growth not only works against the urban poor, but does not have any sizable impact on the rural poor as well. Sectoral classification of the analysis revealed that the secondary sector growth does not have much impact on the poor in both rural and urban areas, and the growth of primary and tertiary sectors in particular leads to poverty reduction in both rural and urban areas. The study has concluded that the urban economic growth fuelled by industrialization is not going to benefit the poor. This reiterates the significance of agricultural sector growth for the overall poverty reduction in the country.

Datt and Ravallion (1998) have shown that agricultural growth did provide benefits to the rural poor and the gain was mostly through wages and prices. Using the data of 24 rounds of National Sample Survey spanning 1958 to 1994, it was shown that increase in the average farm yield helped the poor in the form of higher agricultural wages and lower food prices. It was also revealed that the long-term effects are much larger. It was documented that inequality has increased in the post-reform period in both rural and urban areas, but with a higher rate in urban India. Analysis of data for the period 1993–2005, by dividing it into two periods, revealed that the second period of post-reform (2000–2005) was marked by a higher rate of decline in poverty as compared to the first sub-period (1993–2000). According to Jha (2000), the rise in inequality in the post-reform period was due to the increase in the share of output returns to capital in comparison to labour and changes in the sectoral composition in GDP and sectoral growth rates. Using the NSSO data of the 13th to 55th rounds, the analysis has provided empirical evidence to the hypothesis that the rise in inequality has diminished the poverty-reducing effects of higher growth.

From the literature, it was evident that the results on the relationship between growth, poverty and inequality were varied and not robust. There was also a lack of studies using the general equilibrium framework. Against this backdrop, this chapter attempts to simulate different GDP growth scenarios projected for the future block years 2009/10 and 2019/20 using Computable General Equilibrium models and 2006/07 as the base year. It has also looked into GDP growth impact on sectoral growth, household consumption, wages and income distribution.

Sectors Driving Economic Growth

In the world scene, the position of emerging economies such as India and China has been strengthening steadily over the past two decades. The share of India's GDP in the world, which was at 6% in 2002, is expected to move to 11% by 2025 (Virmani, 2005). The composition of growth is subject to significant changes. The service sector has steadily recorded an impressive growth due to increasing growth in communication, hotels and banking sectors. The industrial growth is led by the construction and capital-intensive manufacturing sectors. Agriculture has not only shown a declining trend in growth but has depicted high volatility in the year-to-year growth caused by highly fluctuating climatic parameters.

India has undergone a structural transformation in agriculture in the post-reform period. Diversification from food grains to other food crops has been quite prominent during the past two decades in Indian agriculture. Experts believe that this transformation is mainly demand-driven. In fact, diversification has helped the growth in agriculture by increasing the overall productivity per unit of land. Gokarn and Gulati (2006) have reported that the productivity of horticultural crops moved up from 7.5 t/ha to 8.9 t/ha during the post reform period till 2002/03, while that of food grains gained a meagre 0.2 t/ha, from 1.4 to 1.6 t/ha during this period. They have also cautioned that the structural rigidities across the regions can prevent its further growth because horticulture requires supportive measures such as a marketing network and infrastructure for processing and the participation of the private sector.

The decadal growth rates for the period ending 2006/07 were computed for the value added at factor cost (at current prices) of the broad sectors and are reported in Table 6.3. Economic growth was found to be mainly driven by four sectors: construction, capital-intensive manufacturing, transport and other services. The growth in the agriculture and allied sector was found to be the lowest, with food crops registering a meagre 5% annual growth at current prices.

Table 6.3. Annual growth rate from 1997/98 to 2006/07 – value-added at factor cost (at current prices). (Author's calculations based on the data extracted from the Directorate of Economics and Statistics, New Delhi.)

Sectors	Growth rate %
Food crops	5.04
Non-food crops	7.57
Dairy, poultry, fisheries and other animal products	7.81
Primary products	11.58
Agri-processing	10.53
Labour-intensive manufacturing	7.58
Petrochemicals	8.90
Capital-intensive manufacturing	14.42
Construction	16.76
Electricity	7.81
Transport	12.16
Other services	13.68

It has been argued that India's reform initiatives are mainly targeted to industry- and trade-related policies and hence these have not helped the agriculture sector. However, due to intersectoral linkages, the benefits of reforms in trade, industry and service sectors have benefited the agriculture sector also to some extent and there has been an increase in agricultural exports from India. The post-reform period is also characterized by a significant decline in the share of public investment in agriculture. The main reason is the high level of subsidy burden and the worsening of the fiscal gap in the union budget. The share of public sector in gross capital formation (GCF) fell from 51.3% in 1980/81 to 25.34% in 1998/99 (Gulati and Bathla, 2002). The share of public investment picked up after 2003, due to the thrust given during the 10th and 11th Five-Year Plans to boost agriculture. The decadal growth in public investment was negative during the decades of 1980–1990 and 1990–2000 and this increased to 15.74% during 2001–2009 (Mahendra Dev, 2012).

Methodology and Approach: Computable General Equilibrium Model

A review of the literature shows that several researchers have built Computable General Equilibrium (CGE) models for India in the past. These models vary in the underlying theoretical framework (i.e. neo-classical or structuralist), focus (short-run versus long-run), treatment of domestic and foreign goods (perfect substitutes versus imperfect substitutes), and details of institutional features of the economy, especially pertaining to the agricultural sector. From amongst them, four studies/models, i.e. Narayana *et al.* (1991), Parikh *et al.* (2002), Polaski *et al.* (2008) and Panda and Ganesh-Kumar (2008), were considered relevant for a detailed review.

After reviewing these models, the model by Panda and Ganesh-Kumar (2008) was chosen for study. It is a static CGE model based on the approach developed by Dervis *et al.* (1982). A distinguishing feature of this approach is that it treats the domestically produced goods and traded goods in a particular sector as imperfect but close substitutes using the Armington specification. The main features of this CGE model are:

- The model is based on a modified Social Accounting Matrix (SAM) for the year 2003/04 prepared by Saluja and Yadav (2006).
- Sectoral outputs are specified as Constant Elasticity of Substitution (CES) production functions with labour and capital as factors that determine the sectoral value added (see Appendix for production structure). Intermediate inputs are considered through Leontief type input–output coefficients obtained from the input–output table embedded in the SAM.
- Sectoral capital and total labour supply are assumed to be fixed. Factors are mobile across sectors, except that mobility of land and capital for agriculture sector farming is restricted within the agriculture sector and the same in the case of non-agriculture sectors. Labour is allowed to be mobile across sectors based on the ratio of wage rate to sectoral price. The labour market closure allows for unemployment of some labour resources when the wage rate is fixed.
- The model distinguishes several types of prices, i.e. import price, export price, domestic market price and producer price. The wedge between these prices arises due to government's tariffs on imports, export subsidies, and indirect taxes/subsidies for domestic goods, and also due to the imperfect substitutability between domestic and traded goods mentioned earlier.
- The model distinguishes ten household expenditure classes, five each in rural and urban areas, which help to capture the distributional impacts of policy alternatives. It is assumed that different household classes save different proportions of their incomes after payment of income taxes in fixed proportions.
- Household incomes depend on the factor income and initial endowment distribution. Households also receive transfers from the government and from abroad. The factor income has to be allocated to the households by income class. This has been done on the basis of initial endowment (factor income) in the SAM. Household's disposable income is net of personal income (direct) after paying tax and making savings. The representative household's preferences are represented through the Linear Expenditure System (LES) of demand, which allows for certain fixed components of consumption to be independent of the changes in income level and is termed as floor-level consumption. The balance consumption varies with respect to net income after adjusting for floor-level consumption.
- Given the commodity prices and incomes, these demand functions define households' real consumption of each commodity. The relevant price here is the 'composite' price, which is the weighted average of market price of the domestically produced good and its import price.
- The model considers several sources of government revenue such as direct taxes, tariff revenue, domestic indirect taxes, non-tax revenue and foreign inflows on government account. On the expenditure

side, it distinguishes government consumption, transfers to households and subsidies for domestic goods and exports.

- National savings are from three sources: private, government and foreign, and the sum total of these determine the aggregate investment in the economy. Sectoral composition of the demand for investment is determined by fixed proportions.
- The domestic market prices clear the domestic market, while the exchange rate clears the foreign exchange market given the level of foreign savings. All transfers from and to the rest of the world are in fixed foreign currency. Foreign savings are derived as the residual difference between foreign payment and receipts.
- The international market is assumed to be large enough to absorb any quantities of goods produced in India and can satisfy any import demand of India. The trading partners are not modelled explicitly, but are addressed as 'Rest of the World' (ROW). The demand by ROW represents India's exports and its supply represents India's imports. Imports, exports and domestic goods are treated as imperfect substitutes through the Armington specification (Dervis *et al.*, 1982). It avoids complete specialization that perfect substitution may entail, and permits cross-hauling, i.e. simultaneous imports and exports in the same sector as observed in reality. The lower Armington[1] elasticity means higher difference between the imported and domestic good and vice versa. With the small country assumption, India faces elastic world supply at fixed world price. Also, the final ratio between imported goods and domestic goods is determined by the cost-minimizing decision of the domestic demanders based on the relative prices of the imported and domestic goods.

Adjustments made in the existing model

The following adjustments were made in the Panda and Ganesh-Kumar (2008) model.

Agricultural sectors

In the existing model high-value-added sectors of agriculture were not explicitly treated as sectors. These were combined with other sectors. In the modified model, the high-value-added sectors were given explicit treatment. The SAM agricultural sectors were disaggregated to obtain these sectors separately.

Factor market disaggregation

The existing SAM of 2003/04 has only two factors, labour and capital. As demonstrated by Polaski *et al.* (2008), it is useful to distinguish different types of labour and capital. In the modified model, the labour was disaggregated into skilled urban, unskilled urban, skilled rural and unskilled rural. The capital was also further divided into land, capital–agriculture and capital–non-agriculture.

Household classification

The existing SAM has five household classes each for rural and urban areas based on the consumption expenditure. The modified SAM aggregated the top two and bottom two household classes to make only three household classes each for rural and urban areas. This was carried out to make the model more meaningful for the future years. There may be households who shift from one group to another and the model is not capable of updating this. The clubbing of finer groups will result in minimal errors.

A household was classified based on three expenditure classes each for rural and urban areas and was based on the following monthly per capita expenditure percentiles:

Category 1: <30%
Category 2: 30–70%
Category 3: >70%.

Numeraire

The model was solved for relative prices and the wholesale price index was set as numeraire.

Dynamics of model

Recursive dynamics

The model suggested by Panda and Ganesh-Kumar (2008) is a static one. The modified model is a recursively dynamic model with which the inter-period changes could be analysed through a series of temporary equilibria. Hence, the parametric assumptions of the model to take it to future years is an important element of the model. The inter-period adjustments consisted of changes in capital accumulation, growth in population and labour supply, changes in total factor productivity, behavioural parameters, and changes in government expenditure and foreign inflow and outflow. With 2006/07 as the base year, we first simulated the model parameters for 2009/10 and keeping this as the base, the results were simulated for 2019/20.

Labour supply growth

India's young population is expected to increase with a declining dependency ratio. According to most population projections, the share of the working age population in the total population will continue to rise for the next 30 years or so. This demographic transition shows a large potential for higher growth through augmented supply of labour.

The total labour endowment was projected by taking into account not only the demographic dividend but also the rural–urban migration effect. Addition to the labour force (in the age group of 15–59 years) has been projected to grow at more than 10 million/year for the first decade and approximately 10 million during the second decade.

Government consumption

Government consumption is likely to grow at 12% per annum during the first block years (2007–2010) due to increased expenditure during the recent meltdown (2008). Thereafter, government consumption is expected to grow at 9% per annum based on the growth rates at constant prices during the previous decade.

Investment

Since the modified model was to be solved at 5-year frequencies, detailed specification of capital was not done. Instead, simple projections based on time trends were used to specify the total stock of capital in the economy and its sectoral composition. Initially, the capital factor (rental value) increased at a growth rate of 9.0%. Considering the high growth rate in capital investment during the previous decade, we expected the same trend to continue for the future blocks. Accordingly, the capital factor was expected to grow at nearly 9.5% for the remaining period.

Since the changes in stock during the period do not follow any trend, we maintained the same proportions as in the base year.

International trade account

The international trade account represents the payments made and received by the government and households with respect to the ROW account in US dollars. For the period, financial year (FY) 2006/07 to FY 2009/10, the financial flows have grown at the compound annual growth rate (CAGR) as per actual. However, considering the trend over the previous decade, it was observed that the growth rate in these payments was varying with period. We therefore assumed continuance of a similar trend but adjusted for the levels already reached by 2009/10 and therefore growth rates for the future periods were adjusted to have a realistic view on their growth.

Government transfers to households

The government transfers to households are expected to grow at the minimum of 2% per annum, representing a small increase in direct spending by government. However, the transfers to households are expected to grow through other social welfare schemes such as the National Rural Employment Guarantee Programme, the Food Subsidy Programme etc., and hence, direct transfers have grown at a marginally higher growth rate than the growth rate in population.

The growth rates of labour supply and population, implemented in the dynamic blocks as parametric changes, are given in Table 6.4.

The population projection was done for rural and urban areas separately for each state and Union Territory by the technical committee headed by the Registrar General of India. The parameters considered were sex ratio at birth, fertility rate, mortality rate, migration, etc. The urban population projection was done using Urban-Rural Growth Differential (URGD) method. It is based on the principle that URGD follows a logistic pattern. For deriving labour participation, the parameters such as population growth in the age group 15–59 and the demographic dividend were considered.

Social accounting matrix

To study the effect of various policy targets on agriculture in the general equilibrium model, a Social Accounting Matrix (SAM) for 2006/07 of India based on the detailed SAM developed by Panda and Ganesh-Kumar (2008) was used. The SAM comprised 32 sectors, seven factors of production and six categories of households.

The 32 sectors comprised 14 sectors from agriculture, one sector of primary products, four sectors of agri-processing, seven sectors of manufacturing and six sectors of services. The SAM was constructed at market prices of the commodities in 2006/07. The sectoral classification is presented in Table 6.5.

Different GDP Growth Scenarios

In this section, the results of alternative growth scenarios are presented along with discussion.

The three scenarios considered for real GDP growth were 6%, 8% and 10%. The factor driving this different GDP growth was the total factor productivity (TFP).

Sectoral changes

The compositional changes (Fig. 6.1) indicated that increasing growth would favour the industry in the long-run.

The growth rates of real GDP have indicated that though the annual growth in agriculture would reach the peak of 6.30% in 2010–2020 under the 10% GDP growth scenario, the share of agriculture in total GDP will come down from 13% to 11% between the 8% and 10% growth scenarios (Fig. 6.1 and Table 6.6). Between the 8% and 10% GDP growth scenarios, the annual growth of the industrial sector would jump from 8.13% to 11.27%, for the services sector this would rise from 8.62% to 10.14%, and the agriculture sector would go up by only 0.86% percentage point (from 5.44% to 6.30%).

Within the agriculture sector, the pulse crops group has depicted the highest annual growth of 6.06%, followed by cotton (5.56%) under the 10% GDP growth scenario (Fig. 6.2). However, on moving from 8% to 10% GDP growth scenario, fruits and vegetables have recorded the highest gain across all the crops, an increase of 1.44 percentage points

Table 6.4. Population and labour growth rate parameters used in the model (based on the Report of the Technical Group on Population Projections constituted by the National Commission on Population – *Population Projections for India and States 2001–2026*).

	Annual growth rate from the base year 2006/07		
Population/labour	2009/10	2019/20	2029/30
Population – rural	0.72	0.82	0.65
Population – urban	3.91	2.38	1.97
Labour – rural	2.57	1.43	0.94
Labour – urban	3.70	2.50	2.26

Table 6.5. Sectors in Social Accounting Matrix (SAM) – 2006/07.

Sl. no.	
Agriculture (14)	
1	Paddy
2	Wheat
3	Other cereals
4	Pulses
5	Sugarcane
6	Oilseeds
7	Cotton
8	Fruits
9	Vegetables
10	Other crops
11	Milk and milk products
12	Poultry and eggs
13	Other animal products
14	Fishing
Primary products (1)	
15	Primary products
Agri-processing (4)	
16	Sugar
17	Vegetable oils
18	Other food products
19	Beverages and tobacco
Manufacturing (7)	
20	Textiles and garments
21	Petroleum products
22	Fertilizers
23	Pesticides
24	Manufacturing-1 (labour intense)
25	Manufacturing-2 (capital intense)
26	Construction
Services (6)	
27	Electricity
28	Water supply
29	Transport services
30	Storage and warehousing
31	Trade
32	Other services

in the annual growth. The coarse cereals have depicted the lowest growth of 2.95% between 2010 and 2020 under the 8% growth scenario.

The production volume of different crops to 2020 has been projected using the real output growth rate and is reported in Table 6.7. Under the 8% GDP growth scenario, rice production would go up from 101 Mt to 153 Mt in 2020, registering an annual growth rate of 4.20%. Wheat production is projected to go up from 83 Mt to 137 Mt with 5.08% annual growth rate in the same period. For both these crops, there is no significant difference between the 6% and 8% growth scenarios and a very marginal growth between the 8% and 10% scenarios.

Alagh (2011) has made projections for the agriculture and allied sector to 2020 based on the UN Alagh model. The projected production of food grains for 2020 is 225 Mt. Our model predicted production of above 300 Mt even at the pessimistic 6% GDP growth. Other cereals supply projections under the alternative assumptions of fertilizer-use and expansion of irrigated area using the partial equilibrium approach were made by Bhalla *et al.* (1999). They have reported cereals production of 281 Mt with technical efficiency improvements. Our model has predicted a very close estimate of 281.89 Mt of cereals by 2020 in the lower growth scenario of 6% GDP growth. The IFPRI's model known as IMPACT, using the base year of 1993, has projected 256 Mt of cereals for 2020. Kumar (1998) has projected cereals supply in 2020 using an econometric approach in the partial equilibrium setting under two scenarios. The first scenario assumed a constant growth in TFP and the second, a declining TFP growth. His estimates were 309 Mt and 269 Mt under these two scenarios, respectively. Our projection under 8% GDP growth is 337 Mt in 2020, which is higher than Kumar's projection. Our projection is based on a general equilibrium approach and hence it has included the feedback effects of first-round impact from the income accounts to the production activities in the second and the subsequent rounds.

Our results have shown that growth in pulse production will be very impressive between 2010 and 2020. It has been documented that due to change in the consumption pattern, pulse consumption per capita is gradually increasing and this requires more production. It is to be mentioned that recently the Government of India through the Department of Agriculture and Cooperation has launched a project called 'Accelerated Pulses Production Programme', under which an area of 1Mha has been identified with

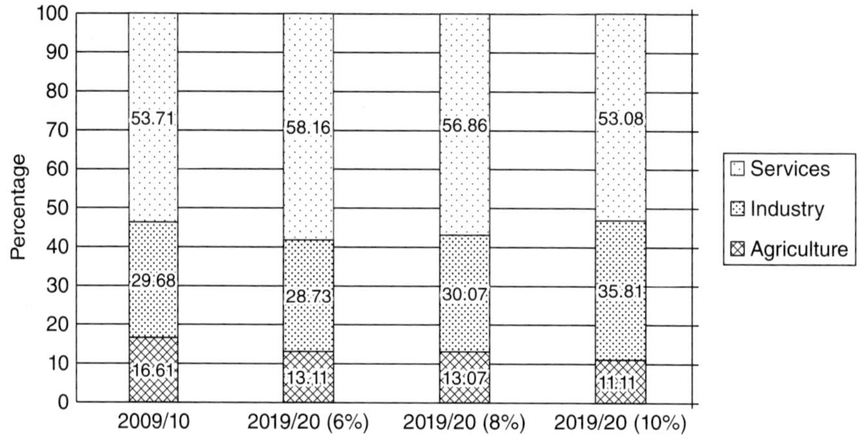

Fig. 6.1. Compositional changes in real GDP in agriculture, industry and services sectors under 6%, 8% and 10% growth (author's calculations).

Table 6.6. Sectoral real value added with different annual growth rates between 2010 and 2020 (author's calculations).

	2009/10 to 2019/20		
Sector	6%	8%	10%
Agriculture	3.45	5.44	6.30
Industry	5.60	8.13	11.27
Services	6.90	8.62	10.14

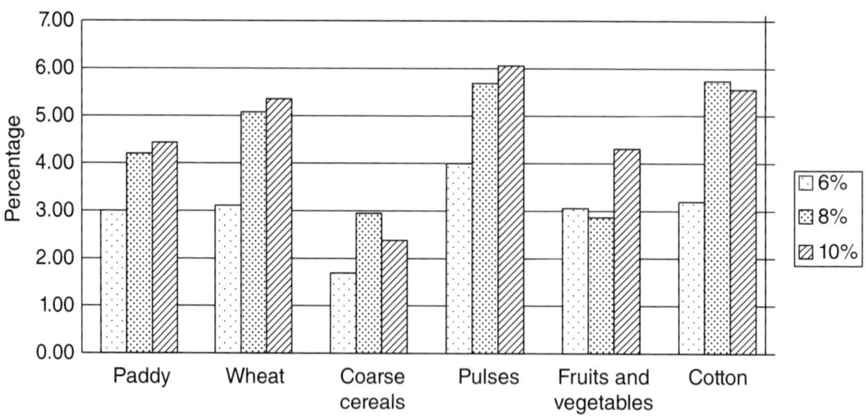

Fig. 6.2. Real output annual growth in the agriculture sector, 2009/10 to 2019/20 (author's calculations).

the objective to increase the production and productivity of pulse crops. Demonstrations on the use of appropriate plant nutrients and plant protection technologies and management practices along with influencing neighbouring farmers to adopt these technologies, are the main features of this programme. Pulses area and production are expected to get a fillip due to this programme, which is consistent with our results.

On perusing the data of the allied agriculture sector that included milk products, poultry, fishery and other animal products, we found that poultry, followed by dairy would register high annual growths of 8.6% and 8.0%, respectively, by 2020 under the 10% GDP growth scenario (Table 6.8).

Rural and urban income distribution

The composition of rural–urban income by household types and income groups is depicted in Table 6.9. A perusal of Table 6.9 reveals that 10% GDP growth would not favour the rural households due to a dip in

Table 6.7. Projected production volume (Mt) of different crops to 2020 under different GDP growth scenarios (author's calculations).

	GDP growth scenarios			
	2010	2020		
Crops	8%	6%	8%	10%
Rice	101.18	131.85	152.67	155.98
Wheat	83.48	108.53	137.02	140.51
Coarse cereals	35.67	41.51	47.7	45.11
Pulses	15.8	22.5	27.48	28.51
Fruits and vegetables	197.88	248.41	262.60	302.11
Cotton	23.08	32.73	40.33	43.44

Table 6.8. Real output annual percentage growth in allied agriculture sector under different GDP growth scenarios (author's calculations).

	2009/10 to 2019/20		
Sectors	6%	8%	10%
Milk and milk products	2.96	6.48	8.00
Poultry	5.74	7.71	8.58
Fishery	3.00	5.91	5.87
Other animal products	4.56	6.81	7.82

Table 6.9. Rural–urban income composition (percentage) by household types and income groups under different GDP growth scenarios (author's calculation).

Household category[a]	2010	2020		
	8%	6%	8%	10%
Rural 1	5.49	5.46	5.47	4.60
Rural 2	14.48	14.92	14.89	12.87
Rural 3	38.01	41.22	40.84	38.95
Urban 1	3.03	2.46	2.43	2.78
Urban 2	9.50	8.16	8.26	9.38
Urban 3	29.49	27.78	28.10	31.42
Total rural	57.98	61.60	61.21	56.42
Total urban	42.02	38.40	38.79	43.58

[a]Rural 1 pertained to households of bottom 30% of income (expenditure) percentile, Rural 2 had households of middle 40% of income (expenditure) percentile and Rural 3 constituted the households of top 30% of income (expenditure) percentile. Urban counterparts followed the same criteria.

agriculture share in total GDP with GDP growth moving from 8% to 10%. The rural share would come down from 61% to 56% (Table 6.9). In 2020, the 10% growth, though it seems to benefit industries, may not really benefit rural industries. In terms of income groups, the 10% growth would mostly benefit urban high-income groups. This analysis has confirmed that the growth is not trickling down. In particular, Rural 2 groups' real income growth would come down on moving from the 8% to the 10% GDP growth scenario during 2010–2020. This is also reflected in the real income per capita figures.

Chhibber and Palanivel (2009) have simulated results based on SAM modelling for income distribution for the year 2009/10 for a pessimistic scenario against the baseline and an optimistic scenario (pre-global financial crisis scenario) of 8% GDP growth in both the years, 2009 and 2010. For the pessimistic (post-crisis) scenario, GDP was fixed at 5.4% for 2009 and 6.5% in 2010. The study has found that during the period of slower growth, the maximum loss would be for the middle and upper-middle income classes of households. The difference in the growth of income for these two classes in rural areas from historical period to 2009/10 is 5.6% and 4.7%, respectively. However, the distribution of income would not favour 'abjectly poor' and 'poor' categories in rural areas. Out of the total income generated in the rural areas, only 2% would reach the 'abjectly poor' and 7.5% to the 'poor' population. Together, the low-income groups in rural areas would get less than 10% of the total income generated in the rural sector. On comparing this with our results, we also found that the middle-income group would be losing out more in the slow growth regime from the base year to 2009/10.

On perusing the results on the real annual income growth rates, it was evident that on moving from the 8% to the 10% growth regime, income was being redistributed from rural to urban areas, particularly across the households of category 1 and category 2 income groups (Fig. 6.3). The real income per capita figures have further confirmed this result (Table 6.10). One can find reason in the fact that agriculture is not being benefited in the increasing GDP growth of 10%. The policies should adopt specific strategies to stimulate agriculture when the economy moves at a higher growth path. This will also help correct urban bias in the income distribution in the process of higher growth. It was interesting to see how the income generation in the industry and services could be tapped to benefit rural India by making use of sectoral linkages.

In another study, Mythili and Harak (2012) have found that in the non-agricultural sector, agri-processing generates the highest output and has income multiplier effects for agriculture. In the wake of increasing

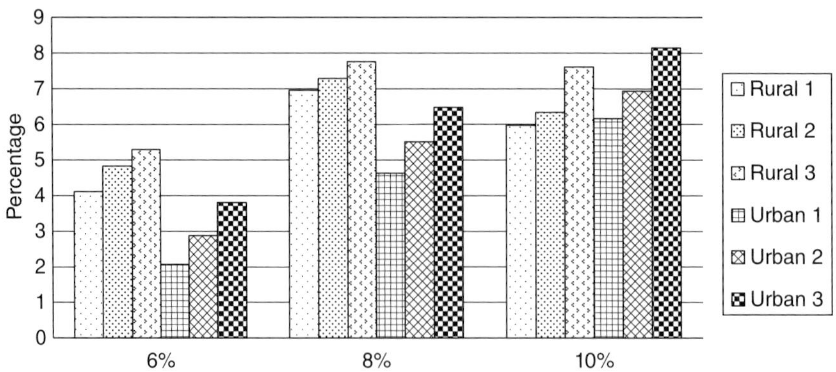

Fig. 6.3. Total real income annual growth rates by household types and income groups, 2009/10 to 2019/20. Based on MPCE, households were classified as follows: Rural 1 pertained to households of the lowest 30% of income (expenditure) percentile, Rural 2 had households of the middle 40% of income (expenditure) percentile and Rural 3 constituted the households of the top 30% of income (expenditure) percentile. Urban counterparts followed the same criteria (author's calculations).

demand for processed foods and changes in the consumption pattern of both rural and urban households, modernization of retail chains and entry of the private sector, the promotion of agri-processing would certainly help the agriculture sector by increasing forward linkages of agriculture with this sector. It was also found that a unit exogenous expenditure in the agri-processing sector among the non-agricultural sectors generates highest income for rural areas.

Wages

The wage rates under different growth scenarios further have corroborated that the dip in rural income in the higher growth scenario is partly due to a wage decline for rural unskilled labour and low growth for rural skilled labour. On moving from 8% to 10% GDP growth, the wage increase would be 25.4% for the rural skilled and 14.3% for the urban skilled labour. The urban unskilled labour has recorded a meagre 7% growth in wages between these two scenarios (Table 6.11).

Conclusions

This study has analysed income distribution across rural and urban India under different GDP growth scenarios using the computable general equilibrium (CGE) model. The study has observed that increasing GDP growth regime beyond 8% does not favour rural households due to the decline in agricultural growth. With GDP growth moving up from 8% to 10%, the rural share in real income will come down from 61% to 56%. Although 10% GDP growth seems to benefit the industrial and services sectors, it will not really benefit the rural industries. The analysis by income groups has revealed that 10% GDP growth will mostly benefit the urban households of Category 3 income-groups.

Table 6.10. Real income per capita – annual income under different GDP growth scenarios (in Rp; author's calculations).

Household category[a]	GDP growth scenario 2010		2020	
	8%	6%	8%	10%
Rural 1	9,035	12,024	16,274	14,298
Rural 2	17,882	24,645	33,214	30,022
Rural 3	62,585	90,784	121,446	121,131
Urban 1	11,689	11,389	15,191	18,206
Urban 2	27,481	28,386	38,819	46,081
Urban 3	113,693	128,874	176,012	205,829

[a]Rural 1 pertained to households of bottom 30% of income (expenditure) percentile, Rural 2 had households of middle 40% of income (expenditure) percentile and Rural 3 constituted the households of top 30% of income (expenditure) percentile. Urban counterparts followed the same criteria.

Table. 6.11. Changes in wages under different GDP growth scenarios (author's calculation).

Labour	2010	2020		
	8%	6%	8%	10%
Rural unskilled	1.00	1.51	1.76	1.68
Rural skilled	1.00	1.30	1.63	1.75
Urban unskilled	1.00	1.05	1.40	1.60
Urban skilled	1.00	1.02	1.39	1.63

The study has also confirmed that the growth is not trickling down. In particular, the growth in real income of rural households of Category 1 and Category 2 income-groups will come down on moving from the 8% to the 10% GDP growth scenario during 2010–2020. This has also been reflected in the real income per capita figures.

A major policy implication drawn from the study is 'how to improve agriculture in the higher GDP growth scenario', as growth in agriculture is going to be one important factor for reducing poverty in rural areas of India. Supplementary measures are needed to tackle the dipping rural poor income with increasing GDP growth.

Acknowledgement

I thank Dr A. Ganesh Kumar, Indira Gandhi Institute of Development Research, for many useful suggestions to the analytical model.

Note

[1] There may be certain demerits in this specification because the gain from the trade will not be fully realized.

References

Alagh, Y.K. (2011) The future of Indian Agriculture. Available at: http://www.iari.res.in/files/18-lecture-Dr-YK-Akagh-21-6-2011.pdf (accessed 23 January 2013).

Bhalla, G.S., Hazell, P. and Kerr, J. (1999) *Prospects for India's cereal supply and demand for 2020*. IFPRI Food, Agriculture and the Environment Discussion Paper 29. International Food Policy Research Institute, Washington, DC.

Chand, R. (2007) *Inequality in Consumption Expenditure in Indian States: 1973–2005*. National Seminar on NSS 61st Round Survey Results. National Sample Survey Organization, Government of India, held at New Delhi, 29–30 October.

Chhibber, A. and Palanivel, T. (2009) India manages global crisis but needs serious reforms for sustained inclusive growth. Paper presented at the tenth annual conference on *Indian Economic Policy Reforms*. Stanford University, 21–23 October.

Datt, G. and Ravallion, M. (1998) Farm productivity and rural poverty in India. *Journal of Development Studies* 34(4), 62–85.

Dervis, K., de Melo, J. and Robinson, S. (1982) *General Equilibrium Models for Development Policy*. Cambridge University Press, Cambridge, UK.

Gokarn, S. and Gulati, G. (2006) India: Country growth analysis. Available at: http://s3.amazonaws.com/zanran_storage/www.adb.org/ContentPages/82470170.pdf (accessed 8 October 2013).

Gulati, A. and Bathla, S. (2002) *Capital Formation in Indian Agriculture: Trends, Composition and Implication for Growth*. Occasional Paper 24. National Bank for Agriculture and Rural Development, Mumbai.

Jha, R. (2000) *Reducing Poverty and Inequality in India: Has Liberalization Helped?* UNU/WIDER Working Paper No. 204. UNU World Institute for Development of Economics Research, Helsinki. November.

Kumar, P. (1998) *Food Demand and Supply Projections for India*. Agricultural Economics Policy Paper 98-01. Indian Agricultural Research Institute, New Delhi.

Mahendra Dev, S. (2012) *A Note on Trends in Public Investment in India*. IGIDR Project Series No. PP-069-SMD2. Indira Gandhi Institute of Development Research, Mumbai.

Mahendra Dev, S. and Ravi, C. (2007) Poverty and inequality: All-India and states, 1983–2005. *Economic and Political Weekly* XLII(6).

Mythili, G. and Harak, N. (2012) Sectoral linkages, multipliers and the role of agriculture. In: Mahendra Dev, S. (ed.) *India Development Report*. Oxford University Press, New Delhi.

Narayana, N.S.S., Parikh, K.S. and Srinivasan, T.N. (1991) *Agriculture, Growth and Redistribution of Income: Policy Analysis with a General Equilibrium Model of India*. Elsevier Science, The Netherlands.

Panda, M. and Ganesh-Kumar, A. (2008) *Trade Liberalization, Poverty and Food Security in India*. IGIDR Working Paper WP-2008-013. Indira Gandhi Institute of Development Research, Mumbai.

Parikh, K.S., Ganesh-Kumar, A. and Darbha, G. (2002) *Growth and Welfare Consequences of a Rise in Minimum Support Prices: An Applied General Equilibrium Analysis for India*. Report submitted to the Ministry of Finance, Government of India, New Delhi.

Polaski, S., Ganesh-Kumar, A., McDonald, S., Panda, M. and Robinson, S. (2008) *India's Trade Policy Choices: Managing Diverse Challenges*. Carnegie Endowment for International Peace, Washington, DC.

Ravallion, M. and Datt, G. (1996) How important to India's poor is the sectoral composition of economic growth? *The World Bank Economic Review* 10(1), 1–25.

Saluja, M.R. and Yadav, B. (2006) *Social Accounting Matrix for India 2003–04*. India Development Foundation, Gurgaon.

Sarkar, S. and Mehta, B.S. (2010) Income inequality in India: Pre- and post-reform periods. *Economic and Political Weekly* XLV(37), 45–55.

Virmani, A. (2005) *Economic Performance, Power Potential and Global Governance: Towards a New International Order*. ICRIER Working Paper No. 150, New Delhi, December.

Appendix

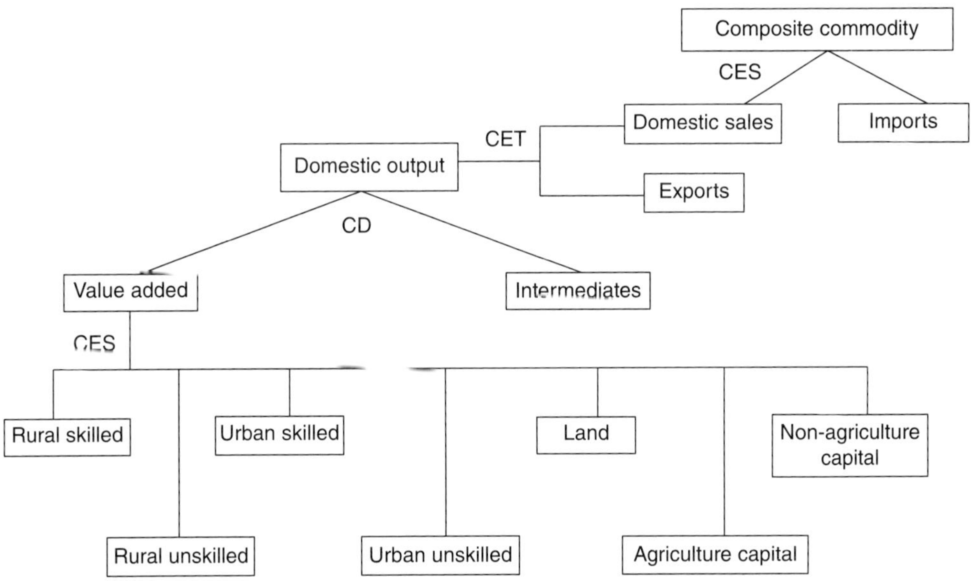

Structural layout of the CGE model.

7 Food Safety Standards for Domestic and International Markets: The Case of Dairy

Anneleen Vandeplas*[1,2] and Mara P. Squicciarini[2]
[1]*European Commission, Brussels, Belgium;* [2]*LICOS, Center for Institutions and Economic Performance, KU Leuven, Belgium*

Introduction

Several chapters in this volume address the question of how demand and supply for key food commodities will evolve in India over the next 20 years, and in particular, whether India will become deficit or surplus in these commodities by 2030. This is a question of high policy relevance to Indian policy makers, who have historically attached major importance to the issue of food self-sufficiency; in other words, of being able to produce all food commodities it needs by itself.[1] However, over the next decades, quantitative considerations of food supply may need to be complemented by qualitative considerations as concerns for food safety and quality[2] are becoming increasingly important. This does not only apply to production for international trade, but also to production for domestic consumption.

In this chapter, we review the relevance of food safety and quality standards for the Indian domestic market and for international trade. We make a distinction between public and private standards, and show how these may reinforce each other. Finally, we present descriptive evidence on food safety practices in dairy production from a primary survey of dairy producers in Andhra Pradesh, a state in south India.

Importance of Quality Standards in the Indian Domestic Market

Domestic markets in Asia are witnessing an increasing demand for food of high quality and safety levels (Pingali, 2007; Ehmke *et al.*, 2008; Minten *et al.*, 2011). Also in India, income growth is spurring concerns over food quality and safety, especially in urban markets (Swinnen, 2007; Birol *et al.*, 2010; Roy *et al.*, 2010). Indian consumers are paying considerable price premiums for tomatoes and rice with superior organoleptic characteristics (Vandeplas and Minten, 2015), and Krishna and Qaim (2008) find that the (stated) willingness-to-pay for residue-free vegetables amongst Indian consumers is more than 50% above common market prices. Birol *et al.* (2010) qualify these results, arguing that a necessary condition for the monetization of such quality premiums is that consumers dispose of credible information regarding quality. This means that Indian consumers are more likely to pay price premiums for

* Corresponding author, e-mail: anneleen.vandeplas@kuleuven.be

  International Trade and Food Security (eds F. Brouwer and P.K. Joshi)

observable food characteristics (e.g. those relating to organoleptic quality) than for food safety characteristics, which are very often unobservable. A study on traditional markets in north India reveals that even retailers are not aware of food safety management practices applied during production of the products they sell (Minten *et al.*, 2012).

Indian newspapers frequently bring stories on pesticide overuse, untreated sewage water-use in irrigation, poor sanitation at food markets and faulty cold storages. A series of scandals in the early 2000s revealed considerably high pesticide-residue levels in bottled water, milk and soft drinks (CSE, 2004; Umali-Deininger and Sur, 2007). In 2013, a large-scale screening of food samples for colour additives revealed that 58% of the investigated 2409 samples exceeded the maximum permissible concentration, and 16% contained non-permitted colours (Dixit *et al.*, 2013). Also, fruits and vegetables have been found to frequently exceed maximum thresholds for heavy metals (Sharma *et al.*, 2009; Finzer *et al.*, 2013). In response, the Government of India is now monitoring pesticide residue levels in food products more closely (GoI, 2011b).

In general, increasing awareness and protests by civil society have urged the Government of India to tackle food safety risks in a more structural way. In an attempt to consolidate the existing food safety regulations, which were previously addressed by a plethora of different acts and laws, Indian policy makers drafted the Food Safety and Standards Act (FSSA) in 2006 (GoI, 2006). To oversee the implementation of this act, the Food Safety and Standards Authority of India (FSSAI) was established in 2011. Under the FSSA, all food traders and manufacturers need to register, and may be subject to random checks by government agencies. The implementation of the FSSA is a matter of the respective state governments.

A major change under the FSSA is that it establishes a system of liability in the food supply chain (Article 27 of FSSA 2006). This means that the act determines which actor in the supply chain bears responsibility when food safety breaks down. If properly enforced, it may have tremendous implications for the organization of food supply chains in India. In particular, manufacturers and packers will be held responsible if a food product issued by them is found unsafe. Wholesalers, distributors and retailers will be held responsible for products supplied after the date of expiry, or if they have stored or handled the product in violation of safety instructions by the manufacturer or the FSSA. In addition, if the manufacturer of a product they sell cannot be identified, they will themselves be responsible for its compliance with applicable safety standards. In particular, the latter issue may have important implications. For example, traders in the wholesale markets will become liable for excess pesticides on agricultural products supplied in bulk by anonymous farmers; milk traders will become liable for any adulterants (e.g. water, oil, skimmed milk powder, soda) added to the milk, even if done by their milk suppliers – unless they can identify the culprit.

These liability regulations apply to formal as well as informal traders – including street hawkers and itinerant vendors. Unsurprisingly, the law has triggered widespread protests, based on claims that poor street vendors will never be able to comply with the food standards which have been imposed.[3] Not surprisingly, this gave rise to major delays in the implementation of reforms. While it was initially planned that the policy would enter into force on 5 August 2012, this deadline has been repeatedly extended.[4]

An interesting political economy issue is revealed from the government's choice to impose food safety and standards only at the level of food traders and manufacturers, not at the level of primary producers of crops, fish, meat or milk. This decision contrasts with the EU regulations for instance, which clearly lay down primary production standards for food. However, regulating agricultural practices of smallholders would likely bring about even more vigorous protests and make implementation of the FSSA even more complex.

A final challenge for implementation of the FSSA was the fact that at the time of introduction of the FSSA, as has been argued by experts, many central and state-level laboratories in India lacked the capacity to carry out the microbiological, chemical and physical tests required to enforce the FSSA and detect food safety hazards (TUV South Asia, 2007; FICCI, 2010).

The Importance of Quality Standards in Trade

Food quality standards are also becoming increasingly important for India's growing participation in international trade. Over the past five decades, more than 150 countries have engaged in multilateral negotiations on the reduction of trade barriers under the aegis of the World Trade Organization (WTO). Increasingly, India is also engaging in bilateral trade agreements. For many products, however, the effects of tariff and quota removal may be strongly mitigated by the existence of sanitary and phytosanitary (SPS) measures or other non-tariff measures (NTM). Especially for food commodities, SPS measures often constitute important barriers to trade between countries. In the wake of successful rounds of tariff reductions, a rapid surge of notifications of new SPS measures to the WTO has been observed over the past two decades (Henson, 2007).

The literature on food quality and safety standards has proliferated, as scholars have started to examine whether such standards should be considered as 'barriers to trade' – for those producers who fail to comply with these standards – or rather as 'catalysts for trade' – for those producers who can comply, and as such gain market share from their competitors (Maertens and Swinnen, 2008; Anders and Caswell, 2009; Munasib and Roy, 2011).

Public versus Private Standards

Public standards are clearly not the only standards of concern in food production and trade. Research has revealed that private standards are often more stringent, more flexible and more agile than public standards (Fulponi, 2006; Smith, 2009; Vandemoortele and Deconinck, 2013). The higher level of private food quality standards may be a response to increasingly stringent public regulations on product liability; on the other hand, they facilitate product differentiation as part of manufacturers' and retailers' competitive strategies.

While private standards have proliferated, especially in developed countries, their impact is increasingly being felt in developing countries as well, among other reasons as a result of the global spread of multinational retailers who frequently impose their international standards upon local markets (Reardon and Berdegue, 2002; Henson and Reardon, 2005).

Also in India, given the growing concerns for food safety, it may not take long before domestic firms pick up private standard-setting as a business strategy. From that perspective, the introduction of the FSSA may only be a prelude of a more widespread transformation of the agri-food system in India. The establishment of liability for food hazards does convey increased responsibility upon food manufacturers and food retailers, which may generate increased incentives for supply chain reorganization. As consumers become more concerned about quality, food manufacturers and retailers may adapt their competitive strategies through an increased focus on quality differentiation.

Standards in the Indian Dairy Sector

We will now look into more detail at the importance of standards in one particular sector, namely the Indian dairy sector. As milk is a perishable commodity, food safety management issues are of high importance in this sector. We start off with a brief description of the dairy sector in India, then discuss the importance of food standards in the domestic dairy market and in dairy trade and finally describe the currently applied food safety practices based on micro-level survey evidence from Andhra Pradesh, a state in South India.

The dairy sector in India

India is the largest milk-producing country in the world with a production of over 120 Mt of milk in 2010 (FAO, 2012). The total value of milk produced in India amounts to more than US$43 billion, which compares to a total value of US$38 billion for rice – India's major food crop (FAO, 2012). This is mainly the result of an impressive growth of over 4% per year recorded over the past four decades, as a result of productivity increases (through adoption of better technologies and better feeding practices), major shifts in demand and important policy changes.

The dairy sector is also important for other reasons in India. Its complementarity with agriculture and its capability to enrich the diet of a largely vegetarian population is well documented. Moreover, as more than half of the Indian labour force is still employed in agriculture (see Ganguly and Pandey, Chapter 2, this volume), the dairy sector is considered as an important source of rural employment and income, and hence as a potential source of pro-poor growth. Dairy production is also often claimed to have positive gender effects, as it provides women with an opportunity to generate income (and thus increases their intra-household bargaining power) while working from home (Sneyers and Vandeplas, 2013).

As far as demand growth is concerned, rising incomes lead to higher consumption and diversified diets: Indian households are gradually increasing their consumption of high-value food products such as fruits and vegetables, meat, fish and dairy products (Pingali, 2007). Recently, fast-food chains and food and non-food industries using dairy ingredients in a wide range of products have also emerged as important sources of dairy demand. Industry experts estimate that demand for milk and milk products will be growing at around 8–10% annually in the coming years (VGBO, 2009). If this is true, and if Indian policy makers want to retain self-sufficiency in milk production, supply growth will need a boost.

The changes in demand for milk do not only affect the quantities demanded, but also quality. With increasing urbanization, Indian households face wider ranges of product choice, and have to rely on markets to buy their daily cup of milk, rather than rearing their own cows or buffaloes. As consumers' awareness will rise with increased exposure to information on food safety, concerns over food safety are expected to increase in the near future as well (Roy *et al.*, 2010). All these factors have influenced the evolution of market demand for milk in India over the past decade, and will continue to exert strong influence on future demand.

Domestic food safety standards in Indian dairy

Public standards

Dairy products produced in India have historically had to comply with India's main law on food safety and quality, the 'Prevention of Food Adulteration (PFA) Act, 1954' (GoI, 2004). This law used to be administered by the Ministry of Agriculture, the Ministry of Food Processing and the Ministry of Health of the Government of India. It covered different aspects of food processing and distribution, with respect to food colour, added preservatives, pesticide residues, packaging and labelling, and prohibition and regulation of sales.

The prohibition and regulation of sales implies that it is strictly prohibited to sell, amongst other items, 'cream which has not been prepared exclusively from milk or which contains less than 25% of milk fat', 'milk which contains any added water', or 'ghee which contains any added matter not exclusively derived from milk fat'. Minimum standards for fat content and non-fat solids content of milk have been specified in the PFA Act of 1954, but microbial count standards have remained conspicuously absent – except for foods intended for infant nutrition, where the microbial count was limited to 40,000 cfu/g (GoI, 2004).[5] Another major shortcoming in the PFA Act seems to be (to date) the absence of maximum thresholds for veterinary drug residues (NDDB, 2009).

Dairy production has also been subject to regulations under the Milk and Milk Products Order, 1973 (amended in 2002), the Standards of Weights and Measures Act (1976) and the Standards of Weights and

Measures (Packaged Commodities) Rule (1977), the Standards of the Bureau of Indian Standards (BIS) and the Infant Milk Substitutes, Feeding Bottles, and Infant Foods (Regulation of Production, Supply and Distribution) Act, 1992.[6] However, none of these regulations has really been able to pin down solid food-safety warranties. For example, none of them lays down specific microbiological standards with which raw milk has to comply. A consolidated FSSA was drafted in 2006. At the level of milk and milk product specifications, however, not much has changed as compared to the earlier regulations. One marked change is that the microbial count threshold for infant foods has been reduced to 10,000 cfu/g, in line with international recommendations (CAC, 1979, 2004).

The establishment of liability along the supply chain in the FSSA may, however, have important implications. At present, the bulk of marketable milk surplus is still collected by the unorganized dairy sector, which hardly pays attention to food quality and safety (NDDB, 2009).[7] With the advent of liability for milk adulteration on milk traders and milk processors, stronger vertical links may need to be established with upstream producers. This may trigger a profound transformation of the dairy sector in India, which to date remains strongly fragmented at all levels.

Private standards

Increasing awareness and strengthening consumer preferences for food quality may spur the emergence of private food standards in the Indian dairy industry. To some extent, quality differentiation already occurs between Indian dairy processors. Interviews with key informants in dairy businesses in Andhra Pradesh have revealed that Indian dairy processors decide to pursue different levels of Hazard Analysis and Critical Control Points (HACCP) and International Organization for Standardization (ISO) certification as a 'competitive' strategy, in an attempt to convince consumers that their milk is of high quality (Squicciarini and Vandeplas, 2010). However, these attempts are as yet mainly confined to the processing factory premises, and quality control up to the farm-level is not very common in the Andhra Pradesh dairy industry.

The potential risks of allowing for safety flaws in food supply chains have been revealed by the 'fake milk' crises in China in 2004 and 2008. In 2004, some milk samples available for retail were found to contain little nutritional value, resulting in malnourishment of babies being fed on infant formula. In 2008, infant formula was found to be contaminated with melamine, a synthetic material that is used to inflate protein counts in milk based on its high nitrogen content. Both crises resulted in the death of several babies in China. Indian newspapers often bring stories on discoveries of 'fake milk' in India as well, but the scandals in India have (fortunately) (so far) never reached the scale of the milk crises in China.[8] Nevertheless, dairy imports by China have risen strongly in response to concerns over safety of domestically produced milk. This could happen to India as well, once Indian consumers become conscious of the potential health hazards to which they are exposed.

If India's production system does not respond appropriately to the increasing demand for quality and safety, even if it manages to keep milk production growth in line with consumption growth, a gap between supply and demand might set in in the high quality segment. This, in turn, may trigger a response by private companies to start catering to this unfilled need. Few companies have taken that step so far, as most are constrained by the atomistic structure of their supply base.

Food safety standards in international trade

We now turn to discussing how food safety standards matter for India in its international trade in dairy products. After several decades of being a net dairy importer, in the second half of the 1990s India became self-sufficient in dairy production and, in 2001, it even became a net exporter. In more recent years, India has started to fluctuate between being a net importer and being a net exporter. In general, imports and exports are still low relative to the production volume and constitute only 0.4% and 0.3% of the global import and export volumes, respectively (FAO, 2012).[9]

Since 2009, India's imports have hiked up, such that India's net dairy exports (in milk equivalents) have been hovering around zero in recent years. Most dairy products are imported by India from the EU. The final destination of India's dairy exports are mainly the neighbouring countries in South Asia and countries in the Middle East.

The current divergence across India's trade partners for imports and exports does not only reflect the differential tariff structures between countries, but also, and to a large extent, differences in international standards applied to dairy products. Low quality and hygiene standards in dairy production have also been identified by the Victorian Government Business Office–India (VGBO, 2009) as a major constraint for the development of the Indian dairy export sector. Another constraint relates to the lack of experience of Indian businesses in the international marketing of dairy products (VGBO, 2009). This is in line with the findings from an Indian dairy company survey by Squicciarini and Vandeplas (2010).

This means that for potential trade partners with stringent food standards, even a significant reduction in tariffs might not lead to increased dairy imports from India in the short to medium term. For instance, the EU imposes strict food safety and quality standards on imported milk products, specifying conditions for primary milk production (e.g. health and hygiene conditions of dairy animals), procurement (e.g. a very short time frame within which milk must be cooled down after milking) and processing, which India's current dairy production system cannot comply with. Still, India is in the process of strengthening its regulatory framework and upgrading its testing facilities in order to facilitate future access of its milk and eggs products to the European market (Sareen, 2003).

Food standards for dairy imports by India

India also imposes SPS barriers upon imported milk. Dairy imports to India – like local dairy products – have to comply with all PFA rules. Once it is implemented, dairy imports will fall under the FSSA.[10]

While, at first sight, India's sanitary requirements do not seem insurmountable by western standards, India has been banning dairy imports from countries maintaining high internal sanitary standards, such as New Zealand, the USA and Australia in the recent past. More specifically, India requires importers to guarantee that the imported dairy product originates from cattle that have not undergone oestrogen treatment in the previous 90 days.[11] As a result of this requirement, Australia and New Zealand were barred from exporting milk to India for several years. Only in 2009, Australia was able to show that, due to a 'voluntary ban' of oestrogen-use, they met the Indian requirements, and trade was re-installed (VGBO, 2009). The ban on New Zealand milk was lifted in 2010.

Similarly, the USA has been suffering from (what they call 'unfounded') sanitary requirements by India, which has limited their exports to India since 2003. US sources claim that India maintains more stringent maximum residue levels (MRL) on dairy imports than on domestic dairy products. Finally, India has also banned dairy imports from China since 1 December 2008 as a result of the melamine adulteration scandal.[12]

Food standards for dairy exports by India

The Export Inspection Council of India (EIC, since 1964) is the government body responsible for monitoring the quality of Indian export products before they cross the border to their final destination. According to the Indian regulations, pre-shipment inspection is compulsory for commodities such as fish and fishery products, milk and milk products, poultry products, egg products, meat and meat products, and honey to be exported. Other products may be certified as well by the EIC, including black pepper destined for the USA and basmati rice for the EU (Sareen, 2007; GoI, 2011c: 84).

Nevertheless, these inspections are not sufficient for the Indian products to gain access to overseas markets. There may be further requirements in addition. For example, in the case of the EU, a valid 'health certificate' delivered by an EU-accredited local government body in India is required for imports of dairy products. This means that the EU will only accept dairy imports from countries that have an EU-accredited government body. The accreditation of a particular government body is usually contingent upon the recognition by the EU of equivalence of the local public food safety inspection system for that particular food product to its own food safety system. In contrast with other Indian trade partners such as Sri Lanka and Singapore, the EU has not accredited the EIC for delivery of valid health certificates for dairy products (GoI, 2011c). This means that, technically speaking, India is not able to export dairy products to the EU as a result of perceived gaps in the Indian public institutional framework for food inspections.

The main reason for the EU not recognizing India's EIC as a competent authority, hence, relates to the substantial 'bridge to cross' (Munasib and Roy, 2011) between India and the EU: the gap between food safety regulations in India and Europe, in particular for dairy products, remains wide. Just to name a few, EU's dairy regulation (EC Regulation No. 853/2004) starts at the farm-level. This means that primary producers of dairy are subject to rules specifying hygienic conditions for animals, milking sheds, milk storage rooms, etc. According to the EU, raw milk must be cooled immediately after milking to not more than 8°C, or processed within 2 h of milking. In addition, it specifies microbial count thresholds for raw milk and requires HACCP compliance by dairy plants.

In contrast, the Milk and Milk Products Order (MMPO) Act (2002) in India, which regulates hygiene standards for milk and milk products, only imposes regulations at the level of dairy plants. India is more lenient when it comes to the allowed timeframe between milking and chilling of the milk: it stipulates that raw milk 'if not collected and brought to the dairy plant within 4 h of milking, be cooled as soon as practicable after procuring, to a temperature of 8°C or lower.' In contrast with the EU, India does not impose any microbial count standards on raw milk, and does not require dairy processing plants to comply with HACCP. However, it does offer subsidies to the processing plants that voluntarily decide to venture into the certification process.[13]

Negotiations on this matter, with regular visits of EU officials to India, have been ongoing,[14] but as long as no final agreement is reached, even Indian companies with the highest level of private food safety and quality standards will not have access to EU dairy markets. For shrimps and basmati rice, on the other hand, the EU does recognize the EIC's competence to deliver valid certification.

In addition to the requirements at the level of the public inspection system, the EU also requires certificates of approval for individual plants as a pre-requisite for exports. In the case of shrimps or basmati rice destined for the EU, these are issued by the EIC. Requirements for dairy processing plants include HACCP compliance, absence of antibiotics or other veterinary drug residues in milk, compliance with specific provisions relating to conditions of packaging and storage and frequent quality checks.[15] Once a processing plant has been approved, it is subject to periodic surveillance by one of the five Export Inspection Agencies (EIA) in Delhi, Calcutta, Kochi, Chennai and Mumbai.[16]

Recognition of a national competent authority is not always a requirement for companies to gain access to foreign export markets. For manufacturers of dairy products to gain access to the US market, for example, foreign food manufacturers register with the US FDA directly, which means that they agree to allow the US FDA to inspect their facilities on a regular basis and suspend their registration if the food they manufacture or distribute is deemed unsafe (see US FDA, 2012). This allows individual plants to export to the USA based on their individual (private) quality standards, even if their country's inspection agency is not recognized as a competent authority. In the new Food Safety Modernization Act, which was signed in 2011, the FDA has expressed its ambitions to increase its reliance on foreign government bodies (and other third parties) for food safety inspections abroad.

For other food imports, such as meat and meat products, food safety is safeguarded by both the FDA and the Food Safety Inspection Service (FSIS). In contrast to FDA, FSIS does require the recognition of a national competent authority for exports to the USA. This recognition is animal type-specific, which means that a country may, for example, be eligible for exports of pork meat to the USA but not for beef.

Food safety practices in dairy production: evidence from Andhra Pradesh

In this section, we describe food safety practices in dairy production in Andhra Pradesh, where 1000 rural households were surveyed. The analysis reveals that food safety practices are based on traditional wisdom, vary across households, and milk farmers receive little guidance, training or inspection by milk traders or by the government. In general, adulteration of milk (including adding water, oil, milk powder or soda) seems to go largely unpunished.

Data collection and categorization of households based on livestock herd size

For the study, a household-level dataset collected during the period April–June 2010 was used. The survey was set up to be representative for the southern half of Andhra Pradesh, covering the Rayalaseema region (in particular, the districts of Kurnool, Cuddapah, Ananthapur and Chittoor) and the southern part of Coastal Andhra (more specifically, the districts of Nellore, Prakasam, Guntur and Krishna) (see Fig. 7.1).

First, the study region was subdivided into four zones based on milk production per rural capita, and buffalo- or cow-based dairy production systems. Within each region, one district was sampled at random. In the selected districts, 50 villages were randomly selected from a district-level list of villages obtained from the Government of Andhra Pradesh (GoAP, 2009). In each village, a census was organized to record the number of female adult dairy animals that each household owned. Households were classified into four categories according to the size of their livestock herd, and based on these, 20 households were sampled in each village according to a stratified random sampling strategy with oversampling of households with larger livestock holdings.

The selected 1000 households were interviewed using a questionnaire that contained detailed modules on dairy production, input markets and output markets, investments, hygienic practices, services offered by buyers, future intentions regarding dairy production, but also all required modules to estimate total household income and total household consumption.

Descriptive statistics

Table 7.1 shows some basic descriptives for the sample households and the rural population of Andhra Pradesh for which the sample is representative (obtained by using proper weighting factors). The head of the average rural household in the population under study was a male, 46 years old and with 3.3 years of education, which means that he did not even finish primary schooling. Almost one-third of the rural households belong to a scheduled caste (SC) or to a scheduled tribe (ST) category, which are both considered to be socially disadvantaged population groups in India.

In 2010, slightly more than half of the population kept at least one dairy animal, but only 3% had more than five dairy animals. The average number of female adult dairy animals was 1.3 when we took all households into consideration, and 2.5 when we only considered the households that had at least one dairy animal. The average herd size increased between 2005 and 2010. About 34.3% of the farmers planned to expand their herds by 2015, 52.3% planned to keep their herd size at par and only 13.7% wanted to reduce their dairy activity. Milk productivity was considerably low, standing at only 3.3 l/dairy animal/day.

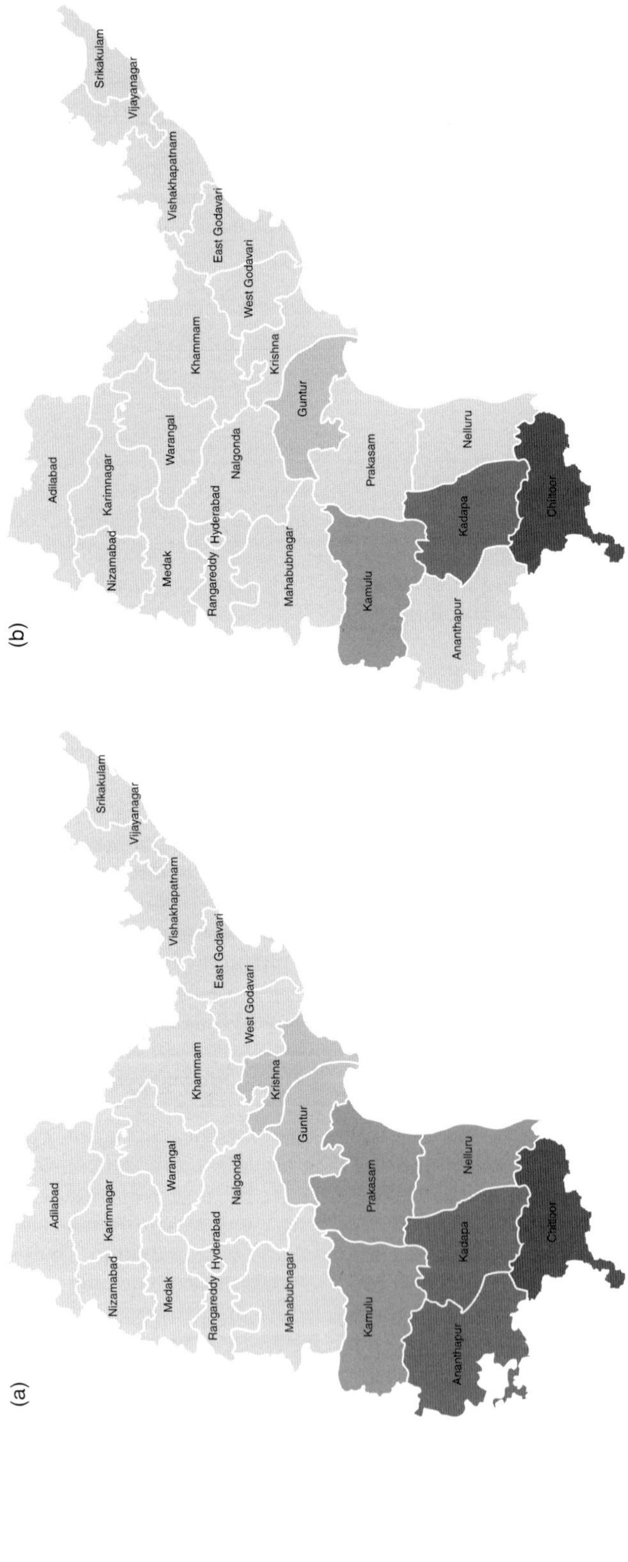

Fig. 7.1. Geographical location of (a) the study area and (b) the sampled districts (Wikimedia Commons, 2008).

Table 7.1. Descriptive statistics of sample households and population under study, 2010 (authors' survey).

	Sample		Population	
Household characteristics	Average	SD	Average	SD
Age of household-head (years)	47.0	11.1	46.3	11.3
Education level of household-head (years)	3.4	5.0	3.3	4.8
SC/ST (%)	24.1		27.7	
Farmers having dairy animals (%)	80.0		51.0	
Farmers producing milk (%)	78.7		49.9	
No. of female adult dairy animals in 2010	2.5	2.6	1.3	1.9
No. of female adult dairy animals in 2005	2.0	3.0	1.1	2.1
No. of female adult dairy animals in 2010 (households with dairy animals)	3.1	2.6	2.5	2.0
No. of female adult dairy animals in 2005 (households with dairy animals)	2.5	3.2	2.2	2.5
Productivity per animal (l/day) (yearly average)	3.4	3.2	3.3	2.7
Cultivated land (acres)	3.8	5.8	2.9	5.0
Cultivated land (households with dairy animal) (acres)	4.3	6.2	4.1	6.3
Non-dairy income per adult equivalent (Rs/year)	32,306.1	54,451.8	31,601.6	50,485.4
Non-dairy income per adult equivalent (US$/year)	807.7	1,361.3	790.0	1,262.1

SC, scheduled caste; SD, standard deviation; ST, scheduled tribe

Since one of the main constraints for dairy production is the availability of green fodder and since access to green fodder is greatly enhanced when households cultivate more land, it should come as no surprise that dairy farmers cultivated on average 3.8 acres of land, which exceeded the population average of 2.9 acres (roughly equivalent to 1.2 ha). Non-dairy income, which comprises total household income excluding income from dairy production, amounted to Rs32,306 (or US$808) per adult equivalent per year on average.

Food safety practices

Table 7.2 shows the prevalence of some practices related to food safety in milk production. The majority of dairy producers milk their animals twice a day. While over 90% of the households wash their hands before milking, very few repeat this action in between milking different cows. This increases the likelihood of transmitting infections between dairy animals. Most households use only water to wash their hands, while less than 5% use soap. Two-thirds of the households consider their dairy animals as clean during milking (without dung on their bodies). Almost all households wash the animal's udder before milking (with water), to remove mud, dust and dung, but less than one-third of the households dry the udder after washing. This increases the probability of contamination of milk, as well as of udder and/or teat infections (mastitis). In more than 30% of the cases additives are used to facilitate milking, mostly oil or ghee. Such additives should be considered as a source of potential contamination of milk, but one that potentially results in higher prices – if milk fat content is raised by the addition of oil or ghee.

The utensils used while milking are mainly of steel or aluminium, which is considered best practice. Still, more than 10% of smaller dairy farmers use plastic utensils, increasing the probability of milk spoilage. Usually utensils are washed with water only (65.1% of the cases); in only 34% of the cases soap is used for washing utensils.

Table 7.3 presents the general milk storage and preservation practices followed by the farmers. Milk remains on the

Table 7.2. Food safety practices in milk production (authors' survey).

Particulars	No. of dairy animals			
	1–2	3–5	>5	Average
How often do you milk your dairy animals per day (%)?				
Once	9.4	7.3	5.6	8.6
Twice	89.9	92.7	93.4	90.9
Do you wash your hands before milking (%)?				
Never	2.2	1.1	0.0	1.7
Before milking	93.8	93.1	94.0	93.6
In between different dairy animals	4.1	5.8	6.0	4.7
Mode of washing hands (%)				
No washing	1.2	1.6	4.0	1.5
Water	94.3	93.5	92.1	94.0
Water and soap	4.4	4.9	3.8	4.5
Dairy animals cleaned during milking (%)	65.4	66.8	55.9	65.2
Udder washed before milking (%)	97.7	99.0	94.1	97.8
Udder dried with cloth/paper (%)	28.5	24.9	21.4	27.0
Additives used on udder before milking (%)	34.1	36.3	32.2	34.6
Type of utensils used for milking (%)				
Plastic	13.9	11.2	7.7	12.7
Aluminium	16.0	18.3	17.7	16.7
Steel	68.3	69.8	72.2	69.0
Other	1.8	0.8	2.4	1.6
Mode of washing utensils (%)				
No washing	1.0	0.2	1.7	0.8
Washing with water and soap	33.0	35.0	40.9	34.0
Washing without soap	66.0	64.8	57.4	65.1

Table 7.3. Milk storage and preservation practices (authors' survey).

Practices	No. of dairy animals			
	1–2	3–5	>5	Average
Time of milk storage on farm after milking (min)	16.6	18.8	16.7	17.3
Location of milk storage before sales (%)				
Inside the house – closed container	71.2	60.4	73.9	68.2
Inside the house – open container	24.7	32.9	24.9	27.1
Outside – closed container	3.0	3.7	0.0	3.0
Outside – open container	1.0	1.6	1.2	1.2
Other	0.0	1.5	0.0	0.4
Type of container to bring milk from farm to buyer				
Plastic	0.9	2.8	0.0	1.4
Aluminium	23.1	21.2	7.5	21.1
Steel	74.1	73.8	92.5	75.7
Other	1.9	2.1	0.0	1.8
Mode of cleaning floor of milk storage (%)				
No cleaning	1.6	0.9	0.0	1.4
Sweeping	63.0	62.3	86.3	63.0
Water	34.8	35.8	13.7	35.0
Water and soap	0.4	0.5	0.0	0.4
Other	0.2	0.5	0.0	0.3

farm for around 17 min on average before the milk producer goes out to sell it, and during this time the milk is usually stored inside the house in a closed container. The floor of the milk storage room is often only swept rather than kept dust-free by water-based cleaning. Most households bring their milk to the milk collection point in steel or aluminium containers. The fact that 60% of the milk producers store their milk for 15 min or less, and that 98% of the milk producers store their milk for 30 min or less on the farm, suggests that most milk producers go out to market their milk right after milking. This is a good practice. The period between milking and cooling is a crucial determinant of microbial content of the milk, because during this period the microbial content grows exponentially.

However, the time of storage at the farm does not cover the complete time before cooling; it does not include the time required for taking the milk to the collection point, performing the market transaction and the time during which the milk is stored at the collection point. Most milk collection points in Andhra Pradesh do not have their own cooling facilities as yet, implying that milk is usually bulked in steel, aluminium, or plastic containers at the collection point, and after each shift (which may last a few hours) these bulk containers are transported to the milk processing plant where the milk is finally cooled down. As a result, it is not uncommon for the time between milking and cooling to take a few hours. This is one of the critical quality bottlenecks of most milk procured within India, which will need to be tackled if milk processors eventually want to produce a dairy product qualifying for exports to the west. However, it is a problem intrinsic to the current milk procurement system, based on a multitude of small dairy farmers, each bringing in very small quantities of milk.

Table 7.4 gives an overview of general hygienic practices in dairy production in the study region. Two-thirds of the farmers wash their animals with water on a daily basis. This percentage is higher for large dairy farmers. Washing the animals here does not only serve the purpose of keeping the dairy animals clean, but it is also a way to refresh the animals in hot weather conditions.

If sick dairy animals are treated with antibiotics, the majority of farmers (around 60%) do not separate the milk from these animals from other milk. Slightly more than one-third of the households throw away milk with antibiotics content. This increases to 60% for large dairy farmers. These figures suggest that milk buyers do not enforce strict antibiotics regulations. Part of this may be a result of the lack of adequate testing equipment. From an operational perspective, it is almost infeasible to test milk for antibiotics content at the supplier-level due to the very small volumes of milk brought in by each supplier.

Hardly any (0.4%) of the sampled dairy farms had an on-farm inspection of milk production and storage practices in the 5 years preceding the survey. When it comes to livestock diseases, foot and mouth disease (FMD) seems to be prevalent in the area, with more than 16% of the households reporting to have had dairy animals suffering from FMD over the 5 years preceding the survey.

Table 7.5 shows the results on the perceived probability of milk adulteration by milk buyers. Only one out of four milk suppliers reported that his milk is checked by the buyer for adulteration upon collection. Almost 40% of the milk suppliers claim that they could add urea, milk powder, caustic soda (to increase the non-fat solids (SNF) content) and vegetable oil (to increase the fat content) without their buyers detecting it. The traditional milkman and the cooperative channel perform better at this level than intermediary traders, who presumably carry less final responsibility for the milk's end product. These findings strengthen the results of a recent survey by the FSSAI, which found more than two-thirds of the analysed milk samples to be adulterated, mainly with water, but also with glucose and, in some cases, even formaldehyde and detergents were found (FSSAI, 2012).

Table 7.4. General hygienic practices in dairy management (authors' survey).

Practices	No. of dairy animals			
	1–2	3–5	>5	Average
Frequency of washing of dairy animals (%)				
Never	0.9	0.8	0.0	0.8
Weekly	29.0	19.3	17.0	25.5
Every other day	5.9	10.9	1.7	7.0
Daily	63.4	67.6	80.5	65.6
More than once daily	0.4	1.3	0.9	0.7
Other	0.4	0.1	0.0	0.3
Main source of water (%)				
Piped water/tap water	53.4	47.5	65.5	52.4
River/stream	19.1	29.7	26.6	22.5
Community ground pump	9.8	6.4	3.4	8.5
Rainwater harvested e.g. on rooftop	0.8	1.8	0.0	1.0
Private ground pump/well	14.0	10.4	3.1	12.3
Other	2.9	4.2	1.4	3.2
Mode of washing dairy animals on a daily basis (%)				
No washing	8.7	4.8	1.7	7.2
Water	56.9	55.0	68.5	57.1
Water and soap/detergent	1.8	2.2	2.4	1.9
Water and scrub	32.1	37.5	26.7	33.3
Other	0.5	0.4	0.6	0.5
When is the dung removed (%)?				
Immediately	12.8	8.6	6.9	11.3
Several times a day	36.2	41.1	48.6	38.3
Once a day	45.5	43.3	41.9	44.7
Less than once a day	5.5	7.0	2.6	5.8
Drainage system for urine disposal (%)	22.6	18.5	25.9	21.6
Milk of dairy animals receiving antibiotics kept separated (%)	36.2	41.2	65.0	39.3
Milk of dairy animals receiving antibiotics thrown away (%)	32.2	37.7	59.7	35.4
Inspections for safety and quality standards (%)	0.7	0.0	0.0	0.4
Incidence of dairy animal diseases in past 5 years (% of households)				
Foot and mouth disease (FMD)	12.3	21.4	34.3	16.2
Haemorrhagic septicaemia	5.1	11.8	16.5	7.6
Black quarter	0.0	1.9	5.1	0.8
Mastitis	0.5	1.0	0.0	0.6
Diarrhoea	1.1	2.0	3.7	1.5

Table 7.5. Detection of milk adulteration (authors' survey).

Particulars	Direct marketing	Traditional milkman	Intermediary trader	Cooperative	Average
Does your milk buyer test for milk adulteration? (% 'Yes')	24.5	32.1	21.8	19.5	23.6
Would your milk buyer notice, if you added [item] to your milk? (% 'Yes for sure' or 'Probably')					
Water	60.6	73.6	53.2	79.2	69.2
Urea	55.4	62.8	42.7	64.6	57.4
Milk powder	58.5	63.4	47.3	71.9	62.8
Vegetable oil	61.9	63.4	46.7	72.0	62.5
Caustic soda	57.6	63.2	46.5	72.0	61.4

Conclusions

Food safety and quality management systems are still in their infancy in India. The current food safety regulations present high risks of food hazards to many consumers and may be considered a risk to public health in India. For example, in the dairy sector, milk adulteration remains common and goes widely unpunished, which means that a poisonous milk crisis is just as likely to happen in India as it used to be in China until a few years back.

The establishment of the FSS Act in 2006 and its implementation in 2014 show that concerns for food safety are on the rise, and that these have reached policy makers. The implications of this changing mind-set are uncertain in the short-term, but once food safety standards are enforced in the way policymakers have conceived them, the potential implications for food supply chains in India will be substantial.

In many countries around the world, the advent of food safety and quality standards has led to significant structural changes in agro-food systems, including the development of stronger vertical links between processors and retailers on the one hand and upstream producers on the other hand. As the agricultural supply base in India is made up of millions of small farmers, such changes will surely have a profound impact on rural livelihoods. While the expectations of such changes may be confronted with strong opposition in India, the potential benefits should not be underestimated. Not only domestic consumers may benefit from improved food safety and quality, there may as well be important implications for international trade.

If India succeeds in aligning its own food standards with international norms as laid down in the Codex Alimentarius, it may surmount current SPS (or other NTM) barriers to exports and gain importantly from new opportunities for trade. Food standards are currently the crucial determinants of trade with high-income countries such as the EU and the USA. Those countries which are currently importing dairy products from India (mainly neighbouring countries in South Asia and countries in the Middle East) may increase their food standards over the next decades as well.

It is important to note that food standards are not always exogenous. Policy makers and research scholars have interpreted the proliferation of NTMs over recent decades as a new form of protection-in-disguise (Sturm, 2006; Swinnen and Vandemoortele, 2011; Beghin and Li, 2013). This would imply that standards are more likely to arise in those commodities where expected trade volumes (in the absence of standards) would be highest. Hence, the endogeneity of food standards may further reinforce their impacts on global trade. As a result, in spite of the fact that NTM issues tend to be disregarded in computable general equilibrium modelling for trade policy analysis, their implications for trade can be substantial and should not be ignored.

Notes

Mara P. Squicciarini is a Doctoral Fellow of the Research Foundation Flanders (FWO). Research for this study was conducted before Anneleen Vandeplas joined the European Commission, while she was a Postdoctoral Fellow of the Research Foundation Flanders (FWO).

[1] To illustrate this, consider the Government of India's declaration of policy objectives, summarized every 5 years in a formal Five-Year Plan. For example, the 7th Five-Year Plan of India (spanning the period 1985–1990) stated explicitly that 'continued fast agricultural growth and self-sufficiency in food must remain a top priority concern of planning in India' (GoI, 1985). The 9th Five-Year Plan (1997–2002) argued that 'Dependence on food aid or largescale imports may entail unacceptable compromises on national security policies. [...] food production is the predominant means of livelihood for a large section of peasant cultivators and agricultural labourers, who are not easily shiftable to other occupations, at least for quite some time. The process of production of food in this predominantly small-holding agricultural economy ensures employment, income as well as food security simultaneously. Last, but not the least, the country

continues to produce most of the times cheaper foodgrains, particularly cereals, as compared to CIF cost of imports. So, except during the years of severe shortfall in production, imports could not be a normal supplement to domestic supplies' (GoI, 1997). Finally, the most recent 12th Five-Year Plan (2012–2017) highlights that 'In the backdrop of the price trends in the international food markets, it would be prudent to plan not only for self-sufficiency in basic food production, but also to maintain a surplus. This surplus can contribute to meeting critical food shortfalls in the neighbouring countries of the region and may strengthen a peaceful climate in the region' (GoI, 2011a).

[2] Food safety and food quality are interrelated concepts. Will and Guenther (2007) define food safety as 'the assurance that food will not cause adverse health effects to the final consumer', and food quality as 'the totality of characteristics of an entity that bears on its ability to satisfy stated and implied needs'. Hence, food quality is a broader concept and includes food safety characteristics.

[3] For a critical discussion of the Food Safety and Standards Bill, see Madhavan and Sanyal (2006) and Govindan (2007).

[4] The implementation of the regulation on the licensing requirement for food business operators was initially postponed to 2014. In 2014, the regulation was implemented, but with an amendment that increased the timeline within which businesses had to comply to 36 months from the moment of publication of the regulation. This means businesses effectively have until 2017 to obtain a licence (FSSAI, 2013, 2014).

[5] The abbreviation cfu/g stands for colony forming units per gram and is an estimate of the concentration of viable bacterial or fungal cells per gram of the sample under consideration.

[6] The Milk and Milk Products Order, 1973 (amended in 2002), aims at monitoring the quality of milk and milk products. It covers processing, distribution and procurement of dairy products, with a main focus on sanitary requirements for dairy plants.

The Standards of Weights and Measures Act (1976), and the Standards of Weights and Measures (Packaged Commodities) Rule (1977) laid down certain conditions for all packaged food (including dairy products), especially with respect to labelling of the food products: the nature of the commodity, the manufacturer, the quantity, the date of manufacture and its shelf-life, as well as the maximum retail price should be mentioned on the package.

Manufacturers of local as well as imported milk powders and condensed/evaporated milk need an 'ISI' label, issued by the Bureau of Indian Standards (BIS), which is the national standards organization under the aegis of Ministry of Consumer Affairs, Food and Public Distribution. The BIS standards cover raw materials permitted and their quality parameters, hygienic conditions under which products are manufactured and packaging and labelling requirements.

Finally, there is an additional legislation on infant foods: the Infant Milk Substitutes, Feeding Bottles, and Infant Foods (Regulation of Production, Supply and Distribution) Act, 1992, issued by the Ministry of Women and Child Development (GoI, 1992).

[7] In 2011, there were 817 dairy companies registered in the organized sector in India: 138 in the cooperative sector, 658 in the private sector and 21 'others' (e.g. public–private partnerships) (DAHD, 2010). Apart from the formal sector, there is still a huge unorganized market in which milkmen or *dudhiyas* operate: vendors who collect milk from local producers and sell it in both urban and non-urban areas. In fact, the latter are handling the major part of total milk production.

[8] The similarity between Indian and Chinese dairy procurement systems and the concomitant food safety hazards have been named as one of the reasons for Fonterra's withdrawal from the Indian dairy market in 2009 (VGBO, 2009).

[9] Nevertheless, India's exports of dairy products have grown significantly in the past decades, especially after 2001, when quantitative restrictions on the exports of dairy products were lifted. Since 2007, there are no longer excise duties to be paid on dairy exports. Moreover, since April 2008, dairy products have been exempted from export subsidies under the 'Focus Market Scheme', a subsidy intended to offset freight costs to selected countries, amounting to 3% of Free On Board (FOB) value of exports.

[10] An additional obstacle for dairy imports is that at the point of entry in India, all imported foods are randomly sampled. Since 2004, the Ministry of Commerce and Industry, Government of India, requires that imports of a certain number of listed 'high risk' food items (such as fats, milk powder, infant milk food and condensed milk) are subject to 100% sampling (except for some exempted fresh and highly perishable products that are, for instance, imported by hotels and restaurants).

[11] This sanitary requirement is based on the precautionary principle, as there is still considerable uncertainty on the short- and long-term effects of synthetic oestrogen on human and environmental health.

[12] Since 2008, this ban has been repeatedly extended. At the time of this writing (April 2015), the current ban is in force at least up to June 2015 (or until further notice) (GoI, 2014).

[13] For example, raw cow milk in the EU has to comply with the following criteria: the plate count (at 30°C) must be less than 100,000/ml. The somatic cell count must be less than 400,000/ml. In comparison, the corresponding standards for milk for liquid consumption in the USA are 100,000/ml and 750,000/ml (FDA, 2011). The US regulations also specify that milk for liquid consumption has to be 'cooled to 10°C or less within 4 hours or less of the commencement of the first milking, and to 7°C or less within 2 hours after the completion of milking.'

[14] In fact, as recent as 2008, the EC carried out a mission to India in order to evaluate the EIC's competence for certifying dairy processing plants, but decided that 'the conditions for the production of milk products for export do not meet the EU requirements as laid down by Commission Decision 2004/438/EC', and that the EIC 'did not ensure that the principles of certification equivalent to those laid down in Council Directive 96/93/EC are followed' (for the full report, see EC, 2008). A major concern of the EU relates to India's monitoring system of residue levels in milk of veterinary drugs, pesticides, aflatoxin and heavy metals, which is why India has recently started to implement a residue monitoring plan for exports to the EU (EIC, 2011).

[15] See EC (2009) for a more detailed overview of EU sanitary requirements for food of animal origin, such as dairy products.

[16] To date, there are 72 dairy processing plants approved for exports. This means that these plants can export dairy products to countries such as Sri Lanka and Singapore. For a full list, see EIC (2012).

References

Anders, S.M. and Caswell, J.A. (2009) Standards as barriers versus standards as catalysts: Assessing the impact of HACCP implementation on US seafood imports. *American Journal of Agricultural Economics* 91(2), 310–321.

Beghin, J.C. and Li, Y. (2013) *The Political Economy of Food Standard Determination: International Evidence from Maximum Residue Limits*. Iowa State University Department of Economics Working Paper No. 13011. Ames, Iowa, USA. Available at: http://www.econ.iastate.edu/sites/default/files/publications/papers/p16181-2013-05-08.pdf (accessed 21 April 2015).

Birol, E., Roy, D. and Torero, M. (2010) *How Safe Is My Food? Assessing the Effect of Information and Credible Certification on Consumer Demand for Food Safety in Developing Countries*. IFPRI Discussion Paper 01029. International Food Policy Research Institute, Washington, DC. Available at: http://www.ifpri.org/publication/how-safe-my-food (accessed 13 November 2012).

CAC (Codex Alimentarius Commission) (1979) Recommended international code of hygienic practice for foods for infants and children. No. CAC/RCP 21-19791. Available at: http://std.gdciq.gov.cn/gssw/JiShuFaGui/CAC/cxp_021e.pdf (accessed 14 November 2012).

CAC (Codex Alimentarius Commission) (2004) Proposed draft revision of the recommended international code of practice for foods for infants and children. Food and Agriculture Organization of the United Nations, and World Health Organization. Available at: ftp://ftp.fao.org/codex/Meetings/CCFH/ccfh37/fh37_04e.pdf (accessed 21 April 2015).

CSE (Centre for Science and Environment) (2004) A Briefing Paper on Pesticide Contamination and Food Safety: Poison vs Nutrition. Available at: http://www.indiaenvironmentportal.org.in/files/poison.pdf (accessed 13 November 2012).

DAHD (Department of Animal Husbandry Dairying & Fisheries) (2010) *Basic Animal Husbandry Statistics 2010*. Government of India, Ministry of Agriculture, New Delhi. Available at: http://dahd.nic.in/dahd/upload/BAHS_2010.pdf (accessed 13 November 2012).

Dixit, S., Khanna, S.K. and Das, M. (2013) All India survey for analyses of colors in sweets and savories: Exposure risk in Indian population. *Journal of Food Science* 78, 642–647.

EC (European Commission) (2008) *Final report of a mission carried out in India from 16 May to 29 May 2008 in order to evaluate the animal and public health controls over the production and processing of dairy products intended for export to the EU*. European Commission Health and Consumers Directorate-General, Directorate F - Food and Veterinary Office. Available at: http://ec.europa.eu/food/fvo/act_getPDF.cfm?PDF_ID=7353 (accessed 13 February 2013).

EC (European Commission) (2009) Regulation (EC) No 853/2004 of the European Parliament and of the Council of 29 April 2004 laying down specific hygiene rules for food of animal origin (OF L 226, 25.6.2004, p. 22) - with Amendments up to 2009. European Parliament and Council. Available at:

http://eur-lex.europa.eu/LexUriServ/LexUriServ.do?uri=CONSLEG:2004R0853:20090420:EN:PDF (accessed 13 November 2012).

Ehmke, M.D., Lusk, J.L. and Tyner, W. (2008) Measuring the relative importance of preferences for country of origin in China, France, Niger, and the United States. *Agricultural Economics* 38(3), 277–285.

EIC (Export Inspection Council) (2011) India Milk Product Residue Monitoring Plan (RMP) for Exports to EU: 2011–2012. Ministry of Commerce and Industry, Government of India, New Delhi. Available at: http://www.eicindia.gov.in/eic/pdf/Archive_2012/RMP%202011-12%20-%20Milk%20Products.pdf (accessed 13 February 2013).

EIC (Export Inspection Council of India) (2012) *List of Approved Units – Milk Products*. Ministry of Commerce and Industry, Government of India. Available at: http://www.eicindia.gov.in/eic/units/milk-appvdunits.htm (accessed 13 November 2012).

FAO (Food and Agriculture Organization) (2012) *FAOSTAT*. Available at: http://faostat.fao.org (accessed 14 November 2012).

FDA (2011) Grade 'A' Pasteurized Milk Ordinance, 2011 Revision. US Department of Health and Human Services, Public Health Service, Food and Drug Administration. Available at: http://www.fda.gov/downloads/Food/GuidanceRegulation/UCM291757.pdf (accessed 13 November 2012).

FICCI (Federation of Indian Chambers of Commerce and Industry) (2010) Bottlenecks in Indian Food Processing Industry-Survey 2010. Available at: http://www.ficci.com/SEDocument/20073/Food-Processing-Bottlenecks-study.pdf (accessed 14 November 2012).

Finzer, L.E., Ajay, V.S., Ali, M.K., Shivashankar, R., Goenka, S., Sharma, P., Pillai, D.S., Khandelwal, S., Tandon, N., Reddy, K.S., Narayan, K.M. and Prabhakaran, D. (2013) Fruit and vegetable purchasing patterns and preferences in South Delhi. *Ecology of Food and Nutrition* 52(1), 1–20.

FSSAI (Food Safety and Standards Authority of India) (2012) Executive Summary on National Survey on Milk Adulteration 2011. Available at: http://fssai.gov.in/Portals/0/Pdf/sample_analysed(02-01-2012).pdf (accessed 20 April 2015).

FSSAI (Food Safety and Standards Authority of India) (2013) Statutory Advisory F.No. 1/1/Enf-1/FSSAI/2012. 5 February 2013. Available at: http://fssai.gov.in/Portals/0/Pdf/Statutory_Advisory%20(05-02-2013).pdf (accessed 20 April 2015).

FSSAI (Food Safety and Standards Authority of India) (2014) F.no. 2-15015/30/2012- Food Safety and Standards (Licensing and Registration of Food Business) Amendment Regulations, 2011. Published in Gazette on 10 June 2014, relating to time extension of Sub-regulation. Available at: http://foodlicensing.fssai.gov.in/PDF/English_Notification(13.06.14).pdf (accessed 20 April 2015).

Fulponi, L. (2006) Private voluntary standards in the food system: The perspective of major food retailers in OECD countries. *Food Policy* 31(1), 1–13.

GoAP (Government of Andhra Pradesh) (2009) Andhra Pradesh Land Record Information. Andhra Pradesh State Unit, National Informatics Centre, Hyderabad. Available at: http://apland.ap.nic.in. (accessed 13 February 2013).

GoI (Government of India) (1985) *7th Five Year Plan*, Vol. 1. Planning Commission, New Delhi. Available at: http://planningcommission.nic.in/plans/planrel/fiveyr/7th/vol1/7vol1.html (accessed 13 February 2013).

GoI (Government of India) (1992) The Infant Milk Substitutes, Feeding Bottles and Infant Foods (Regulation of Production, Supply and Distribution) Act, 1992. Available at: http://wcd.nic.in/infantmilkpact1.pdf (accessed 13 February 2013).

GoI (Government of India) (1997) *9th Five Year Plan*, Vol. 2. Planning Commission, New Delhi. Available at: http://planningcommission.nic.in/plans/planrel/fiveyr/9th/vol2/v2c4-3.htm (accessed 13 February 2013).

GoI (Government of India) (2004) The Prevention of Food Adulteration Act, 1954 & The Prevention of Food Adulteration Rules, 1955 (as on 01.10.2004). Available at: http://dbtbiosafety.nic.in/act/PFA%20Acts%20and%20Rules.pdf (accessed 13 November 2012).

GoI (Government of India) (2006) *Food Safety and Standards Act, 2006*. Ministry of Law and Justice, Legislative Department, New Delhi. Available at: http://www.fssai.gov.in/Portals/0/Pdf/FOOD-ACT.pdf (accessed 14 November 2012).

GoI (Government of India) (2011a) *Faster, Sustainable, and More Inclusive Growth. An approach to the Twelfth Five Year Plan*. Planning Commission, New Delhi. Available at: http://planningcommission.nic.in/plans/planrel/12appdrft/appraoch_12plan.pdf (accessed 13 February 2013).

GoI (Government of India) (2011b) *Monitoring of Pesticide Residues at National Level: Annual Report 2010–11*. Department of Agriculture and Cooperation, Ministry of Agriculture, New Delhi. Available at: http://agricoop.nic.in/dacdivision/MPRNL.pdf (accessed 13 November 2012).

GoI (Government of India) (2011c) *Annual Report 2010–2011*. Department of Commerce, Ministry of Commerce & Industry, New Delhi. Available at: http://commerce.nic.in/publications/contents-PDF-2010-11.htm (accessed 13 February 2013).

GoI (Government of India) (2014) *Notification No. 84 (RE-2013)/2009/2014 of 23.06.2014 on Prohibition on import of milk and milk products from China*. Ministry of Commerce & Industry, Department of Commerce, New Delhi. Available at: http://dgft.gov.in/exim/2000/NOT/NOT13/not8413.htm (accessed 20 April 2015).

Govindan, A. (2007) *India Food and Agricultural Import Regulations and Standards. India's Food Safety Standards Act - Status and Outlook 2007* (No. GAIN Report IN7033). Global Agricultural Information Network, USDA Foreign Agricultural Service, New Delhi.

Henson, S. (2007) The role of public and private standards in regulating international food markets. *Journal of International Agricultural Trade and Development* 4(1), 63–81.

Henson, S. and Reardon, T. (2005) Private agri-food standards: Implications for food policy and the agri-food system. *Food Policy* 30(3), 241–253.

Krishna, V.V. and Qaim, M. (2008) Consumer attitudes toward GM food and pesticide residues in India. *Applied Economic Perspectives and Policy* 30(2), 233–251.

Madhavan, M.R. and Sanyal, K. (2006) PRS Legislative Brief Food Safety and Standards Bill, 2005. India Together: The News in Proportion. Available at: http://www.indiatogether.org/2006/feb/law-foodsafe.htm (accessed 13 November 2012).

Maertens, M. and Swinnen, J. (2008) Standards as barriers and catalysts for trade, growth and poverty reduction. *Journal of International Agricultural Trade and Development* 4(1), 47–61.

Minten, B., Murshid, K.A.S. and Reardon, T. (2011) *The Quiet Revolution in Agrifood Value Chains in Asia: The Case of Increasing Quality in Rice Markets in Bangladesh*. IFPRI Discussion Paper 01141, International Food Policy Research Institute, Washington, DC.

Minten, B., Vandeplas, A. and Swinnen, J. (2012) Regulations, brokers, and interlinkages: The institutional organization of wholesale markets in India. *Journal of Development Studies* 48(7), 864–886.

Munasib, A. and Roy, D. (2011) *Sanitary and Phytosanitary Standards as Bridge to Cross*. IFPRI Discussion Paper 01029. International Food Policy Research Institute, Washington, DC. Available at: http://www.ifpri.org/publication/sanitary-and-phytosanitary-standards-bridge-cross (accessed 13 November 2012).

NDDB (National Dairy Development Board) (2009) *Base Working Paper on Strategy and Action Plan for Ensuring Safety of Milk and Milk Products*. Food Safety and Standards Authority of India, Expert Group on Milk and Milk Products, Anand, Gujarat, India, p. 41.

OECD (Organization for Economic Co-operation and Development) (2001) *Agricultural policies in emerging and transition economies: special focus on non-tariff measures*. OECD, Paris.

Pingali, P. (2007) Westernization of Asian diets and the transformation of food systems: Implications for research and policy. *Food Policy* 32(3), 281–298.

Reardon, T. and Berdegue, J.A. (2002) The rapid rise of supermarkets in Latin America: Challenges and opportunities for development. *Development Policy Review* 20(4), 371–388.

Roy, D., Birol, E., Deffner, K. and Karandikar, B. (2010) Developing country consumers' demand for food safety and quality: Is Mumbai ready for certified and organic fruits? In: Bennett, J. and Birol, E. (eds) *Choice Experiments in Developing Countries: Implementation, Challenges and Policy Implications*. Edward-Elgar Publishing, Cheltenham, UK, pp. 261–277.

Sareen, S. (2003) *Food Safety in Food Security and Food Trade: Case Study: India Responds to International Food Safety Requirements*. IFPRI Focus 10, Brief 11 of 17. International Food Policy Research Institute, Washington, DC. Available at: http://www.ifpri.org/publication/case-study-india-responds-international-food-safety-requirements (accessed 14 November 2012).

Sareen, S. (2007) Certification scheme for dairy products export. In: Gupta, P.R. (ed.) *Dairy India 2007*. Thomson Press (India) Limited, New Delhi, pp. 87–89.

Sharma, D. (2005) Food Safety Bill May Hurt Hawkers. India Together: The News in Proportion. No. 10. February. Available at: http://indiatogether.org/2005/feb/dsh-safefood.htm (accessed 13 November 2012).

Sharma, R.K., Agrawal, M. and Marshall, F.M. (2009) Heavy metals in vegetables collected from production and market sites of a tropical urban area of India. *Food and Chemical Toxicology* 47(3), 583–591.

Smith, G. (2009) *Interaction of Public and Private Standards in the Food Chain*. OECD Food, Agriculture and Fisheries Working Paper No. 15. Organization for Economic Co-operation and Development, Paris.

Sneyers, A. and Vandeplas, A. (2013) *Girl Power in Agricultural Production: How Much Does it Yield? A Case-Study on the Dairy Sector in India*. LICOS Discussion Paper 2013/341, LICOS, KU Leuven, Belgium.

Squicciarini, M. and Vandeplas, A. (2010) *The Dairy Sector of Andhra Pradesh: Findings from a Company Survey* (No. TAPSIM deliverable D5.3). LICOS, KU Leuven, Belgium.

Sturm, D.M. (2006) Product standards, trade disputes, and protectionism. *Canadian Journal of Economics*, 39(2), 564–581.

Swinnen, J. (ed.) (2007) *Global Supply Chains, Standards and the Poor*. CABI Publishing, Wallingford, UK.

Swinnen, J. and Vandemoortele, T. (2011) Trade and the political economy of food standards. *Journal of Agricultural Economics* 62(2), 259–280.

TUV South Asia (2007) *Comprehensive Study of Food Regulations and Standards, Food Testing Infrastructure*. Codex Resource System, HACCP/ISO/Organic Products Certification/Accreditation System. Part I: Executive Summary. Ministry of Food Processing Industries, Government of India, New Delhi.

Umali-Deininger, D. and Sur, M. (2007) Food safety in a globalizing world: Opportunities and challenges for India. *Agricultural Economics* 37, 135–147.

US FDA (US Department of Health and Human Services, Food and Drug Administration) (2012) Guidance for Industry: Questions and Answers Regarding Food Facility Registration, 6th edn. Available at: http://www.fda.gov/food/guidanceregulation/ucm331959.htm (accessed 21 April 2015).

Vandemoortele, T. and Deconinck, K. (2013) When are private standards more stringent than public standards? *American Journal of Agricultural Economics* 96(1), 154–171.

Vandeplas, A. and Minten, B. (2015) Food Quality in Domestic Markets of Developing Economies: A Comparative Study of Two Countries. *Agricultural Economics* 14(5), 617–628.

VGBO (Victorian Government Business Office-India) (2009) *Market Opportunities in Indian Dairy Value Chain*. A research study for the VGBO. Department of Business and Innovation, Australia.

Wikimedia Commons (2008) Andhra Pradesh Districts map. Available at: https://upload.wikimedia.org/wikipedia/commons/2/29/Andhra_Pradesh_districts_map.svg (accessed 13 November 2012).

Will, M. and Guenther, D. (2007) *Food Quality and Safety Standards as Required by EU Law and the Private Industry. With special reference to the MEDA countries' exports of fresh and processed fruit & vegetables, herbs & spices*, 2nd edn. Division 45 – Agriculture, Fisheries, and Food. Deutsche Gesellschaft fuer Technische Zusammenarbeit (GTZ), Germany.

8 India's Poultry Sector: Trade Prospects

Rajesh Mehta,[1]* R.G. Nambiar[2] and P.K. Joshi[3]†
[1]Formerly of Research and Information System for Developing Countries, New Delhi; [2]FLAME University, Pune, Maharashtra; [3]International Food Policy Research Institute, New Delhi, India

Introduction

Today, poultry is one of the fastest growing segments of Indian agriculture and the poultry industry is growing at the rate of 8–10% per annum. India is also the third largest producer of eggs and the sixth largest producer of chicken meat in the world and has the potential to emerge as the world leader in poultry. India is also an exporter of poultry meat and eggs to the European Union (EU), the Middle East and neighbouring South Asian countries. As of today, Indian poultry exports constitute only a fraction of the total world trade. The domestic poultry industry is definitely price competitive, and India is gifted with natural resources that favour the growth of poultry. However, these price advantages are being offset by the countless tariff and non-tariff barriers prevalent in export destination country markets. The implementation of Uruguay Round talks should have mitigated some of these trade barriers, but their implementation is taking a lot of time. Nevertheless, there has been a gradual opening up of the trade in agricultural goods since 1991. The vital questions are: What are the trade opportunities for India's poultry products? Is there scope for expansion in India's poultry trade? This chapter has assessed the potential of such trade opportunities for India.

World Poultry Industry – A Profile

Global production trends

In Table 8.1, the yearly production statistics are presented for both poultry meat and total meat production in the world. A glance at the data reveals two facts: (i) poultry meat production represents approximately one-third of the total meat production; and (ii) production of poultry meat has grown faster than of other meats. For instance, the production of poultry meat has recorded an annual growth of 3.5% as against 1.9% growth for meat as a whole.

Worldwide, chicken is the dominant source of poultry meat, i.e. around 86–87% is chicken meat. The remaining 13–14% is duck, turkey and geese meat together. The production of chicken meat jumped from 58.6 Mt in 2000 to 86.2 Mt in 2010, i.e. 30 Mt

* Rajesh Mehta passed away at the age of 65 years on 12 May 2015, during the process of editing this book. We deeply regret the loss to his colleagues, friends and family

† Corresponding author, e-mail: p.joshi@cgiar.org

Table 8.1. World production of total meat, poultry meat and eggs, 2000–2010 (FAO, 2012).

Year	Total meat production (Mt)	Poultry meat production (Mt)	Share of poultry meat in total meat (%)	Production of hen eggs (Mt)
2000	234.2	68.6	29.29	55.3
2001	236.9	71.0	29.98	56.5
2002	243.8	73.8	30.28	57.8
2003	248.5	75.1	30.25	58.8
2004	254.2	78.t	30.76	60.0
2005	260.2	80.8	30.94	61.3
2006	267.7	83.0	30.99	62.6
2007	274.0	88.0	32.08	64.5
2008	280.1	92.5	33.03	66.8
2009	285.3	94.2	33.02	68.0
2010	292.8	98.0	33.50	68.9

were added over the 10-year period. Also, almost all the increase in poultry meat production pertained to the increase in chicken meat production.

Egg production is another component of the poultry industry. Global egg production rose from 55 Mt in 2000 to 68.9 Mt in 2010 (Table 8.1).

Leading producers of chicken meat

The top three producers of chicken meat in the world are Brazil, China and the USA – these three together produce nearly 45% of the world poultry production (Table 8.2). Next come Mexico (2.68%), the Russian Federation (2.53%) and India (2.53%), followed by Indonesia (1.65%), Iran (1.65%), Argentina (1.59%) and South Africa (1.47%).

China leads the world with 44% of global egg production followed by the USA with 8.5% and India with a 5.3% share, then come Japan (3.9%), Mexico (3.7%) and the Russian Federation (3.5%). Indonesia, Ukraine and Thailand have 3% or less share each in world egg production. The combined egg production of these ten countries was 49 Mt (75% of global production) in 2010, as depicted in Table 8.2.

International trade scenario in poultry

Table 8.3 presents annual data on international trade in poultry meat for the period 2000–2009. A perusal of Table 8.3 reveals two things: (i) the volume of international trade in poultry has been on the rise during the past 10 years: the volume of exports in 2009 amounted to 14.2 Mt as against 8.7 Mt in 2000, i.e. the volume of poultry trade has almost doubled during 2000–2009; and (ii) the proportion of poultry production entering international trade, though small, has depicted an ascending trend. In 2000, this proportion was 12.8% and it increased to 15.1% in 2009.

Brazil is the leading exporter of poultry meat – the country accounted for more than one-fifth of the world exports in 2010. Next in the league come the USA and The Netherlands with a relative share roughly of 15% each. Germany and France follow next with a share of 7–8% each, followed by Poland with 4%. The other countries, namely Belgium, the UK, Hungary and China, have shares of 2–3% each. Note that, though India is a very large producer of poultry, it has only a marginal presence in its international trade. India's share in world exports of poultry was less than 1% and it ranked 39th in terms of export share in 2010.

Among the importers, Germany tops the list with 11% of the world total. Both Hong Kong and the UK are also major importers with a share of 7–8% each in the world totals. Next in the importing league come The Netherlands, Saudi Arabia, Japan and France, with shares varying between 5% and 6%, followed by the Russian Federation, China and Belgium with a share of 4% each. India's share in the world imports is negligible.

Table 8.2. Top ten countries' production of chicken meat and hen eggs, 2010 (FAO, 2012).

Chicken meat				Hen eggs			
Rank	Country	Production (Mt)	Share (%)	Rank	Country	Production (Mt)	Share (%)
1	USA	16.97	19.72	1	China	28.00	44.04
2	China	11.85	13.77	2	USA	5.41	8.51
3	Brazil	10.69	12.42	3	India	3.41	5.37
4	Mexico	2.68	3.12	4	Japan	2.51	3.96
5	Russian Federation	2.53	2.94	5	Mexico	2.38	3.75
6	India	2.30	2.67	6	Russian Federation	2.27	3.58
7	Indonesia	1.65	1.92	7	Brazil	2.08	3.28
8	Iran	1.65	1.92	8	Indonesia	1.37	2.17
9	Argentina	1.60	1.86	9	Ukraine	1.02	1.60
10	South Africa	1.47	1.71	10	Thailand	0.98	1.54

Table 8.3. Poultry meat: trade, production and their ratio, 2000–2009 (FAO, 2012).

Year	Poultry trade (export) (Mt)	Poultry meat production (Mt)	Trade/ production (%)
2000	8.78	68.59	12.8
2001	9.58	71.00	13.5
2002	9.63	73.85	13.0
2003	10.12	75.18	13.5
2004	9.72	78.19	12.4
2005	10.96	80.84	13.6
2006	11.10	82.98	13.4
2007	12.62	87.91	14.4
2008	13.92	92.52	15.1
2009	14.21	94.20	15.1

Poultry Industry in India

Structure

A point worth emphasizing about Indian poultry is that it is self-sufficient in terms of technology and management of its poultry system. For instance, the country has developed genetically superior breeds[1] (pure line and grandparents) capable of: (i) adjusting to local conditions; and (ii) giving higher production. To support this expanding industry, the country has built over time a strong base, which comprises production of world-class medicines and vaccines, Specific Pathogen Eggs (SPE), farm and hatchery automation systems, pelleted feed,[2] eggs and poultry processing, a nationwide network of disease diagnostic laboratories, etc. Backyard poultry, though still important if we look at the number of birds under it, is less significant in terms of production level because of low productivity. While agricultural cultivation is still a major occupation, many farmers are taking to poultry to earn extra income.

It is a well-known fact that Indians prefer fresh chicken, which is generally sold as live birds and slaughtered in the neighbourhood shops Hence, around 95% of the poultry market is a wet market and only 5% is processed.[3] However, consumption trends are slowly changing with time. Today, with changing life-styles, the Indian consumer is accepting both dressed chicken and fresh chilled chicken. This changing consumption pattern provides a lot of scope for processing.

Integrated farming, also known as contract farming that has taken place in other countries, has also begun to take hold in India. Its operation has been negligible in the past two decades. However, in more recent times, large-scale vertical integration is slowly catching up. Pioneered first by Venkateshware Hatcheries (VH) Group[4] and more recently by the Suguna, the model is gaining popularity among poultry farmers, especially in the western and southern parts of India. The model is gaining wide acceptance because of the fact that the integrator takes most of the risks – provides day-old chicks, feed, medicines and supervision, all of which lead to higher productivity – while the farmers are assured of considerable earnings.

Based on Livestock Census 2007,[5] India has around 617 million chickens, while numbers of all other birds, except ducks, are negligible. Livestock Census also has data pertaining to size-wise distribution of poultry farms (i.e. number of birds per farm). As per the 2007 census, there are about 10,000 layer farms and 47,000 broiler farms in the country. About 65% of broilers are reared in small-scale (3000–10,000 birds) and medium-scale (10,000–50,000) farms. The remaining 35% of broilers are reared in large-scale farms with numbers varying from 50,000 to 400,000 birds. This distribution of total number of birds in different farm sizes is shown in Table 8.4. The share of birds reared in large farms is 26% of total birds.

Trends in poultry production in India

India's poultry industry has constantly been on the rise, as shown by Fig. 8.1. The data show:

1. A sharp jump in egg production. In 2009/10, India produced almost 60 billion eggs compared to 21 billion in 1990/91, i.e. almost a 200% rise over the 20-year period.
2. A steep rise in poultry meat production. India produced 2 Mt of poultry meat in 2009/10 as against 0.5 Mt in 1990/91, i.e. a three-fold increase in production.
3. Note the upward trend is felt much stronger after 2000/01 – poultry meat production rose from 0.9 Mt to 2 Mt between 2000/01 and 2009/10, and egg production increased from 30 billion to about 60 billion eggs.

This upward trend in poultry production can be attributed to various factors.

1. The productivity level: because of the genetic base, the productivity level (feed conversion ratio or FCR) of broilers and layers in India is comparable to that found in developed countries. For instance, the FCR for broilers is 1.8 kg of feed for 1 kg of meat, which is more or less the same as that in developed countries. Similarly for layers, the FCR is 130 g of feed per egg, which again is comparable to the developed world.
2. The production of poultry meat/eggs from broilers/layers is taking a shorter time period.
3. The surge in production depends, apart from productivity, on the size of poultry production also. As per the Livestock Census 2007, the total number of poultry birds was 648.8 million in the country. Out of this, 617.7 million were chicken: 325 million reared in poultry farms (154 million layers and 198 million broilers) and the rest in backyard poultries.
4. India is one country in the developing world with a self-reliant technology capable of producing every essential input – vaccines, farm and hatchery automation systems, feeds, etc.

Table 8.4. Indian chicken broilers: distribution of flocks by farm size (Livestock Census, 2007, Department of Animal Husbandry, Dairy and Fisheries, Government of India, New Delhi).

Broiler farm size (no.)	Percentage of birds in total
0–250	0.15
250–750	1.14
750–2,000	10.13
2,000–5,000	11.06
5,000–10,000	22.15
10,000–25,000	19.43
25,000–50,000	3.99
50,000–100,000	2.60
100,000–200,000	2.65
200,000+	26.71

Relative importance of poultry in agriculture

The rising ascendency of poultry can be gauged by looking at its contributions to different aspects of the economy. One such aspect that can be gauged easily is its share in the agriculture sector: the share rose to 3.4% in 2008/09 from 2.72% in 2004/05.

Another way to gauge its importance is by looking at the relative share of poultry meat vis-à-vis other meats, namely buffalo meat, mutton and pig meat in production/consumption. In Table 8.5, these data are presented for three years, viz. 1990, 2000

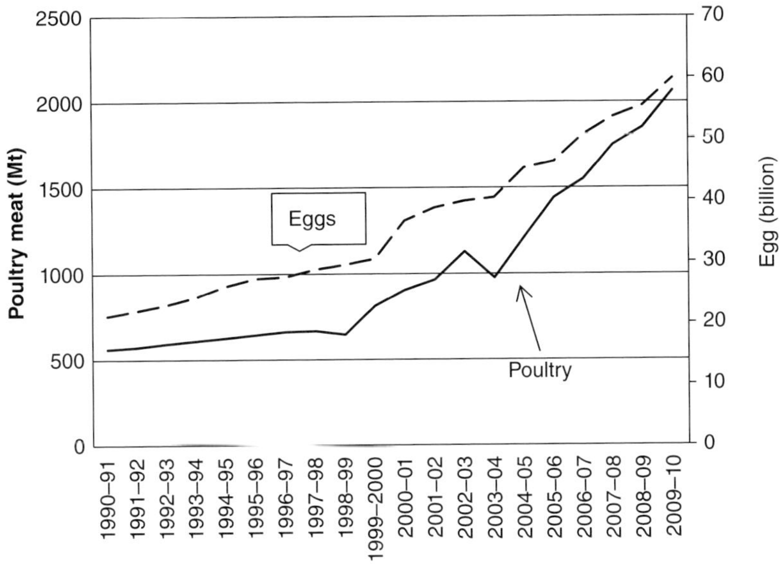

Fig. 8.1. Indian poultry: meat and egg production, 1990/91 to 2009/10 (eggs: GoI, 2010; poultry meat: FAO, 2012).

Table 8.5. India: production of different types of meat in 1990, 2000 and 2010 (FAO, 2012).

		Production (thousand tons)			% distribution		
Type of meat		1990	2000	2010	1990	2000	2010
Buffalo		1078	1255	1462	28.17	28.25	23.32
Cattle		1036	981	1086	27.06	22.07	17.32
Poultry	Chicken	531	864	2300	13.87	19.43	36.67
	Duck	30	40	38	0.78	0.90	0.60
Goat		430	469	586	11.23	10.55	9.35
Meats not elsewhere specified		127	147	175	3.32	3.31	2.79
Pig		413	456	332	10.79	10.48	5.30
Sheep		181	221	289	4.73	4.97	4.61
Total		3828	4445	6272	99.95	99.96	99.97

and 2010. A glance at the data shows that poultry meat has outpaced its other major competitors, namely buffalo meat, cattle meat, goat and sheep meat and others. Prior to 2000, both buffalo and cattle meats topped the list, but in 2010 poultry (and that again chicken meat) outpaced both these meats and now tops the list. High mutton prices, religious restriction on beef and pork and the limited availability of fish outside coastal regions have all contributed to make poultry meat the most preferred meat in India.

India's Trade in Poultry

Tables 8.6 and 8.7 present the export/import data of poultry meat and eggs for the years 1990, 2000 and 2009. Several points emerge from the data displayed in these tables. First, India has only a marginal position in international trade of poultry meat. In 2009, out of a total of 14.2 Mt of poultry meat exported across the world, only 0.002 Mt originated from India. Second, the import of poultry meat to India is very small. In 2009, poultry meat imports to India amounted to

Table 8.6. World and India: export and import of poultry meat in 1990, 2000 and 2009 (FAO, 2012).

	Export (t)			Import (t)		
Year	World	India	India/World %	World	India	India/World %
1990	2,677,081	228	0.009	2,649,508	0	0
2000	8,780,674	259	0.003	7,741,460	1	1.29
2009	14,212,076	1,656	0.012	12,701,540	61	4.80

Table 8.7. World and India: Export and import of eggs in 1990, 2000 and 2009 (FAO, 2012).

	Export (t)				Imports (t)			
	India		India in World %		India		India in World %	
Year	Hen eggs, in shell	Eggs – liquid, dried	Hen eggs, in shell	Eggs – liquid, dried	Hen eggs, in shell	Eggs – liquid, dried	Hen eggs, in shell	Eggs – liquid, dried
1990	1,524	23	0.18	0.02	0	0	0	0
2000	11,344	5,496	1.20	3.17	23	0	0.003	0
2009	44,180	7,286	2.69	2.25	66	12	0.004	0.004

only a meagre figure of 61 t as compared to world totals of 12.7 Mt. Third, India's main exports of poultry are eggs and egg powder. In 2009, India exported 44,180 t of hen eggs (2.6% of world totals) and 7286 t of liquid and dried eggs (2.25% of world exports). If we examine the data carefully, we can see India's share continuously rising in the first half of the 2000s decade, but diminishing thereafter.

The major destinations of India's poultry exports are its neighbouring Asian countries, the EU countries and the Middle East. Among the Asian countries, the most important are Pakistan, Maldives, Sri Lanka, Indonesia and Afghanistan; they absorbed about 40–45% of India's poultry exports. Among the EU countries, the most important are Germany, Denmark and The Netherlands; the three together accounting for 20% of India's poultry exports. Among the Middle East countries, the most important are Saudi Arabia, Bahrain and UAE; these countries together accounted for 8–10% of poultry exports from India.

As per UN COMTRADE data, a major source of India's imports of poultry is Brazil – nearly half of its imports are from Brazil. India also imports from EU countries, the three main countries being France, Germany and the UK; their combined share in India's poultry imports was 46% in 2010. On the other hand, a major export item from India is birds' eggs in shell. India is a major supplier to the world market for this item with a share of 11.7%. India has a significant presence in the export of birds' eggs, other than in-shell, and egg yolks (fresh, dried, cooked by steaming or by boiling in water, moulded, frozen or otherwise preserved), the share of egg yolks dried (HS040811) in global exports being 3.2%. In the rest of the items, however, India's presence is negligible.

The main imports to India are live fowls of species *Gallus domesticus* and ducks, weighing not more than 185 g. In 2009, out of the total poultry imports of US$5.4 million, the cited items amounted to US$4.9 million. Again, one can notice that India is a smaller market for imports than the rest of the world.

What kind of barriers exist in India against imports of poultry? A part of the answer follows from the tariff rates applicable to the imports of poultry products. For example, there are two items, namely cuts of edible offal of species *Gallus domesticus* fresh/chilled (HS code: 020713) and cuts of edible offal of species *Gallus domesticus* frozen (HS code: 020714) that attracted 100% import duty in 2012/13. Barring these two items, all the rest have a 30% import duty. These rates have not changed for the past two decades.

Price Competitiveness of India's Poultry Products

Table 8.8 reports the producer prices of hen eggs in shell, chicken live and chicken meat for the top five countries with the highest share in international trade. The prices in India have also been added at the end of the table. Such absolute price comparison is one of the ways of gauging competitiveness of India's poultry products.

The analysis of these data has shown that Indian prices for 'hen eggs in shell' were generally lower than those in Brazil, the USA, Germany, France and The Netherlands. Hence, in a scenario of agricultural trade liberalization, one can expect India to have a definite advantage in the case of egg export.

In the case of poultry meat, India is not at all price competitive. This may be due to the fact that the price of India's poultry meat does not differ over different cut-pieces. Most of Indian poultry meat is bought in the wet market. Chickens are slaughtered, dressed, and sold to the consumers in their presence. All the cut-pieces of chicken meat are sold to the consumer at a fixed price per bird. On the other hand, in western countries, the prices of different cut-pieces of chicken meat are significantly different. Chicken meat is sold in the markets in processed and/or frozen forms. The price of fillet (breast or other clean) meat is very high as compared to that of leg quarters,[6] because consumers in western countries prefer fillet meat over other cut-pieces. The price of leg-quarter is negligible. In fact, they throw leg-quarter in the dustbin at the time of dressing. On the other hand, India's preference of poultry meat is for leg pieces. The preference of Indian consumers for leg pieces (*Tangri*) can be seen clearly in restaurants. Hence, in a liberalized trade regime, there can be a win–win situation because India can export fillet meat and the western countries can export legs (including leg quarters) to India. This is an underassumption that India's landed price of breast meat in the EU and other western countries will be lower than that of their local prices, and similarly, the landed price of poultry legs from the EU and other western countries to Indian markets would be lower than the corresponding domestic prices in India.

Table 8.8. The producer price (US$/t) of poultry products of India, selected EU and other countries for years 2005 and 2009 (FAO, 2012).

	Hen eggs in shell		Chicken live (by weight)		Chicken meat	
Country	2005	2009	2005	2009	2005	2009
Brazil	672	916	752	1022	977	1327
USA	489	922	970	1008	1316	1381
Netherlands	686	1402	870	1094	1115	1402
Germany	1318	2408				
France	880	1597	1045	1449	1491	1906
India	580	726	1198	1503	1597	2004

Liberalization of India's Imports

Liberalization of India's tariffs

In this section, the potential effects of a complete elimination of tariffs by India on poultry meat imports are assessed. Currently, India levies an 87% tariff on the import of poultry meat. If this is completely removed, then what would be the effect of India's imports of poultry meat and, as a consequence, on its domestic prices and output?

Tariff elimination in all probability is expected to reduce the domestic price of poultry meat. The exact magnitude of the fall in price will depend on the response of imports to tariff cut. Due to lack of information on the magnitude of effects, the AGLINK-COSIMO model[7] was used. As per the model's estimates, the difference in prices before and after tariff reduction worked out to be Rs52,940 per 1000 t, which as a proportion of pre-liberalization price depicts a decline by 39%.[8] This in turn can be expected to stimulate imports. Consequently, India's imports of poultry meat are expected to increase from a level of 0.061 t to 1.519 Mt.

Thus, according to this scenario, the resulting effects are likely to be disastrous to the domestic poultry industry – production will tend to fall by 29%. This would affect the smaller farmers more than the larger ones because their profitability is lower.

Liberalization of maize imports

The two main feeds used by Indian poultry farmers are maize and soya. The availability of both of these at reasonable prices is very important because feed costs alone account for 50–70% of the total costs. Any change in the prices of these feeds will affect the growth of the poultry industry. The domestic production of maize has fallen considerably and has often led to shortages. Importing maize is one way of tackling this domestic shortage.

To capture the impact of this, we again resorted to the AGLINK-COSIMO model. The model results have shown no significant change in the domestic price of maize as the maize price in India is lower than the world price of coarse grains.

Conclusions

1. Poultry's share in India's exports and imports is negligible. In 2009, out of 14.2 Mt of poultry meat exports across the world, only 0.002 Mt originated from India. Within poultry, egg and egg powder are the most important export commodities. The major destinations of India's exports are the neighbouring Asian countries, the EU and the Gulf countries.
2. The Indian poultry sector is protected by applied (and bound) tariffs of 30–100%. Poultry meat, the most sensitive product, is protected by a tariff of 100%.
3. India is definitely price competitive in eggs, but not in poultry meat. Hence, one can expect India to have an edge in international trade of eggs, but not in poultry meat.
4. If India lowers import duty on poultry meat from the existing 87% to zero, then it will push up imports by 1.519 Mt. However, it will prove disastrous to the domestic poultry industry.

Notes

[1] Though imports of pure line have been open since 1985; however, due to the development of genetically superior breeds indigenously, there have been no imports of pure-line birds.
[2] Earlier, compound feed was not available. Now a number of feed mills have started up.
[3] There are around a dozen processing units for broilers and three units for egg processing.
[4] Venkateshware Hatcheries (VH) Group is a conglomerate, which produces 85% of day-old chicks (DoCs) of chicken layers and 73% of DoCs of chicken broilers in the country.
[5] The Livestock Census is taken every 6 years in India. The livestock census is for the year 2007. The data are collected at the state level because agriculture and livestock fall under the state subject.
[6] As per the supermarket, Sainsbury, the price of fresh Basics British Chicken Legs was £1.78/kg as compared to the price of fresh British Chicken Breast Fillet Portions of £12.49/kg. http://www.mysupermarket.co.uk/#/grocery-categories/fresh_chicken_pieces_in_sainsburys.html (accessed 21 August 2012).
[7] For details, see among others OECD (2007).
[8] See Mehta and Nambiar (2008) for details.

References

FAO (2012) FAO Statistical Yearbook. Available at: http://faostat.fao.org/site/291/default.aspx (accessed 3 July 2012).

GoI (2010) Animal Husbandry Statistics 2010. Government of India, New Delhi. Available at: http://dahd.nic.in/dahd/upload/BAHS_2010.pdf (accessed 21 August 2012).

Mehta, R. and Nambiar, R.G. (2008) The poultry industry in India. In: *Poultry in the 21st Century: Avian Influenza and Beyond*. FAO Animal Production and Health. Food and Agriculture Organization (FAO) of United Nations, Rome.

OECD (2007) Working Paper on Agricultural Policies and Markets 2007. Documentation of the AGLINK-COSIMO model, AGR/CA/APM(2006)16/Final. Available at: http://www.oecd.org/dataoecd/54/11/45172745.pdf (accessed 21 August 2012).

9 Employment Guarantee Programme and Income Distribution

G. Mythili*
Indira Gandhi Institute of Development Research, Mumbai, India

Introduction

The Government of India has launched many schemes for the generation of additional income for the poor. One such scheme which has received the attention of many economists and political analysts is the Mahatma Gandhi National Rural Employment Guarantee Scheme (MGNREGS). It is basically an employment generation programme for the rural poor households with the policy of direct transfer of money to them through provision of public works. The scheme stems from the enactment of the National Rural Employment Guarantee Act (NREGA) notified in September 2005, which was later renamed as the Mahatma Gandhi National Rural Employment Guarantee Act (MGNREGA). It intends to enhance the livelihood security in rural areas by providing at least 100 days of guaranteed wage employment in a financial year to every rural household whose adult member volunteers to do unskilled manual work.[1]

MGNREGS was launched in 2006 initially in 200 districts of India and was extended to cover 130 more districts in 2007. As per the act, employment shall be provided within a radius of 5 km of the village where the applicant resides at the time of applying. In cases where employment is provided outside such radius, it must be provided within the Block, and transport allowance and daily living allowance shall be paid in accordance with programme rules. The programme may also provide for the training and up-gradation of the skills of unskilled labourers.

The works permissible under the scheme include water conservation and water harvesting, drought proofing (afforestation and tree plantation), irrigation canals, renovation of traditional water bodies, tank desilting, land development, flood control and protection works including drainage in waterlogged areas, rural connectivity to provide all-weather access and any other work notified by the central/state government.

Various employment guarantee programmes have been launched by the Government of India since the 1980s to achieve multiple objectives, which included providing employment opportunities during lean agricultural seasons and incidence of natural disasters like floods, droughts, etc., and creating rural infrastructure which supports economic activities in rural areas. These were the National Rural Employment Programme (1980), Rural Landless Employment Guarantee Programme (1983), *Jawahar Rozgar Yojana*

* Corresponding author, e-mail: mythili@igidr.ac.in

(1989), Employment Assurance Scheme (1993), *Jawahar Gram Samridhi Yojana* (1999), *Sampoorna Grameen Rozgar Yojana* (2001) and Food for Work Programme (2004). Unlike the earlier wage employment programmes that were allocation-based, MGNREGS is demand-driven.

This chapter is devoted to analysing the impact of MGNREGS on income distribution across rural households in India.

An Overview of Past Studies

Several studies have analysed the impact of employment guarantee schemes (EGS) on the basis of partial equilibrium framework. The empirical studies on rural public work programmes have regarded these schemes as income insurance to the agrarian labour force due to the presence of seasonality (Basu, 2011), for building long-term capital assets (Basu, 1981) and as influencing rural–urban migration (Ravallion, 1991). Some of the notable studies are by Kamath (2010), Basu (2011), Mukherjee and Sinha (2011), Mahendra Dev (2011) and Jha and Gaiha (2012).

With the help of a theoretical model, Mukherjee and Sinha (2011) analysed the impact of MGNREGS on the labour market and the income of poor households in rural areas. They concluded that the income of the rural poor can be enhanced by creating work opportunities and market access and not merely through the launching of income-generating schemes like MGNREGS.

Analysing the impact of MGNREGS on the welfare of women and children using the secondary data of states and focus group discussions, Mahendra Dev (2011) concluded that there are significant regional variations in the functioning of MGNREGS across the country. It is working much better in states such as Rajasthan, Andhra Pradesh and Himachal Pradesh and not working well in states such as Bihar, Jharkhand and Uttar Pradesh. The author has also noted that the achievements are still short of its potential.

The macroeconomic impacts of MGNREGS in terms of output enhancement through the 'multiplier and accelerator' effect, if implemented properly, have been looked into by Kamath (2010). Since rural population has a higher propensity to consume, it will have a multiplier effect on output and the increasing output would stimulate private investment through an 'accelerator effect'. The author has pointed out the need of increased mobilization and awareness generation about MGNREGS to depart from the line of most of the existing poorly-implemented public work schemes.

A relatively important but neglected element of real income transfers, net of opportunity cost of time, due to MGNREGS has been studied by Jha *et al.* (2012). Using the primary household data for three states, the study reported that net transfers of MGNREGS are quite modest and its poverty-reduction potential is also limited. Basu *et al.* (2005) raised the issue of contestability effect of MGNREGS, i.e. whether the alternative source of employment in public works would raise the private wages to retain them in agriculture.

The impact of MGNREGS based on a general equilibrium approach has been discussed by Khan and Saluja (2007). Using a village survey, they concluded that even though it has many beneficial effects, the major concern was leakage and corruption. The authors have recommended a more decentralized administrative mechanism for the scheme to make it more effective. Based on the panel data of rural households, Ravi and Engler (2009) concluded that MGNREGS has a significant impact in alleviating rural poverty as it improves food security, increases probability of holding savings and reduces anxiety level of the participating poor households.

Sathe (1991) reported the positive effect of EGS on agricultural and rural non-agricultural activities. Basu (2011) maintained that the institution of an employment guarantee programme increases the permanent wage but displaces some permanent agricultural labourers into casual labourers.

An equally important aspect of an employment guarantee scheme is the creation of productive assets or infrastructure that will have a bearing on agricultural productivity. The effects of projects such as public irrigation or flood control measures in vulnerable areas created through EGS will be taken into consideration in output and labour-hiring decisions (Basu, 2011) since these projects reduce the output risk of weather fluctuations.

Examining the labour and output market responses to a programme such as MGNREGS, Basu (2011) determined the optimum compensation to the programme employees. Considering the seasonality in agricultural production and the institution of permanent labour contracts, the author has concluded that the technological changes and productivity increases in EGS programmes seem to benefit the labourers more than the direct increase in wages or relief programmes. The author has further noted that if the elasticity of EGS input with respect to permanent labourers is high, a specific subsidy targeted towards the hiring of permanent labourers best serves the dual objectives of increased agricultural productivity and better welfare of the labourers.

We have come across a study where the scenario of the impact of MGNREGS was analysed using a general equilibrium modelling approach with the use of social accounting matrix (SAM) for the year 2003/04. Kumar and Panda (2009) found that MGNREGS increases the private consumption by 0.3% and the Gross Domestic Product (GDP) by 0.5%. They also found that if the scheme fully covers the bottom 70% of rural population, then unemployment will be reduced by 0.5%. So far, MGNREGS has not been studied in the dynamic setting using the general equilibrium modelling approach. This study has attempted to fill this gap.

Some Statistics

With budgetary allocation of Rs11,300 crore (×10 million) in 2006/07 under the umbrella of MGNREGS, it is easily the largest rights-based social protection initiative in the world (Farrington *et al.*, 2007). As per Schedule I of the Act, the public works carried out under MGNREGS have 'creation of sustainable rural assets' as one of the main objectives.

The expenditure incurred by the government and employment provided to rural households therein during the period 2006/07 to 2010/11 reported in Table 9.1 reveals that during this 4-year period the government expenditure on MGNREGS increased almost 5 times and its employment provision by about 2.5 times. The wage and non-wage expenditure was in the ratio of 70:30.

Modelling Aspects

The model used in the study followed the neo-classical approach and assumed the labour market to be functioning efficiently with the flexible wage rate determined as a market clearing price between labour endowment and factor employment in the commodity sectors. Unemployment was assumed to be voluntary (exogenous). The government provides employment guarantee under the MGNREGS and incurs expenditure for both wage bill and non-wage components. The non-wage component can be a maximum 40% of the total expenditure. The employment to the rural unemployed

Table 9.1. Expenditure break-up and employment to households under MGNREGS, 2006/07 to 2010/11 (Hanumantha Rao, 2012).

Year	Total expenditure (Rs million)	% of total expenditure		Number of households provided employment (in million)
		Wage bill	Non-wage components	
2006/07	88,233	66.21	33.79	21.016
2007/08	158,568	67.72	32.28	33.909
2008/09	272,501	66.79	33.21	45.115
2009/10	379,097	69.53	30.47	52.585
2010/11	393,772	57.65	42.35	54.95

labourers results in additional income generation for households, thereby increasing the household expenditure on consumption goods and creating additional demand for commodities.

In the base year, the SAM reflected the full employment scenario in the economy. However, to map the employment generated under the MGNREGS during the later periods, the total labour supply in the economy has to be modified to account for the unemployment. Under the scheme, the unemployed unskilled labour in the rural households is provided employment on demand. The unskilled labour endowments of rural households were therefore updated for the subsequent period to reflect the increased labour supply in the economy. A supply function was introduced for the MGNREGS labour. The wage rate under MGNREGS is set at a level lower than the market wage rate of unskilled labour in the rural areas (say at 70% of the market wage rate) so that it does not lead to shifting of labour from the normal agricultural/economic activity.

The non-wage component in the government expenditure on MGNREGS has been introduced in line with the existing norms. MGNREGS also creates additional capital and this has to be accounted for in the model. The mobility of capital has been restricted to within the agricultural sector only. This has implications for the model in terms of both government expenditure and employment of rural households having unskilled labour endowment. In the model, it has been assumed that only the unemployed unskilled labour belonging to the category[2] 1 and category 2 income-classes of rural households would seek employment under MGNREGS; these two categories of households constitute 70% of the rural population.

The materials used for the programme are expected to generate demand for various commodities like tools and machinery, cement, transportation, petroleum products, electricity, etc. We modified the base equation of the model initially to introduce unemployment and then for movement of unemployed labour for employment under MGNREGS. The modification was as given below.

Equation for labour employment under MGNREGS

$$ELSEGS * ((WGEGS/PF)^{**}\varepsilon)) = LSEGSQ \quad (9.1)$$

where *ELSEGS* is the labour entitled for employment under MGNREGS, *LSEGSQ* is the rural labour employed under MGNREGS, *WGEGS* is the MGNREGS wage rate, which is always less than the market wage rate, *PF* is the market wage rate and ε is the labour elasticity for MGNREGS.

In other words, it refers to the percentage change in labour employment under MGNREGS due to 1% change in the wage of MGNREGS relative to market wage.

Equation for labour demand under MGNREGS

$$LDEGSQ(\text{fl}) = CVR(\text{fl}) * LSEGSQ(\text{fl}) \quad (9.2)$$

where *LDEGSQ* is the labour demand under MGNREGS, fl is factor labour and *CVR* is the coverage factor indicating the eligible household category for employment under MGNREGS; it has the value '1' for rural unskilled labourer and '0' for others.

It has been assumed that MGNREGS will not compete with the open labour market and hence it will not enter directly into the production activities. Therefore, all those eligible labourers who offer themselves for work under MGNREGS will have to be absorbed under the scheme. This means additional labour demand has to be created by the public works. The additional demand for capital goods created due to launching of public works under MGNREGS has also been allocated to a host of manufacturing sectors in the model.

Impact of MGNREGS

Impact on GDP

The business-as-usual (BAU) scenario of 8% GDP growth has been compared with the

MGNREGS scenario basically because all the parametric assumptions were similar to this GDP growth. It was also found that MGNREGS had increased the GDP growth only marginally, from 8.0% to 8.04%. It should be noted that the total resources of the government were kept constant and the government resources were diverted to the scheme. On the other hand, the income of rural poor was enhanced with the introduction of MGNREGS and as a result had created more demand for various commodities, which had an impact on many sectors of the economy.

As for composition of real GDP, MGNREGS has brought down the share of agriculture in GDP in the long run, between 2010 and 2020, from 13.0% to 11.8% (Fig. 9.1). This could be in view of the shortage of labour availability for agriculture due to the presence of MGNREGS, though the rule on paper says that MGNREGS will absorb labour mainly during the off-season when the labour has no other work opportunity.

Impact on industries

It is evident from Fig. 9.2 that MGNREGS gives a boost to industry. The sectors such as construction have seen a considerable increase in output under MGNREGS. Real income share between rural and urban areas in this scenario has shown that between 2010 and 2020, real income has been reallocated from rural to urban as compared to BAU (Fig. 9.3).

Impact on household income

The income distribution across rural and urban areas by household groups (Fig. 9.4) showed that, though initially the impact of MGNREGS was positive for the rural poor, it will not benefit the rural poor in the long-run because the reallocation of resources to MGNREGS has effectively withdrawn resources from the productive sectors and hence has affected the wage income.

As expected, between BAU and MGNREGS scenarios, the income has not shown much variation across the urban households as MGNREGS is basically a rural programme (Fig. 9.5). However, MGNREGS can suppress the migration of labour from rural to urban areas and it would affect the rural–urban aggregate income distribution. The annual growth rates for the current consumption (Fig. 9.6) have depicted a picture similar to the one for income.

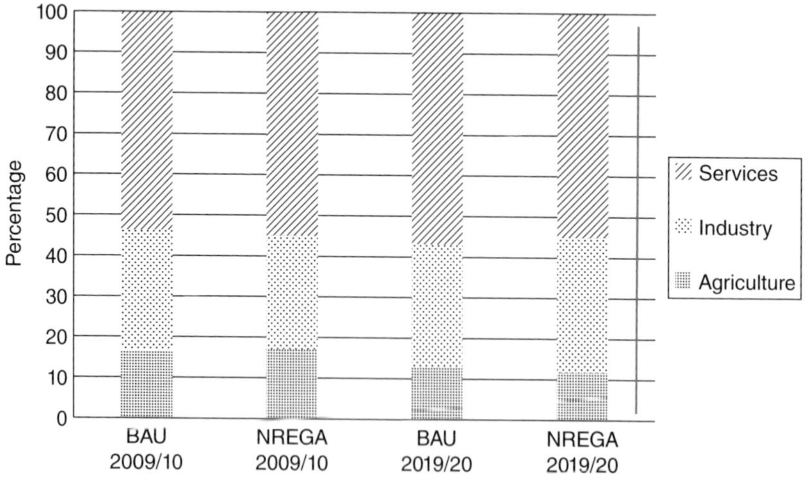

Fig. 9.1. Composition of real GDP across different sectors in India (BAU, 'business-as-usual' scenario; author's calculations).

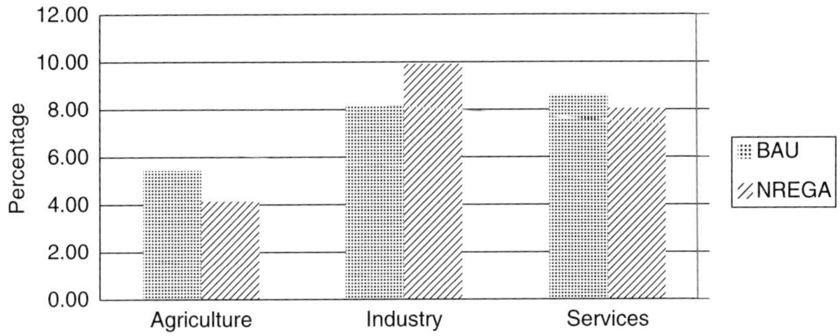

Fig. 9.2. Real GDP sectoral annual growth rates between 2010 and 2020 (BAU, 'business-as-usual' scenario; author's calculations).

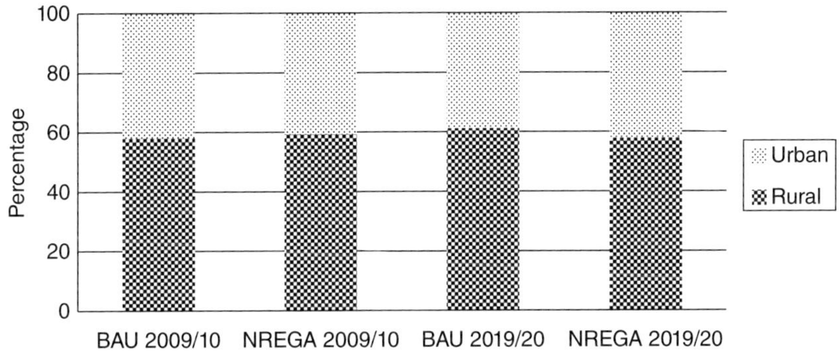

Fig. 9.3. Real income composition across rural and urban areas (BAU, 'business-as-usual' scenario; author's calculations).

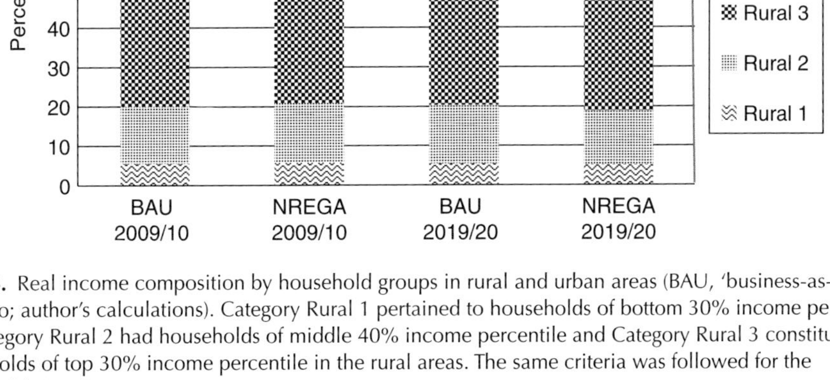

Fig. 9.4. Real income composition by household groups in rural and urban areas (BAU, 'business-as-usual' scenario; author's calculations). Category Rural 1 pertained to households of bottom 30% income percentile, category Rural 2 had households of middle 40% income percentile and Category Rural 3 constituted households of top 30% income percentile in the rural areas. The same criteria was followed for the households in urban areas.

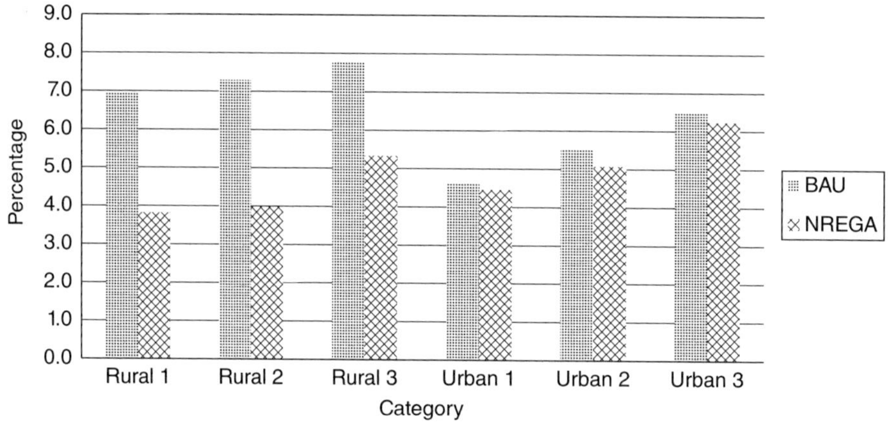

Fig. 9.5. Real income annual growth rates by household groups between 2010 and 2020 (BAU, 'business-as-usual' scenario; categories are as per Fig. 9.4; author's calculations).

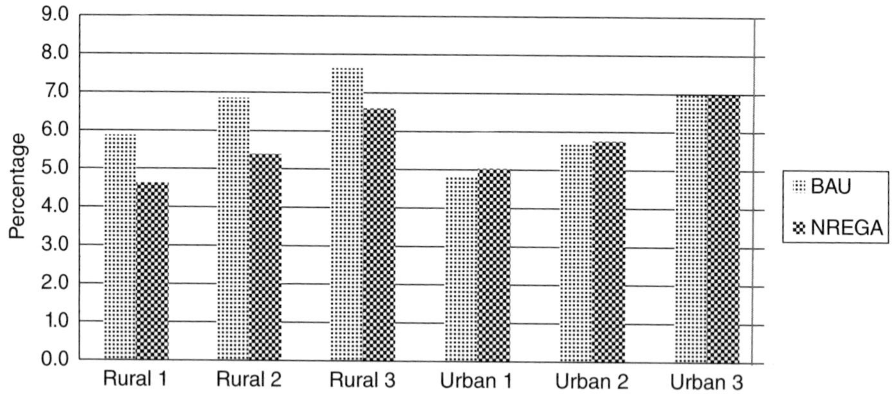

Fig. 9.6. Current consumption annual growth rates between 2010 and 2020 (BAU = 'Business-as-usual' scenario; author's calculations). Category Rural 1 pertained to households of bottom 30% income percentile, category Rural 2 had households of middle 40% income percentile and Category Rural 3 constituted households of top 30% income percentile in the rural areas. The same criteria was followed for the households in urban areas.

Impact on agricultural labour supply

A debate is going on whether MGNREGS pushes up the wage for agricultural labour due to the fact that labour could be diverted from agriculture to MGNREGS and hence could cause labour scarcity in agriculture. However, the results provided in Table 9.2 did not support this argument in the long-run, as shown by the 2019/20 results. The results of annual real income per capita (Table 9.3) show that the per capita figures for 2019/20 are lower for MGNREGS than under the baseline scenario. This has substantiated the earlier results that MGNREGS is not likely to benefit the rural poor in the long-run.

Conclusions

It is a known fact that during the time of world economic slowdown, the government's counter-measures such as fiscal expansion

Table 9.2. Projected labour wage rates to 2009/10 based on 2019/20 (author's calculations).

		2009/10		2019/20	
Type of labour	2006/07	BAU	MGNREGS	BAU	MGNREGS
Rural unskilled	1.00	1.12	1.16	1.97	1.68
Rural skilled	1.00	1.10	1.11	1.79	1.65
Urban unskilled	1.00	1.03	1.03	144	1.42
Urban skilled	1.00	1.02	1.07	1.42	1.49

BAU, Business as usual scenario

Table 9.3. Annual real income per capita (Rs/year) across different rural and urban areas (author's calculations).

	2009/10		2019/20	
Household type[a]	BAU	MGNREGS	BAU	MGNREGS
Rural 1	9,035	10,252	16,274	14,279
Rural 2	17,883	20,462	33,215	28,973
Rural 3	62,585	68,681	121,447	108,749
Urban 1	11,688	12,409	15,191	15,937
Urban 2	27,481	29,379	38,820	39,858
Urban 3	113,694	119,657	176,013	179,371

BAU, Business as usual scenario

[a]Category Rural 1 pertained to household of bottom 30% income percentile, Rural 2 households of middle 40% income percentile and Rural 3 constituted households of top 3% income percentile in the rural areas. The same criteria was followed for the households in urban areas.

bring some relief to the developing economy, but the poorer household classes do not benefit from such measures and may even be worse off. With the implementation of schemes such as MGNREGS, this can be overcome. These schemes benefit the rural poor, whose real income also rises. As a result, there is expansion in the domestic output and consequently an increase in GDP. This highlights the need for implementation of well-designed targeted anti-poverty programmes.

The area of contention for MGNREGS seems to be that the wages of MGNREGS look reasonably good, relative to market wages, and this may cause upward wage pressure. However, if this translates into high prices, then the real wage does not change. Tracing the food price crises in 2008, and the inflation rates in rural areas, there was a substantial hike in inflation from less than 1% in 2000/01 to more than 12% in 2008/09.[3]

The present study has adopted the general equilibrium approach to find the impact of MGNREGS. It has revealed that MGNREGS is likely to have a negative impact on agriculture in the long-run. It has only helped industry, though in the early period, between 2007 and 2010, agriculture and services improved marginally; between 2010 and 2020, the study has shown a reduction in share of agriculture in total GDP from 13.0% to 11.8%. This could be due to government resources being diverted to MGNREGS from erstwhile productive sectors.

The real income in rural India has gone down partly due to lower agricultural growth and partly due to lower market wage as compared to the BAU scenario during 2010–2020. The overall picture is that MGNREGS has contributed to the growth of industry and has provided a big fillip to the industries such as manufacturing, both labour and capital intensive, and the construction sector. The market wages of the unskilled labour in the rural areas are not increasing due to MGNREGS, against the expectation that it would push up agricultural wages. It has been confirmed that MGNREGS supplements only the off-season employment and it does not draw agricultural labour away from farming. The real income per capita has also supported the result that MGNREGS pushes up the income of urban poor, and not of rural poor, in the long-run, because of higher growth of the manufacturing and construction sectors under the MGNREGS regime.

The policy implication is not only that MGNREGS may not be sustained in the long-run given the limited resources of the government, but MGNREGS will also not continue to provide benefits to the rural poor, as was intended originally.

Acknowledgement

I would like to thank Nitin Harak, Doctoral Student of Indira Gandhi Institute of Development Research, for assisting me in the early stages of preparation of this Chapter.

Notes

[1] The entitlement of 100 days of wage employment can be shared within a household, i.e. more than one member of the household can be employed (simultaneously or at different times).

[2] Category Rural 1 pertained to households of the bottom 30% income percentile category, Rural 2 had households of the middle 40% income percentile and Rural 3 constituted households of the top 30% income percentile in the rural areas. The same criteria was followed for the households in urban areas.

[3] Labour Bureau, Government of India. Available at http://www.labourbureau.nic.in/indtab.pdf (accessed 16 April 2011).

References

Basu, K. (1981) Food for work programmes: Beyond roads that get washed away. *Economic and Political Weekly*, January 3–10.

Basu, A., Chau, N. and Kanbur, R. (2005) The National Rural Employment Guarantee Act of India, 2005. Available at: http://www.arts.cornell.edu/poverty/kanbur/egaoxfordcompanion.pdf (accessed 16 October 2012).

Basu, A.K. (2011) *Impact of Rural Employment Guarantee Schemes on Seasonal Labor Markets: Optimum Compensation and Workers' Welfare*. Discussion Paper No. 5701. Institute for the Study of Labor (IZA), Bonn.

Farrington, J., Holmes, R. and Slater, R. (2007) *Linking Social Protection and the Productive Sectors*. Briefing Paper 28, Overseas Development Institute, London.

Hanumantha Rao, K. (ed.) (2012) *Rural Development Statistics 2011–12*. National Institute of Rural Development, Ministry of Rural Development, Government of India, New Delhi.

Jha, R. and Gaiha, R. (2012) *India's Rural Employment Guarantee Scheme as it is – Interpreting the Official Report*. ASARC Working Paper No. 2012-12. Australian National University, Australia South Asia Research Centre.

Jha, R., Gaiha, R. and Pandey, M.K. (2012) Net transfer benefits under India's rural employment guarantee scheme. *Journal of Policy Modelling* 34, 296–311.

Kamath, R. (2010) National Rural Employment Guarantee Act: An effctive safety net? *IIMB Management Review* 22, 42–55.

Khan, A.U. and Saluja, M.R. (2007) *The National Rural Employment Guarantee Act*. Working Paper 0701. India Development Foundation, New Delhi.

Kumar, G. and Panda, M. (2009) Global economic shocks and Indian policy response: An analysis using a CGE model. In: Parikh, K.S. (ed.) *Macro Modelling for the Eleventh Five Year Plan of India*. Academic Foundation, New Delhi.

Mahendra Dev, S. (2011) *NREGS and Child Well-being*. Working Paper No. WP 2011-004. Indira Gandhi Institute of Development Research, Mumbai.

Mukherjee, D. and Sinha, U.B. (2011) *Understanding NREGA: A Simple Theory and Some Facts*. Working Paper No. 196. Centre for Development Economics, Delhi School of Economics, Delhi.

Ravallion, M. (1991) Market responses to anti-hunger policies: Effects on wages prices and employment. In: Drèze, J. and Sen, A. (eds) *The Political Economy of Hunger*, Vol. 2, *Famine Prevention*. Clarendon Press, Oxford, UK.

Ravi, S. and Engler, M. (2009) Workfare in low income countries: An effective way to fight poverty? *The case of India's NREGS*. Available at: http://www.isid.ac.in/~pu/conference/dec_09_conf/Papers/ShamikaRavi.pdf (accessed 23 February 2011).

Sathe, M.D. (1991) *Rural Employment and Employment Guarantee Scheme in Maharashtra*. Development Group, Vijayanagar, Pune, Maharashtra.

10 India's Price Support Policies and Global Food Prices

Gerdien Meijerink[1]* and P.K. Joshi[2]

[1]*CPB Netherlands Bureau for Economic Policy Analysis, The Hague, The Netherlands;* [2]*International Food Policy Research Institute, New Delhi, India*

Introduction

The agricultural price policy of India aims to achieve the twin objectives of assuring remunerative prices to the farmers and providing food grains to the consumers at reasonable prices (GoI, 2012). The price policy was designed during the green revolution period to ensure farmers received minimum support prices (MSPs) and assured procurement of their produce. The policy of subsidizing food grains was developed during the period of relative shortage of agricultural products, particularly of food grains. Several policy instruments and complementary policies have been implemented to achieve these twin objectives (see also MoSPI, 2012). India also has a targeted public distribution system (TPDS) to improve the food security of India's poor. In addition, India pursues an active trade policy to shield the domestic food market from the impact of global prices.

During 2007–2008 and also in 2011, the global food prices climbed to new heights and showed a high degree of volatility. In this chapter, we have discussed the impact of global price spikes in relation to India's agricultural price policy. We have argued that the higher global wheat and rice prices put pressure on India's domestic prices. India has successfully shielded its domestic markets from the global price volatility through export bans. However, the policies did not succeed in avoiding price rise in the domestic market. When farmers could not profit from the higher global prices because of the export bans, they lobbied successfully for higher MSPs. These triggered a higher supply response from the farmers. The buffer stocks of major staples, viz. wheat and rice, increased consistently because of India's buffer stock policy, which allows farmers to sell their produce to the government at MSPs. In principle, the larger buffer stocks should be good news, as more wheat and rice are available for the poor in India. However, due to inefficiencies in India's TPDS, the stocks are not being distributed sufficiently and properly. Much of India's rice and wheat reserves were stored in inappropriate places because stocks exceeded the storage capacity, leading to rotted stocks unsuitable for human consumption.

The chapter has been organized into six sections. After this introduction, we have analysed the global rice and wheat prices and their transmission to domestic prices. The third section provides the impact of increasing minimum support prices for rice and wheat on increasing their stock levels in India. In section

* Corresponding author, e-mail: g.w.meijerink@cpb.nl

four, the impact of India's agricultural price policies on domestic prices of rice and wheat has been discussed. In section five, the prospects of existing price policy have been assessed under a trend scenario for 2010 to 2030. Section six discusses major findings and gives conclusions.

Global and Regional Food Prices of Rice and Wheat

Global food prices started rising around 2006. The rice prices reached a peak in 2008, while wheat prices peaked in 2008 and 2011 (Figs 10.1 and 10.2, respectively). We referred to Meijerink *et al.* (2011) for an explanation for the peaks in international prices. India responded to these increases in price with various trade policies to protect Indian consumers from rising rice and wheat prices.

A ban on futures trading in wheat and rice was implemented from February 2007 to May 2009 by the Government of India (*Times of India*, 2009). In 2008, the government suspended futures trading in soybean oil, potatoes, rubber and chickpeas to stem soaring inflation at a time when global food and energy prices were rising steeply (WSJ, 2008). Furthermore, since 2007, the Indian government has permitted the states to impose stock limits on the private trade of wheat under the Essential Commodities Act. In addition, the government mandated in 2008 the large food companies and trading companies operating in India to declare their stock levels (Singh, 2012).

An export ban on wheat was imposed from February 2007 to 9 September, 2011. However, exceptions were made on humanitarian grounds for exports to some countries like Nepal, Bangladesh and Afghanistan. The government also permitted an export quota of 650,000 t of wheat products, although not whole grain, during a marketing year. However, Indian wheat exports were

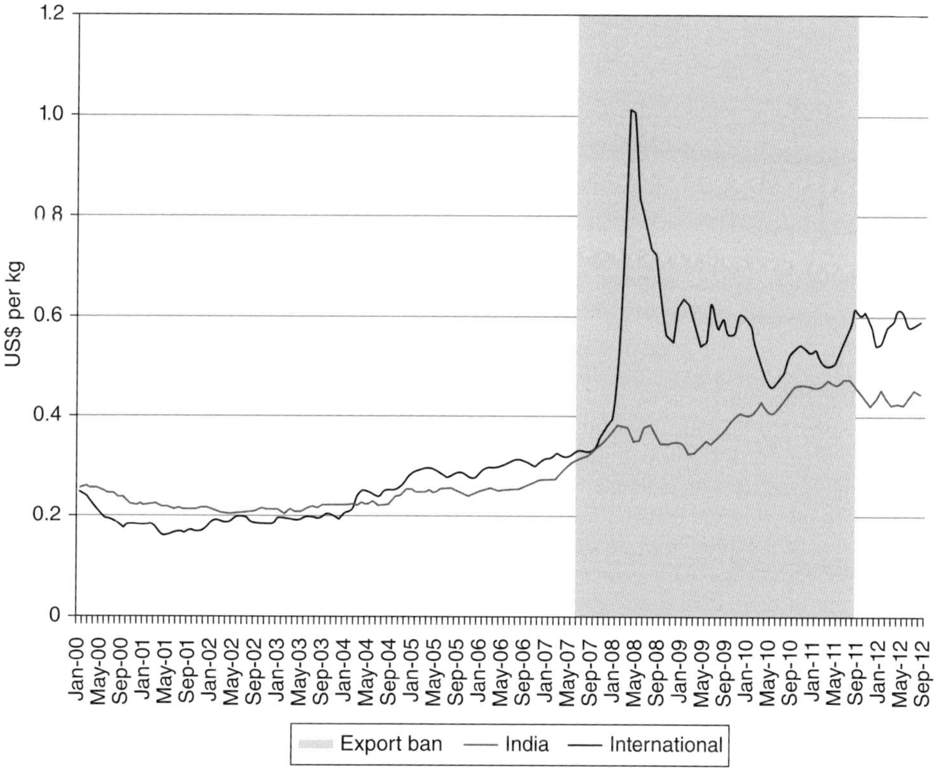

Fig. 10.1. International and Indian rice prices, 2000–2012 (US$ per kg) (GIEWS, 2012; IMF, 2012).

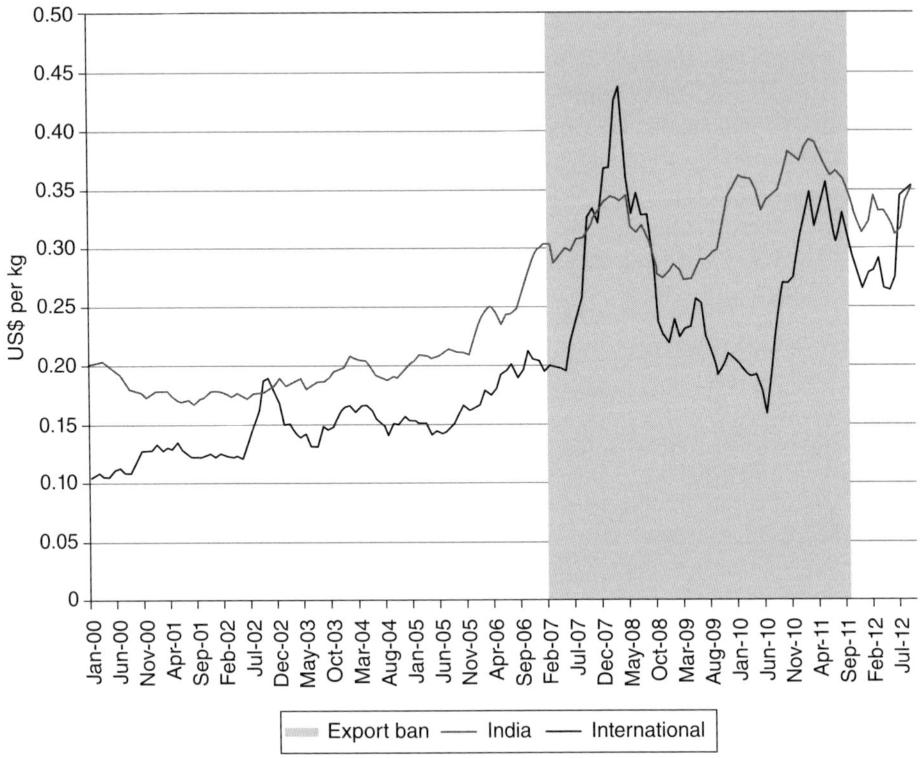

Fig. 10.2. International and Indian wheat prices 2000–2012 (US$ per kg) (GIEWS, 2012; IMF, 2012).

already low due to uncompetitive prices and quality issues. When wheat stocks rose to new highs in 2012, the Indian government decided to allow wheat exports at prices below the MSP. This was, however, in violation of the World Trade Organization rules (DNA, 2012). An export ban on non-basmati rice was also imposed from September 2007 to 9 September 2011. Again, exceptions were made for some countries on humanitarian grounds. The exports of basmati rice continued to be allowed subject to a minimum export price of US$900 per ton. The ban was later lifted because of massive rice production, 'more-than-sufficient' rice stocks and relatively weak domestic prices (Singh, 2012).

In addition, the government lowered the duty on wheat imports to zero in September 2006, which was extended indefinitely in October 2007. As imported wheat works out to be relatively more expensive than local wheat after accounting for the costs on shipping, clearance and inland transport, the imports of wheat became too small. In March 2008, the Indian government removed the import duty on rice. The zero duty on rice was effective through 31 March 2012 (Singh, 2012).

These policies could not prevent the domestic rice and wheat prices from rising. Figures 10.3 and 10.4 show the prices of rice and wheat, respectively, in the Chennai, Delhi, Mumbai and Patna markets. These prices are the averages of retail and wholesale prices. The prices have depicted an increasing trend, even after the export ban.

There are several reasons behind the increases in domestic prices of wheat and rice, but they are mainly associated with India's food grain management (Basu, 2011; Nair and Eapen, 2011). First, the continuous increase in MSPs inflated the open market food grain prices (see Box 10.1 for an explanation of MSPs). The procurement of food grains at a higher MSP required the government to charge higher prices for food grains sold in

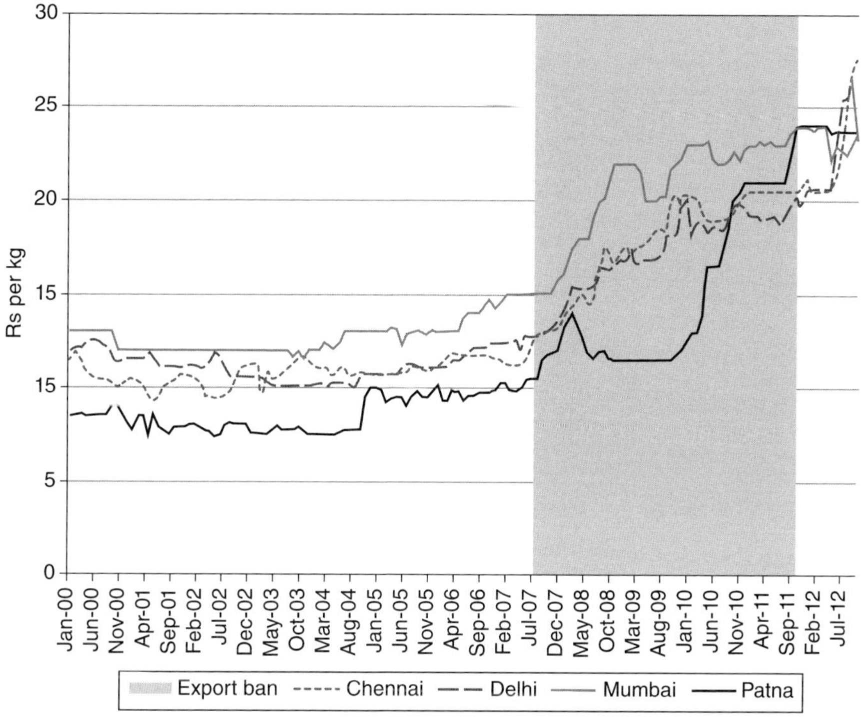

Fig. 10.3. Rice prices in Chennai, Delhi, Mumbai and Patna markets, 2000–2012 (Rs per kg) (GIEWS, 2012).

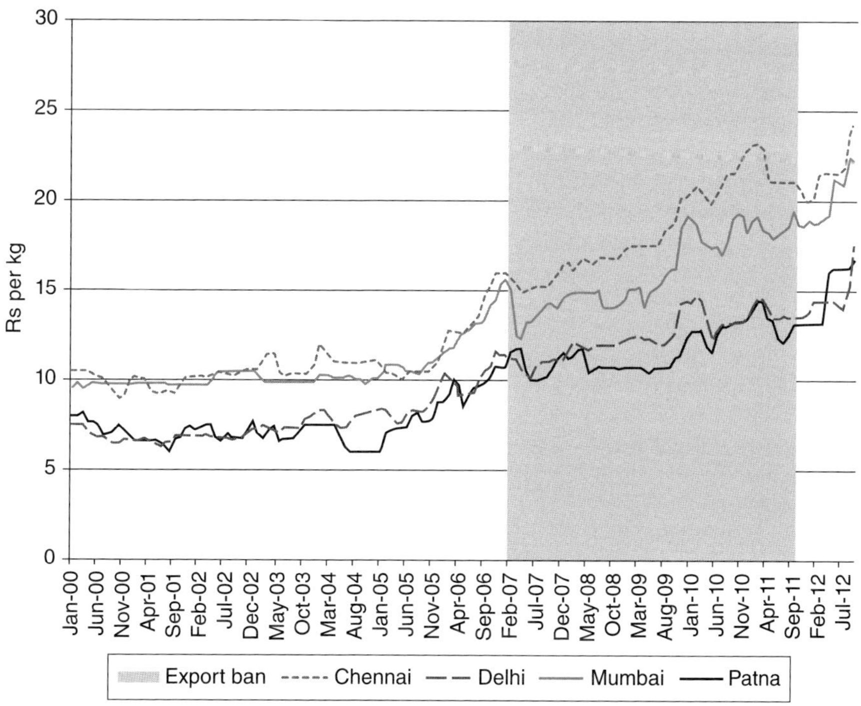

Fig. 10.4. Wheat prices in Chennai, Delhi, Mumbai and Patna markets, 2000–2012 (Rs per kg) (GIEWS, 2012).

Box 10.1. Minimum Support Prices

For major agricultural products, the Government of India announces each year their minimum support prices (MSPs), which are fixed taking into account the recommendations of the Commission for Agricultural Costs and Prices (CACP). The CACP recommends MSPs for 24 important crops. Besides taking into account the cost of production, the CACP considers such other factors as demand–supply gap, price situation, global availability, intercrop price parity and terms of trade between agriculture and non-agriculture sectors.

The government intervention takes place when prices of the relevant commodities fall below the MSP, resulting in procurement at the MSP by the Food Corporation of India (FCI) for cereals, and other agencies for non-cereal crops. Producers of these crops are assured that the state will purchase all supply for sale should the market price fall below the MSP.

With the exception of some states, the FCI purchases the relevant commodities at the procurement prices set by the government and sells them at the Central Issue Price, which is also fixed by the government.

The MSP is viewed as a form of market intervention scheme by the central government and as one of the supportive measures (safety nets) to the agricultural producers; it insulates producers against fluctuations in market prices. In addition, it allows the government to create an incentive structure for farmers to allocate resources towards the desired crops.

Source: GoI (2009)

the open market. Second, the high levels of food grain procurement by the Indian government deprived the private sector of sufficient grains for meeting the requirements of ordinary consumers, thereby putting additional pressure on food grain prices. Finally, the Indian government failed to stabilize food grain (especially rice and wheat) prices through the sales of buffer stock grains in the open market.

The increase in MSPs has thus been a key element in rising domestic prices. Figure 10.5 shows the MSP for rice and its average market price, based on the prices in four markets depicted in Fig. 10.3. Figure 10.6 shows the MSP for wheat and its average market price based on Fig. 10.4. The MSPs and domestic market prices of rice are very closely related. The rising MSP could be a reflection of the strong lobby power of farmers for a higher price. For instance, the government rejected the recommendation of the Commission for Agricultural Costs and Prices (CACP) in November 2012 to keep the MSP of wheat for 2012/13 unchanged at Rs12.85 per kg. The CACP had recommended the MSP for wheat at Rs11.20 per kg in 2010/11 because the cost of production was only Rs8.26 per kg. The Government of India did not accept this recommendation and asked the CACP 'to review the recommendations in view of increasing cost of inputs such as fertilizer and diesel' (*Business Standard*, 2012).

An export ban can also contribute to the pressure on MSPs. In November 2012, CACP advised stabilizing MSPs and recommended that due to the export ban on wheat, the government could increase its MSP by 10% (*Business Line*, 2012a). Figures 10.5 and 10.6 support this: MSPs as well as domestic prices started rising considerably after imposing the export ban. The logic behind this is that under an export ban, Indian wheat and rice producers can no longer profit from high world prices, and demand higher MSPs to compensate for their loss.

Impact of India's Agricultural Price Policies on Domestic Rice and Wheat Prices

One of the goals of India's agricultural price policies is to maintain stable prices. It is clear from the figures that India has not been successful in preventing the rise in domestic rice and wheat prices, which we will discuss later. However, it was successful in shielding India's domestic market from the importing price volatility. Figures 10.1 and 10.2 have confirmed this; India's domestic rice and wheat prices have not shown the sharp peaks as observed in the international prices in 2008 and 2011.

Rapsomanikis and Mugera (2011) investigated price transmission and volatility

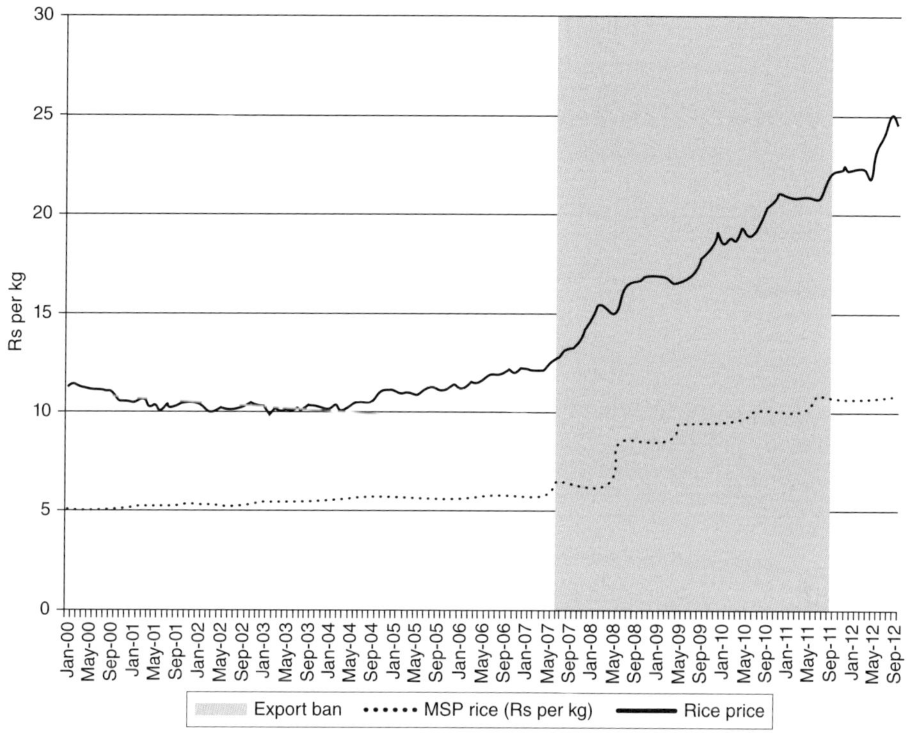

Fig. 10.5. Minimum support prices and domestic prices for rice, 2000–2012 (Rs per kg) (GIEWS, 2012; MoA, 2012). Note: the rice price was averaged over the four selected markets and averaged for wholesale and retail prices.

spill-overs of global prices to developing countries' food markets. They have found evidence that the global and domestic prices for wheat, rice and maize are co-integrated; international prices are transmitted to domestic markets but with a lag. In their analysis of the Indian rice market, they found that the world prices and Indian market prices are endogenous, i.e. both the Indian and the world prices adjust to their long-run equilibrium relatively rapidly, correcting about 16% of the divergence each month.

We replicated the analysis carried out by Rapsomanikis and Mugera (2011) for rice and wheat till September 2012, instead of December 2010, capturing some additional information. We differentiated between the Indian markets of Delhi, Chennai, Mumbai and Patna and constructed an average price for India. We used the average of wholesale and retail prices for rice and wheat as provided by the Department of Consumer Affairs (2012). The GIEWS website (GIEWS, 2012) provides the same prices. For international prices, we used the figures provided by the IMF (2012).

We first tested for stationarity by the Augmented Dickey-Fuller (ADF) test and the Phillips-Perron (PP) test. The difference between these tests is that the PP allows for fairly mild assumptions concerning the distribution of the errors in that they can be weakly dependent and heterogeneously distributed. The tests are based on a random walk and the fact that a random walk has a unit root. If the variable in question follows a random walk, it is non-stationary. If estimations are done with non-stationary data and residuals, then regressions may be spurious. The main method for inducing stationarity is to take the first difference of the time series data.

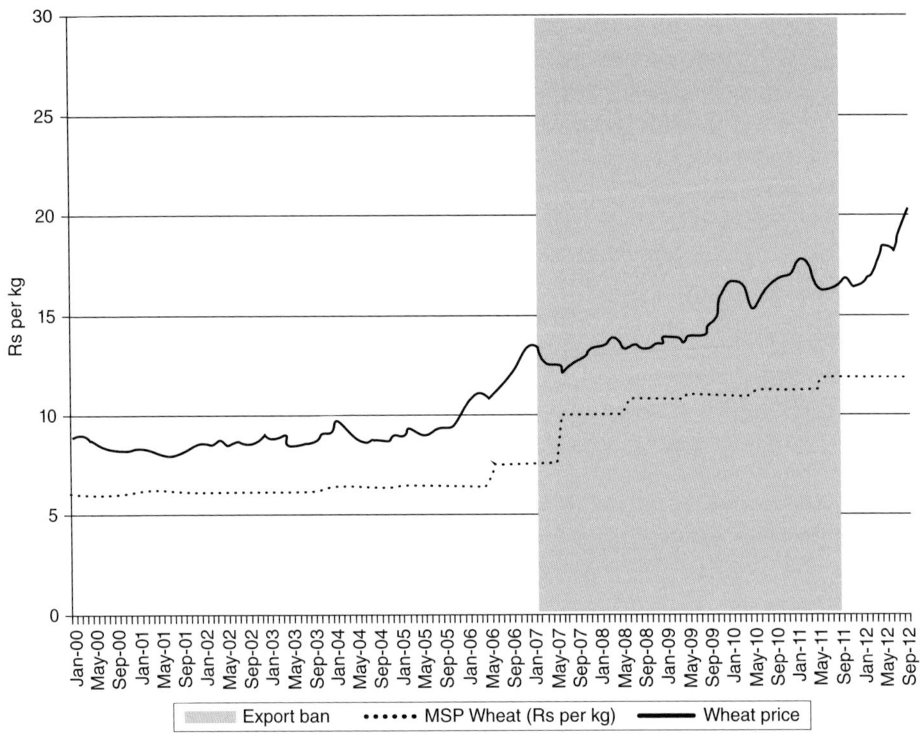

Fig. 10.6. Minimum support prices and domestic prices for wheat, 2000–2012 (Rs per kg) (GIEWS, 2012; MoA, 2012). Note: the wheat price was averaged over the four selected markets and averaged for wholesale and retail prices.

The results of ADF and PP tests are given in Table 10.1. The different test values occasionally lead to different conclusions about whether we can reject the null hypothesis (H_0), such as for international wheat price and Chennai rice prices. As the ADF test does not have a strong power, we preferred to rely on the PP test values. Also, because the sample size was not very large (only 115 observations), we checked the ADF p-value. For the international wheat price, the ADF test p-value was 0.027, thus we could not reject the H_0 at the 5% level. For the Chennai rice price, the PP test value was smaller than the critical value, leading us to reject the H_0. We performed the Kwiatkowski-Phillips-Schmidt-Shin (KPSS) test to check all values and for the Chennai rice price the test value (0.451) was well above the 1% critical value of 0.216. In the case of the KPSS test, the null hypothesis was different: it assumed stationarity of the variable of interest. Therefore, our conclusion was that we cannot reject the H_0 of non-stationarity for all price series.

The test values for changes in prices (Δp_t) showed that we could reject the H_0 for the first differences of prices, which means that the first differenced price series became stationary (i.e. the price series were all integrated of order 1).

To test whether the wheat and rice prices in India are driven by the international prices, we used the Engle-Granger test to find co-integration. The Engle-Granger test is based on the ADF test for stationarity. The two non-stationary variables containing a unit root (i.e. I(1) variables) are co-integrated if the error-term is stationary (i.e. I(0)). We ran a 'co-integrating regression' selecting the various domestic market prices as dependent, including the international price as independent.

Table 10.1. Test values for ADF and PP tests for non-stationarity in prices of rice and wheat.

	P_t		Δp_t	
Market	ADF	PP	ADF	PP
Rice				
Chennai	−3.30	−3.72	−11.68	−11.68
Delhi	−2.74	−2.84	−11.02	−11.03
India	−3.24	−3.20	−5.25	−8.86
International	−3.05	−3.08	−5.90	−6.61
Mumbai	−2.59	−2.55	−9.41	−10.69
Patna	−2.41	−2.33	−10.41	−10.38
Wheat				
Chennai	−3.07	−2.44	−9.10	−8.91
Delhi	3.32	−2.31	−6.08	−8.67
India	−2.91	−2.16	−8.19	−7.96
International	−3.64	−2.59	−9.36	−8.96
Mumbai	−3.22	−2.79	−8.52	−8.91
Patna	−2.75	−2.44	−11.32	−10.98

With constant and trend, the 5% and 1% critical values for ADF test are −2.88 and −3.48, respectively. For the PP test, these are −2.92 and −3.44, respectively.

The results of the ADF test on the residuals are presented in Table 10.2. Because the sample was not large (only 147 observations), we have also presented the p-values. We could find statistical significance for none of the rice or wheat markets, implying that neither wheat nor rice markets are influenced by the world market.

We therefore concluded that India's trade policy was, in fact, successful in shielding its domestic rice and wheat markets from global price volatility. We carried out the analysis for the whole period because the period during which the export ban was imposed was too short for statistical analysis of this kind. These results are illustrated by the figures of prices for wheat and rice for the four Indian markets and the international price (see Fig. 10.3 for rice and Fig. 10.4 for wheat). It can be observed that different prices have a similar path in the long-run, but in the short-term they are often on opposing paths.

This supports the findings of Dasgupta *et al.* (2011), who have found that although international wheat prices are a strong driver of domestic quarterly wheat prices in India, there are, however, significant wedges between international and domestic wheat prices. Their results revealed that while domestic wheat prices were more stable than international prices, especially in an era of significant trade wedge (export ban) after 2006, there were unexpected opposite movements of domestic wheat prices, counter to international trends in recent quarters, and the fit is therefore relatively poor.

Table 10.2. Results of Engle-Granger tests for co-integration (independent variable: international price).

	Engle-Granger test	
Market	Test statistic	p-value
Rice		
Delhi	−2.1445	0.7067
Chennai	−2.3759	0.5878
Mumbai	−2.2474	0.6557
Patna	−2.8611	0.3287
India	−2.3767	0.5873
Wheat		
Chennai	−3.4761	0.1048
Delhi	−3.4761	0.1048
Mumbai	−2.6665	0.4292
Patna	−2.4069	0.5709
India	−2.6951	0.4138

The prices are logged values.

Impact of Increasing MSPs on Rising Food-grain Stocks

It seems that the higher MSPs have led to increasing food grain stocks in India (see also Kotwal *et al.*, 2011). The private traders and the government both procure food grains from the market, usually at MSP. However, if MSPs for rice and wheat increase and remain higher than the market prices, producers will: (i) produce more rice and wheat; and (ii) sell more to the government. The assured procurement provides a guarantee to the farmers for the purchase of rice and wheat, therefore the Food Corporation of India (FCI) is obliged to procure all wheat and rice offered by the farmers for sale (see Box 10.1). The fact that the government buys most of the rice and wheat produced has led to the exit of private traders, thus forcing the government to procure even more. Due to cumbersome administrative procedures, it is difficult for the government to unload the excess food grain stocks quickly to effectively counter the price rise. Figure 10.7 shows the rising stocks of wheat and rice since 2000.

Maintaining sufficient buffer stocks is an important objective of India's food policy because it enables:

- supplying food grains to TPDS and other welfare schemes;
- meeting emergency situations arising out of unexpected crop failures or natural disasters; and
- moderating food grain prices through market interventions.

India targets a total buffer stock of rice and wheat ranging between 16.2 Mt in April and October and 26.9 Mt in July. In addition, India aims to maintain 3 Mt of wheat and 2 Mt of rice as strategic reserves (Kapur, 2012) (see Fig. 10.7). In recent years, the actual stocks exceeded the buffer norms with a large margin. The Indian Economic Survey has criticized the Indian government for creating 'artificial shortages' through the 'policy of stocking grain well above the buffer norms' (MoF, 2013, p. 191).

The peak in stocks during 2000 to 2003 raised considerable concerns in India at that time, as it necessitated substantial amounts of food credit and food subsidies to finance the stocks. The stock levels duly dropped, and from 2004 to 2008 the stocks remained

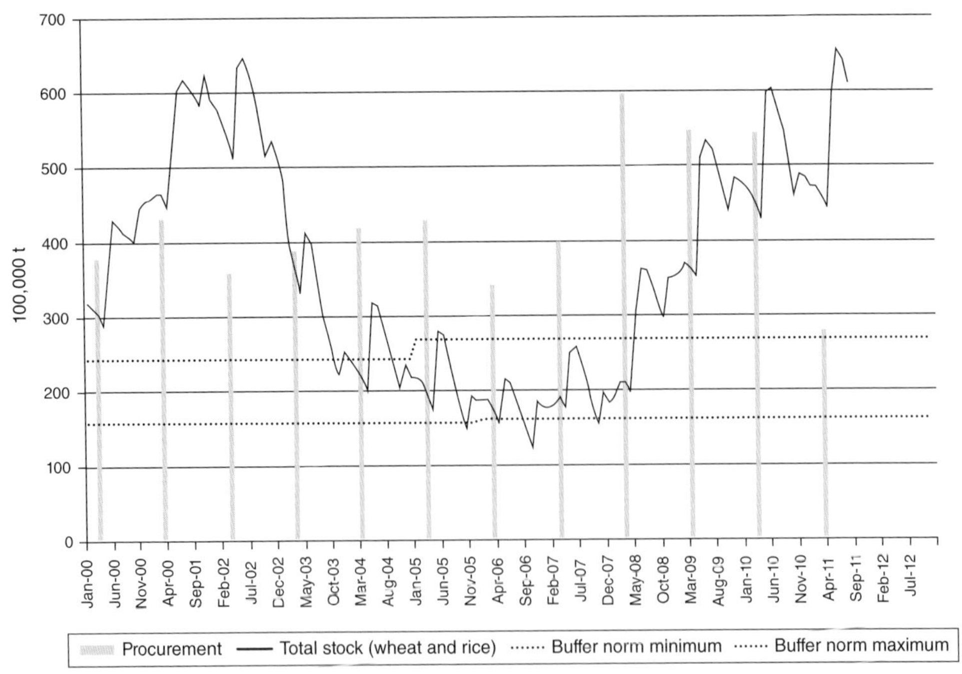

Fig. 10.7. Stocks of rice and wheat – procurement and buffer norms, 2000–2010 (Food Corporation of India, 2012a, b, c). Note: data for 2012 not completely available.

relatively stable, not including seasonal fluctuations. The stocks of wheat and rice started increasing again after 2008, when world wheat and rice prices started rising. In the agricultural year of July 2008–June 2009, a record volume of food grains were produced, estimated at 233.87 Mt (*Thaindian News*, 2009).

Most of the stocks are stored in the central pools located in different states of India. In 2012, the total covered storage space had a capacity of 52.85 Mt, which is clearly insufficient given that buffer stocks reached over 60 Mt in 2012 (Fig. 10.7). The FCI therefore announced at the end of 2012 that it planned to increase India's total covered storage capacity for food grains to about 71 Mt by constructing new warehouses (godowns) in 19 states (*Business Line*, 2012b).

In addition, the government released an additional 5 Mt of food grains to the 'below poverty line' (BPL) families and allowed sale under the Open Market Sale Scheme of 3 Mt of wheat in the summer of 2012. It also decided to allow wheat exports at prices below the MSP (DNA, 2012).

Despite overflowing stocks, there is considerable evidence that rice and wheat were not reaching the poor, who should be the real beneficiaries of the TPDS. The public distribution system is riddled with inefficiencies and there are reports of high leakages, which deprive the poor of access to the food grain (Jha *et al.*, 2007) (see also Box 10.2 and the discussion section, this chapter).

Figures 10.9 and 10.10 show the relationships between rice and wheat stocks and Indian rice and wheat prices from 2000 to mid-2012.

We tested whether MSPs have indeed pushed up stocks (i.e. whether there is a statistically significant connection between the rice and wheat MSPs and their buffer stocks). For this, we examined the existence of a co-integrated combination of the two series through a Vector Error Correction Model (VECM), which can be written as:

$$\Delta y_t = \mu_t + \Pi y_{t-1} + \Sigma_{t=1}^{p-1} \Gamma_t \Delta y_{t-1} + \epsilon_t \quad (10.1)$$

where $\Pi = \Sigma_{t=1}^{p} A_t - I$ and $\Gamma_t = -\Sigma_{j=t+1}^{p} A_j$

There are typically three steps in this analysis:

1. Test to determine the number of co-integrating vectors, the co-integrating rank of the system (Johansen co-integration test).
2. Estimate a VECM with the appropriate rank, but subject to no further restrictions.
3. Probe the interpretation of the co-integrating vectors as equilibrium conditions by means of restrictions on the elements of these vectors. We have not provided the results of step 3 here.

When we performed step 1 for the MSP for rice and rice stocks for the whole period, the Johanssen test indicated that there was no co-integration.[1] On taking the period January 2006 to September 2012, the Johanssen test indicated co-integration between MSP values and stock levels for rice (Table 10.3).

We then estimated a VECM for rice, using rank 1 (resulting from the Johanssen test). The error correction term of the equation for stocks was found significant,[2] which means that rice stocks do adjust to the changes in MSP, indicating that MSPs for rice indeed push up rice stock levels but not vice versa. The same applies for the VECM results for wheat (Table 10.4); after August 2006 MSPs pushed up wheat stocks.

To conclude we may say that higher MSPs have indeed led to increasing stocks, thus adding a 'push factor' to the TPDS. Because the distribution side was left unchanged, the high MSPs indirectly resulted in piling up of rice and wheat stocks. This could have been a win–win situation, with rice and wheat producers earning a better income and poor consumers obtaining more rice and wheat. However, large parts of these additional stocks of rice and wheat became unsuitable for human consumption due to the lack of proper storage facilities, leaving the poor consumers unable to benefit from the additional harvests.

Box 10.2. India's Targeted Public Distribution System

The FCI procures and maintains the buffer stocks of food grains, especially of rice and wheat. The responsibility for distribution of food grains to the beneficiaries rests with the state governments through the Targeted Public Distribution System (TPDS) and various welfare schemes for poverty alleviation. In 1997, the TPDS replaced the earlier public distribution system with differential prices for the 'below poverty line' (BPL) population and for the 'above the poverty line' (APL) population.

The TPDS ensures distribution to consumers of essential commodities: currently wheat, rice, coarse grains, sugar and kerosene. It is operated under the joint responsibility of the central and state governments; the central government is responsible for the procurement, storage, transportation and allocation of stocks, while the state governments are responsible for identification of beneficiaries, issuing of ration cards and distribution of food grains to them through 'fair price shops' set up for this purpose.

Until 1997, the annual cost of the TPDS was less than 0.5% of GDP, a cost of almost US$2 billion or Rs107 billion (World Bank, 2011). In 2011/12, these costs increased to almost 1% of GDP (Rs600.57 billion) (Kapur, 2012). The costs on the food subsidy programme are shown in Fig. 10.8.

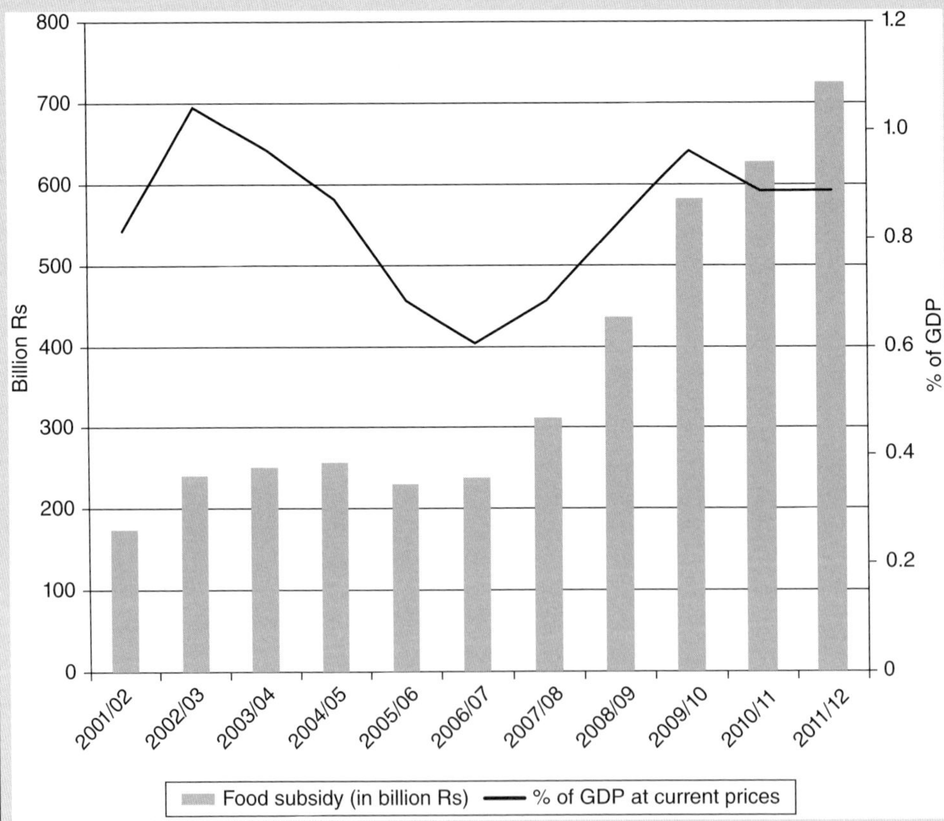

Fig. 10.8. Subsidy on food grains for distribution under the TPDS (2001–2012) (in billion Rs and as percentage of GDP) (Department of Food and Public Distribution, cited in Kapur, 2012; MoF, 2013).

In 2012–2013, under the TPDS about 26.42 Mt of rice and 23.52 Mt of wheat were to be distributed to different categories of poor (Department of Food and Public Distribution, 2012). These categories are: APL, BPL and AAY (*Antyodaya Anna Yojana*, for the poorest of the poor families).

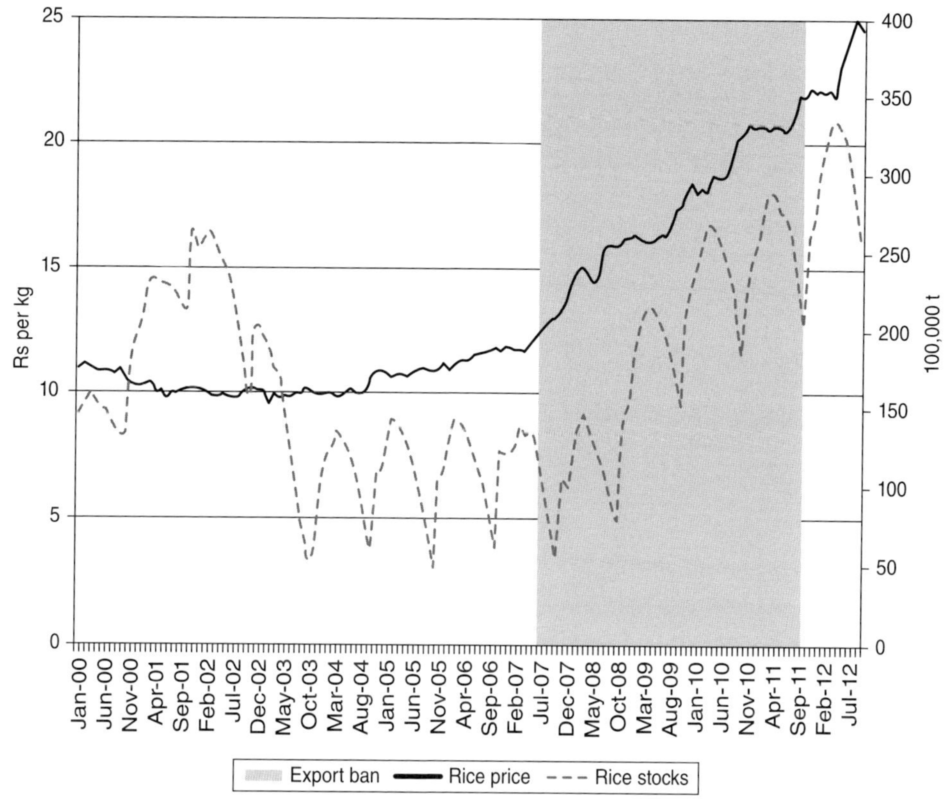

Fig. 10.9. Rice stocks (in 0.1 Mt: right-hand scale) and Indian rice price (average of retail and wholesale prices in Rs per kg: left-hand scale) (Food Corporation of India, 2012a; GIEWS, 2012).

A Baseline Trend to 2030

The key question in policy makers' minds is: 'will the system be sustainable if continuous rising (MSP) prices push up stock levels and associated costs?' In the coming decades, both production and consumption are expected to increase. The baseline scenario (see Woltjer and Rutten, Chapter 5, this volume) has revealed that demand for rice will only be met in 2030 and little import of rice will be necessary. This also means that domestic rice prices are expected to increase, unless imports of cheap rice are possible.

On the other hand, import of cheap rice may put Indian rice farmers at a disadvantage. If international rice prices are high, or fluctuate widely, imposing export restrictions in the case of rice may be rational to shield India's domestic rice market. For wheat, however, India will produce a surplus, thus easing domestic prices. In this case, it is not likely that an export ban will be necessary, especially not with high world prices. In both cases, if the current system of MSP and the obligation of India to buy from farmers is upheld, rice and wheat stock capacity will need to increase substantially.

The projected economic growth in India will mean that more poor will be lifted out of poverty, thus there will no longer be a need for food aid through the TPDS. The distribution of rice and wheat from buffer stocks will thus lessen, increasing the likelihood of surplus stocks if the logistics of the TPDS remain inefficient.

Discussion and Conclusions

This study has analysed the impact of rising global wheat and rice prices on the Indian

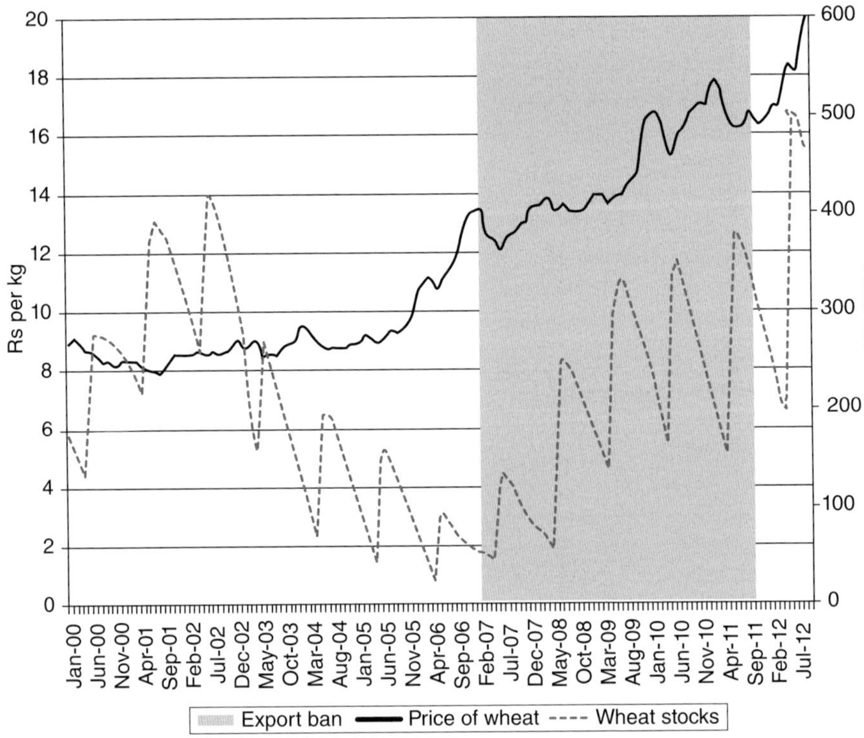

Fig. 10.10. Wheat stocks (in 0.1 Mt: right-hand scale) and Indian wheat price (average of retail and wholesale prices in Rs per kg: left-hand scale) (Food Corporation of India, 2012a; GIEWS, 2012).

Table 10.3. Results of Johanssen test (corrected for sample size (df = 67)) for rice.

Rank	Trace test	p-value
0	21.580	0.005
1	1.205	0.282
Eigenvalue	0.222	0.015

Number of equations = 2
Lag order = 1 (based on AIC (Akaike criterion) and BIC (Schwarz Bayesian criterion))
Estimation period: 2006:01 – 2012:09 (T = 81 months)
Case 3: Unrestricted constant
log likelihood = –61.431 (including constant term: –291.299)

Table 10.4. Results of Johanssen test (corrected for sample size (df = 67)) for wheat.

Rank	Trace test	p-value
0	0.004	0.004
1	4.070	0.049
Eigenvalue	0.222	0.015

Number of equations = 2
Lag order = 2
Estimation period: 2006:08 – 2012:08 (T = 73 months)
Case 3: Unrestricted constant
log likelihood = –231.98 (including constant term: –439.145)

price policy system. The Indian government is heavily involved in India's wheat and rice markets, two of the main staple crops in India. Its goal is two-fold: ensuring minimum prices for producers on the one hand and providing subsidized food to India's poor on the other hand. When international prices started to increase, the Indian government imposed an export restriction on wheat from February 2007 to September 2011 and on all rice varieties, except basmati, from September 2007 to September 2011. It also lowered import duties on wheat. These trade policies were aimed at shielding the domestic market from the price increases in the world market.

The analysis presented in this chapter shows that India has been successful in shielding the domestic rice and wheat prices from the global price volatility. India's wheat and rice prices have been found relatively more stable compared to those of the world market, especially of rice. However, this policy has come at some cost. When world prices increased and the Indian government imposed export restrictions, Indian producers were no longer able to profit from these international high prices. The Indian government therefore was obliged to raise MSPs to conciliate producers.

These higher domestic rice and wheat prices led to higher supply of these commodities. Because the Indian government purchases all the rice and wheat that producers sell, stocks also increased, which is shown by our analysis. In the past years, India's foodgrain stocks have risen to such levels that stocking capacity (warehouses, etc.) has become a constraint, leading to losses of rice and wheat stocks. At the same time, due to an inefficient distribution system, it did not lead to more availability of rice and wheat for India's poor.

The long-term (baseline) projections (see Woltjer and Rutten, Chapter 5, this volume) show that for wheat, India is expected to produce a surplus even in 2030. But, for rice, the picture is somewhat different, with supply lagging behind demand until 2030. Demand for rice and wheat will increase with growth in GDP and population, but the composition of demand is expected to change. With increasing incomes and more urban populations, the Indian population is expected to reduce the share of food in expenditures. The share of rice is expected to decrease most. This has consequences for India's TPDS, which, given these developments, can be expected to play a relatively smaller role by 2030. Therefore there is a risk of a future mismatch between a growing supply with MSPs and a relatively reduced demand for the TPDS.

The study has concluded that although the price policy has shielded domestic rice and wheat prices from global volatility, it has come at a cost in terms of rising MSPs and increasing stocks. The objectives of the price policy, made public from time to time, are multiple and often conflicting. The price policy aims to achieve simultaneously: (i) reasonable prices for producers; (ii) reasonable prices of grains released for public distribution systems; and (iii) reasonable prices in the open market. There is no clarity on the priority among these multiple objectives or the trade-offs involved.

In addition, any attempt to use high prices to encourage agricultural production can lead to distortions in the price structure. Kapila (2006) therefore has argued that India's price policy misses its goal because the aggregate supply response of agriculture is very weak. Kumar and Joshi (Chapter 4, this volume) have estimated the supply response elasticity for rice as only 0.2357 and for wheat as 0.2164.

India's TPDS has also received criticism. A recent World Bank report has found that the cost of TPDS is 1% of GDP, covers up to 23% of households, but that the effect on poverty reduction is low due to high leakages and diversion of grains from the TPDS. Only 41% of the grains released by government reached households in 2004/05 (World Bank, 2011). Recent reports by the Office of Supreme Court Commissioners of India (OSCCoI, 2012) have mentioned various problems with the implementation of the food schemes. See also Kattumuri (2011) for a recent overview of studies.

Finally, some authors have pointed out that price policy is a weak instrument for income transfers. A study by Shutes *et al.* (2012) has analysed the poverty impacts of high food prices under different trade regimes in India.

The authors argue that when global prices for rice and wheat increase, domestic wheat and rice production becomes relatively cheap, which induces the consumers to switch to domestic wheat and rice, thus increasing demand. This, in turn, will put pressure on domestic prices. However, high global prices also induce the producers to increase their supply, especially when exports are limited, as they were in India. Such a situation will put a downward pressure on prices. It depends on the elasticities of demand and supply as to what the net effect will be, but there will always be a 'dampening effect' of price policy on income through these second-round effects.

In September 2013, India adopted the National Food Security Bill (NFSB), which puts the 'Right to Food' approach at centre stage. As per this act, 67% of the people will be given subsidized food at almost one-tenth of the economic cost, which will add significantly to the cost to the exchequer. The NFSB will impact all agricultural policies, including trade and marketing policies, by making them more restrictive and state-controlled. Gulati (2013) expects that the NFSB will lead to even greater government intervention in cereal markets as well as increasing restrictions to cereal trade. Although it is too early to know what the implications of the NFSB will be, this suggests that the conclusions from our analysis remain relevant, or will even become more pronounced.

Notes

[1] Results available from authors upon request.

[2] Full details available from authors upon request.

References

Basu, K. (2011) India's foodgrains policy: An economic theory perspective. *Economic & Political Weekly* 46(5), 37–45.

Business Line (2012a) Farm prices panel against raising MSP of wheat next year. 4 September 2012, New Delhi edition. Available at: http://bit.ly/Rg0sC4 (accessed 16 January 2013).

Business Line (2012b) Covered foodgrain storage capacity to be increased. 17 October 2012, New Delhi edition. Available at: http://bit.ly/UpAmAm (accessed 6 December 2012).

Business Standard (2012) Foodgrain pricing panel asked to rework wheat MSP. 30 November 2012, New Delhi edition. Available at: http://bit.ly/UpsH52 (accessed 6 December 2012).

Dasgupta, D., Dubey, R.N. and Sathish, R. (2011) *Domestic Wheat Price Formation and Food Inflation in India*. Working Paper Series. Department of Economic Affairs, Ministry of Finance, Government of India, New Delhi.

Department of Consumer Affairs (2012) Price monitoring information system. Available at: http://fcainfoweb.nic.in (accessed 30 November 2012).

Department of Food and Public Distribution (2012) Monthly allocation of foodgrains for AAY, BPL and APL categories. Available at: http://bit.ly/TaNsij (accessed 5 October 2012).

DNA (*Daily News & Analysis*) (2012) Wheat export ban lifted to settle cabinet feud. 5 July. Available at: http://bit.ly/O7oWN9 (accessed 2 December 2012).

Food Corporation of India (2012a) Month-wise foodgrains stock in central pool (as on 1st day of the month). Available at: http://bit.ly/RkHGvo (accessed 1 October 2012).

Food Corporation of India (2012b) State-wise procurement of rice and wheat in major rice and wheat producing states. Available at: http://fciweb.nic.in/stocks/index (accessed 6 December 2012).

Food Corporation of India (2012c) Buffer norms and strategic reserve. Available at: http://fciweb.nic.in/stocks/view/8 (accessed 11 March 2012).

GIEWS (2012) Food price data and analysis tool. Available at: http://bit.ly/9mgQIT (accessed 1 October 2012).

GoI (Government of India) (2009) *Economic Survey 2008–2009*. Ministry of Finance, New Delhi.

GoI (Government of India) (2012) *Economic Survey 2011–2012: Agriculture and Food*. New Delhi. Available at: http://www.indiabudget.nic.in (accessed 16 January 2013).

Gulati, A. (2013) India's policy options to ensure agri-markets contribute to food security. In: *Farm Policy Reforms and the Future of the Multilateral Trading System: What Could Evolving Policies Mean for Global Regulatory Frameworks?* ICTSD Conference, 24 September 2013, Geneva.

IMF (International Monetary Fund) (2012) IMF primary commodity prices. Available at: http://bit.ly/nLjNDZ (accessed 30 November 2012).

Jha, S., Srinivasan, P.V. and Landes, M. (2007) *Indian Wheat and Rice Sector Policies and the Implications of Reform*. Economic Research Report Number 41. United Stated Department of Agriculture Economic Research Service, Washington, DC.

Kapila, U. (2006) Developments in Indian agriculture. In: Kapila, U. (ed.) *Indian Economy Since Independence*, 17th edn. Academic Foundation, New Delhi, pp. 207–246.

Kapur, A. (2012) *Targeted Public Distribution System, Government of India Budget Briefs 2011–12*. Government of India Budget Briefs. Accountability Initiative, Centre for Policy Research, New Delhi.

Kattumuri, R. (2011) *Food Security and the Targeted Public Distribution System in India*. Asia Research Centre Working Paper. London School of Economics, London.

Kotwal, A., Murugkar, M. and Ramaswami, B. (2011) The political economy of food security in India. In: *Globalization and Public Sector Reforms in China and India*. Asia Research Centre, Copenhagen Business School, Copenhagen, Denmark.

Meijerink, G., van Berkum, S., Shutes, K. and Solano, H.G. (2011) *Price and Prejudice: Why Are Food Prices so High?* LEI, Wageningen UR, The Netherlands.

MoA (Ministry of Agriculture) (2012) Minimum support prices of various agricultural commodities (Table 11.5.1). Available at: http://bit.ly/ZDe1l9 or http://cacp.dacnet.nic.in (accessed 11 March 2012).

MoF (Ministry of Finance) (2013) *Economic Survey 2012–2013*. Government of India, New Delhi.

MoSPI (Ministry of Statistics and Programme Implementation) (2012) *Draft Manual on Agricultural Prices and Marketing*. Government of India, New Delhi.

Nair, S. and Eapen, L.M. (2011) Wheat price inflation in recent times: causes, lessons and new perspectives. *Economic & Political Weekly* 46(36), 58–65.

OSCCoI (Office of Supreme Court Commissioners of India) (2012) Reports submitted to the Supreme Court Commissioners by its State Advisers. Supreme Court Commissioners of India. Available at: http://bit.ly/XH9kZ8 (accessed 30 November 2012).

Rapsomanikis, G. and Mugera, H. (2011) Price transmission and volatility spillovers in food markets of developing countries. In: Piot-Lepetit, I. and M'Barek, R. (eds) *Methods to Analyse Agricultural Commodity Price Volatility*. Springer, Dordrecht, The Netherlands, pp. 165–179.

Shutes, L.J., Ganesh-Kumar, A. and Meijerink, G. (2012) *Fluctuating Staple Prices and Household Poverty in India*. Research memorandum. LEI, Wageningen UR, The Netherlands.

Singh, S. (2012) *India. Grain and Feed Annual 2012*. GAIN Report. United Stated Department of Agriculture Foreign Agricultural Service, Washington, DC.

Thaindian News (2009) Farm production could fall by 15-20 percent by Mukherjee. 27 August 2009 edition. Available at: http://tinyurl.com/yzo86k7 (accessed 30 November 2012).

Times of India (2009) Ban on wheat futures removed. 16 May 2009 edition. Available at: http://bit.ly/13t8UK0 (accessed 30 November 2012).

World Bank (2011) *Social Protection for a Changing India*, Vols I and II. The World Bank, Washington, DC.

WSJ (*Wall Street Journal*) (2008) India bans futures trading in four more commodities. 8 May 2008 edition. Available at: http://tinyurl.com/m4gz288 (accessed 30 November 2012).

11 Biofuel Commitments in India and International Trade

Geert Woltjer* and Edward Smeets
LEI Wageningen UR, The Hague, The Netherlands

Introduction

The production and use of biofuels has increased rapidly during the past several years. Global ethanol production has increased from 48 billion l in 2005 to 113 billion l in 2012 and biodiesel production has increased from 5.3 billion l to 28 billion l (FAO and OECD, 2012). As of today, more than 50 countries across the world, including India, have implemented biofuel policies (Sorda *et al.*, 2010). These policies typically consist of subsidies on biofuel use or production or biofuel blending mandates.

The recent rise in use of biofuels is driven by the concerns over energy security, climate change and rising fossil fuel prices as well as the additional demand for agricultural commodities. Consequently, rising farm incomes form important benefits of these biofuel policies. Furthermore, biofuels are often seen as a stimulant for rural development and employment.

In recent years it has become clear that the use of current first-generation biofuels, which are made from conventional starch-, oil- and sugar-containing crops, such as wheat, maize, rapeseed, palm fruit, soybean and sugarcane, potentially have various disadvantages. First, food security may be negatively affected by the higher prices of agricultural commodities. Second, the greenhouse gas balance is not as beneficial as was initially assumed, partly as a result of the loss of natural vegetation due to indirect land-use change (ILUC). These ILUC effects also reduce biodiversity. Finally, various studies have suggested that biofuel production can negatively affect the socio-economic conditions in the rural areas of developing countries, for example, through the insidious dissipation of indigenous land-use rights.

In 2001, India implemented a pilot programme aimed at realizing 5% ethanol blending (E5) and launched a National Mission on Biodiesel in 2003 to achieve 20% biodiesel blends (B20) by 2011/2012 (Pohit *et al.*, 2011). In 2009, the Government of India approved the National Policy on Biofuels that includes an indicative 20% blending target by 2017, both for biodiesel and bioethanol (GoI, 2009). The objectives of this policy are to reduce the dependency on imports of fossil oil, reduce greenhouse gas emissions, promote rural development and generate employment opportunities. A pre-requisite thereby is that biofuels may not be produced at the cost of food crops. For this reason, the production of biodiesel from non-edible oilseeds is promoted only on waste, degraded

* Corresponding author, e-mail: Geert.Woltjer@wur.nl

and marginal lands. At this moment, the production of *Jatropha* is not commercially viable, except for some heavily subsidized projects, and it is expected that the blending target for biodiesel for 2017 will not be realized (Pohit *et al.*, 2011; USDA, 2012). In the case of ethanol the main feedstock is molasses, which is a by-product of sugarcane processing. However, it is expected that the 20% blending target for ethanol by 2017 cannot be realized using only molasses (Pohit *et al.*, 2011; Raju *et al.*, 2012; USDA, 2012).

Several studies have been conducted on the status of biofuel use in India and its impact on land use, food production and the environment (Pohit *et al.*, 2011; Ravindranath *et al.*, 2011; Schaldach *et al.*, 2011). Several case studies have also been carried out in this regard (Mahapatra and Mitchell, 1999; Agoramoorthy *et al.*, 2009; Findlater and Kandlikar, 2011; Sasmal *et al.*, 2012). This chapter adds to these studies by investigating the consequences of the National Biofuel Policy of India and of biofuel policies in other countries on poverty, welfare, land use, trade, food security, etc. in India to the year 2020 using a global economic model.

Modelling of Biofuel Policies

The impact of biofuel policies is mostly investigated using the MAGNET general equilibrium model framework (i.e. Modular Applied GeNeral Equilibrium Tool, Woltjer and Kuiper, 2014), which is an extended version of the Global Trade Analysis Project (GTAP) model (Hertel, 1997). For the simulations in TAPSIM (Trade, Agricultural Policies and Structural Changes in India's Agrifood System; Implications for National and Global Markets), the MAGNET model was modified so that it could take into account the production of ethanol from molasses, the intensification of crops and livestock production, the use of by-products of biofuel production as animal feed, and also the substitution possibilities for different feedstocks of biofuel production.

Introducing New Sectors in MAGNET Model

Several new sectors were added to the MAGNET model to generate a model suitable to analyse the impact of biofuel policies, namely:

- In India, the production of ethanol from molasses, which is a by-product of sugar production, is a split off from the sugar industry. In most other countries, biofuels are produced from the conventional agricultural crops, which are already considered in the MAGNET model. Growing sweet sorghum is also a potentially feasible option for marginal lands in India, though the yields are likely to be low (Ravindranath *et al.*, 2011) and, therefore, the production of ethanol from only molasses and sugarcane was considered.
- The production of animal feed was separated from the sector 'other feed and food', which includes, for example, canned fish. It was needed to account for the impact of high-value by-products of biofuel production. For maize and wheat ethanol, these are distiller's dried grains (DDG) and the main by-product of biodiesel production which is the oilcake obtained from crushing of oilseeds in the vegetable oil sector.
- The vegetable oil sector was split into two sub-sectors, one that produces relatively cheap crude vegetable oils, which are used for biodiesel production, and the other that produces relatively expensive refined and processed vegetable oils, which are used in the food processing industry and also in cosmetics.
- Finally, a key issue was the impact of biofuels on the intensification of agriculture, i.e. on crop yields. One of the main ingredients of this intensification is the increase in fertilizer use. In the GTAP database, fertilizer has been included in the chemical sector. For this reason, the fertilizer sector had to be split from the chemical sector.

Substitution in Production

The next step involved the modelling of substitution between fossil fuels and different biofuels, between biofuel by-products and other feeds for livestock, and between different inputs for biofuel production. The MAGNET model has a flexible constant elasticity of substitution (CES) nesting structure for production. This flexibility creates the opportunity to change the substitution possibilities between inputs in case it is relevant for a specific policy scenario. For all sectors that are not discussed below, a standard GTAP production structure with a substitution possibility between energy and capital was used (Fig. 11.1). The value-added as well as all intermediate inputs had fixed technical coefficients. Within the value-added inputs, there was a non-capital value-added nest and a capital-energy nest. Within the capital-energy nest, capital and energy can be substituted. Within the energy nest, different types of energy can be substituted. The elasticity of substitution between capital and energy was set at 0.5, and between different types of energy at 1.0, following the energy variant of GTAP, GTAP-E. For the non-capital value-added nest, the standard GTAP substitution elasticities were used.

In the petroleum sector, crude oil is converted into conventional fossil fuels and ethanol or biodiesel is blended with these petroleum products. For this reason, the first CES level of the petroleum sector concerns the blending and substitution of biofuels and fossil fuels. The share of biofuels in fossil fuels is exogenously determined, based on the biofuel policies in various countries (Sorda *et al.*, 2010). The elasticity of substitution between fossil fuels and biofuels is thereby set very high (50), while a substitution possibility between biodiesel and ethanol is assumed to be relatively small (3). The production structure of the fossil fuel sector has the standard GTAP configuration: a CES nest for value-added inputs and fixed coefficients for the intermediate inputs.

The ethanol and biodiesel sectors convert the biofuel feedstock into biofuels. These sectors follow the standard production structure with one extra nest, viz. the feedstock nest (Fig. 11.1). Biodiesel is produced from vegetable oil only, and therefore no substitution is possible. For ethanol, the substitution

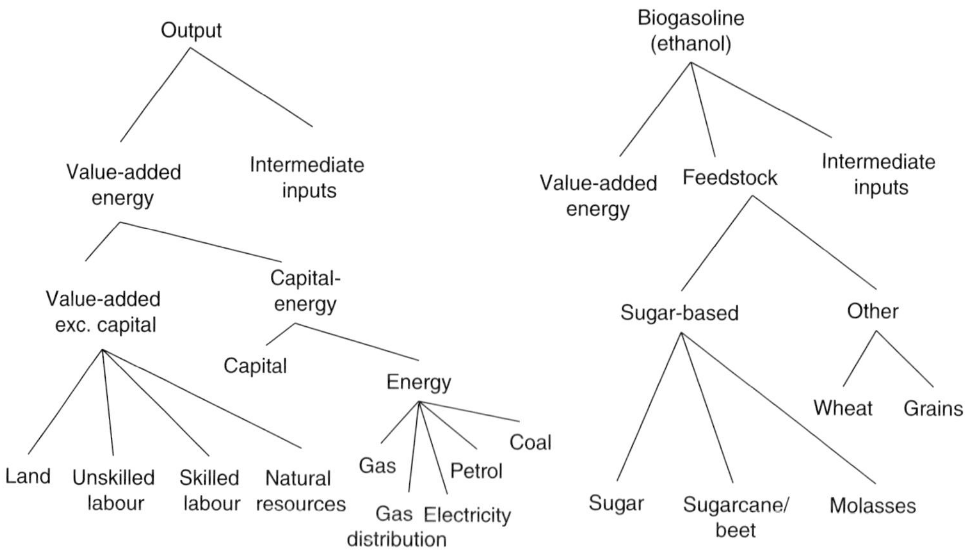

Fig. 11.1. The standard CES production structure (left) and the CES production structure of the ethanol sector (right).

possibilities are much larger as substitution is possible between sugar-based feedstock and other feedstock with a substitution elasticity of 5. Much easier is the substitution between sugar-based feedstock (sugarcane, sugarbeet and molasses; substitution elasticity of 50) and between starch-based feedstock (wheat and grain; substitution elasticity of 20). These large elasticities were chosen to prevent very large changes in prices when large changes in biofuel shares as a consequence of biofuel targets were required, while the fixing of biofuel shares in the model (see below) guaranteed that these large elasticities would not influence the substitution between biofuels and fossil fuels.

The intensification of crop production requires that the use of cropland can be replaced by higher use of fertilizers, i.e. when you apply more fertilizer, less land per kilogram of crop is required. The structure of crop-producing sectors is based on the standard production structure, but the value-added nest was split into a land-fertilizer nest and a standard value-added nest that excluded land. The substitution between land and fertilizer is possible only within the land-fertilizer nest. For the substitution between fertilizer-land and value-added nests, a substitution elasticity of 0.1 was used, which is the default value in the MAGNET model, and the substitution elasticity between land and fertilizer was set at 0.8.

Modelling Biofuel Targets

In most countries, biofuel policies are formulated as a target share of biofuels in the fuels used for road transport. The data on biofuel production, consumption and share in road transport in the base year of the MAGNET model (2007) were taken from the International Energy Agency (IEA, 2011). The value of biofuel production in each country in the GTAP database was calculated by multiplying these numbers with the price of the biofuels. The price of biofuels was based on cost structures of the ethanol and biodiesel production per feedstock and the price of feedstock was calculated from the required quantities and the average price of these quantities in the model.

The biofuel is blended with fossil fuels in the petroleum sector and the mixed fuel is sold to the users of transport fuels. This mandatory blending is budget-neutral from the government point of view. To achieve this, a CGE model involves implementing two policies. First, the biofuel share of transport fuel is specified and made exogenous such that it can be set at a certain target. An endogenous subsidy is modelled to achieve the required biofuel share. Second, to ensure that this incentive instrument is budget-neutral, the biofuels subsidy is financed by an end-user tax on petrol consumption, implying that the petrol user pays for the cost involved for using biofuel. This implicit subsidy is in line with the reality, although in some countries tax exemptions for biofuel are also implemented.

Scenarios

Three scenarios have been introduced in the MAGNET economic model that can be compared with a baseline in which the share of biofuels used in transport fuels remains constant at the level in 2007 in all the countries. In the first scenario, only biofuel policies outside India have been considered (called Non-India biofuels). This implied a biofuel share in transport by 2020 of 5% for the EU and South-east Asia, 10% for the USA, Indonesia, the rest of southern Asia, 15% for China and 25% for Brazil. In the second scenario, a biofuel share of 20% for India by 2020 has been assumed (called India biofuels). In the third scenario, the other two scenarios have been combined, i.e. both India and the rest of the world fulfil their biofuel commitments (called Global biofuels).

Biofuel Production and Feedstock Demand

The average worldwide biofuel share in transport fuels is 6.0%, 1.4% and 6.7% for the Non-India biofuels, the India biofuels and the Global biofuels scenarios, respectively. The biofuel share in the baseline scenario is 0.7%. Most of the production of biofuels will take place in the regions where the demand for biofuels is generated and their trade is limited.

The production of biofuels requires extra feedstock, as shown in Table 11.1. The India biofuels scenario results in an increase in the use of sugarcane largely and molasses partially. The increase in use of molasses outside India is the result of small shares in 2007 of ethanol from molasses in southern America and South-east Asia in combination with the high substitution elasticity with sugarcane.

The increase in demand of feedstock used for biofuel production would generate a substantial price effect by 2020 (Table 11.2). The biofuel policy in India would result in a 27% higher sugarcane/beet price and 11% higher molasses price in 2020. The biofuel policies in the rest of the world would especially affect the price of coarse grains (+18%) and vegetable oils (+19%). The price of wheat would be much less influenced by the biofuel policies in India and elsewhere.

Table 11.1. The volume of use of agricultural commodities for biofuel production by 2020 in the three scenarios relative to the baseline scenario (in million constant 2007 US$).

	Scenario		
Agricultural commodity	Non-India biofuels	India biofuels	Global biofuels
Wheat	2,475	2	2,542
Coarse grains	90,656	54	90,273
Sugarcane/beet	13,627	11,143	24,941
Molasses	1,255	2,072	3,508
Vegetable oils	58,436	38	59,732

Table 11.2. The percentage change in global price of feedstock input for biofuels by 2020 in the three scenarios relative to the baseline scenario.

	Scenario		
Feedstock	Non-India biofuels	India biofuels	Global biofuels
Wheat	5	1	6
Coarse grains	18	0	21
Sugarcane/beet	13	27	38
Molasses	–1	11	12
Vegetable oils	19	0	21

The increase in prices of agricultural commodities reduces the non-biofuel demand for agricultural commodities. Therefore, the net effect of the extra biofuel production on agricultural production would be less than the increase in demand for biofuels, as can be seen by comparing the values in Table 11.1 and Table 11.3. One can even see that for the commodities that are used in small amounts, i.e. wheat and molasses, the production volume is less in the biofuels scenarios than in the baseline scenario. For molasses, this is a little confusing, because in the production of ethanol from sugarcane some molasses is implicitly produced that is directly converted into ethanol, which is not mentioned separately in the statistics.

Further, Table 11.4 shows that the total effect of biofuel policies on production of agricultural commodities would be significant. In the Global biofuels scenario, the global production of sugarcane would increase by 40%, the production of coarse grains by

Table 11.3. The volume of global use of agricultural commodities by 2020 in three different scenarios relative to the baseline scenario (in million constant 2007 US$).

	Scenario		
Agricultural commodity	Non-India biofuels	India biofuels	Global biofuels
Wheat	–988	28	–919
Coarse grains	86,047	264	84,558
Sugarcane/beet	13,480	11,727	25,141
Molasses	–325	–82	–401
Vegetable oils	54,028	90	55,014

Table 11.4. The percentage change in volume of global production of agricultural commodities by 2020 in three scenarios relative to the baseline scenario.

	Scenario		
Agricultural commodity	Non-India biofuels	India biofuels	Global biofuels
Wheat	–0.5	0.0	–0.5
Coarse grains	31.2	0.1	30.7
Sugarcane/beet	20.7	18.0	38.7
Molasses	–1.5	–0.4	–1.9
Vegetable oils	47.5	0.1	48.3

30% and the production of vegetable oils by almost 50%. The production of molasses would remain relatively constant in all three scenarios, but this is the consequence of not counting molasses production that is directly converted into ethanol.

Land Use and Intensification

As a consequence of biofuel policies, the use of agricultural land would change. Table 11.5 shows that the total global area of agricultural land would increase 1.9% in the case of the Global biofuels scenario. In particular, the area of cropland used for biofuel feedstock crops would increase, resulting in a 4.5% increase in the global area of cropland. The (limited) increase in use of land for livestock is caused by the substitution away from crops as animal feed towards roughage as a consequence of higher crops prices. The biofuel policy in India may increase land use of the main biofuel feedstock, sugarcane, by about 8% compared with the Non-India biofuels scenario.

The increase in biofuel use would also result in increased crop yields (Table 11.6). The increase in crop yields would be the highest in the regions with ambitious bio fuel policies and where the demand for crops would increase most, i.e. for vegetable oils in the EU, for coarse grains in the USA (Non-India), for sugarcane and sugarbeet almost everywhere in the case of the global biofuels directive, and a small increase for wheat in the EU.

Table 11.5. The percentage change in global land use by 2020 in three scenarios relative to the baseline scenario.

	Scenario		
Agricultural commodity	Non-India biofuels	India biofuels	Global biofuels
Wheat	–2.3	–0.2	–2.8
Coarse grains	11.2	–0.1	9.7
Sugarcane/beet	13.0	9.7	21.1
Oilseeds	13.1	–0.4	11.6
All crops	4.5	0.0	3.8
Livestock	0.2	0.2	0.8
Primary agriculture	1.8	0.1	1.9

The increase in crop yields is driven by the increase in demand for crops for biofuel production and the increase in land prices. Table 11.7 shows the likely changes in prices of agricultural land under different scenarios. As a consequence of the relatively low elasticities of substitution between different types of land, we see occurrence of relatively large price differences. Especially in India, the price of land used for sugarcane production would increase rapidly, by 150% or more. To what extent these effects would be correct is an empirical question, whereby we must be aware that the price is for effective land units, so if expansion of land requires the use of less suitable land, this also implies an increase in the need for land.

Table 11.8 depicts the changes in the use of land for different crops by 2020 under three different scenarios. In the India biofuels scenario, the area under sugarcane cultivation in India would expand by 68%. The impact of the Global biofuels scenario on land use for production of other biofuel feedstock in India would be less, but still substantial (4–13%). This means that part of the additional production of biofuel feedstock needed to meet the biofuel blend mandates outside India directly or indirectly would come from India. The effects of the Global biofuels scenario would be strongest for the feedstock types used in the regions with aggressive biofuel policies (e.g. oilseeds in the EU, coarse grains in the USA).

The increase in crop yields and land prices could also result in an increase in the use of fertilizers per hectare and, to a certain extent, to a higher use of capital and labour per hectare. In Table 11.9 we can see this clearly for sugarcane production. The biofuel policy in India could result in a 112% increase in the use of fertilizers per hectare. However, more capital and labour would also be required, among other reasons to improve the irrigation of sugarcane.

Table 11.10 shows the results for intensification and input-use of crop production

Table 11.6. The percentage change in crop yields per hectare by 2020 in three scenarios relative to the baseline scenario.

		Scenario		
Crop	Location	Non-India biofuels	India biofuels	Global biofuels
Wheat	India	1.6	1.2	2.9
	EU27	4.4	0.2	4.7
	Rest of world	0.7	0.2	1.3
Coarse grains	India	1.0	0.4	1.3
	EU27	2.8	0.1	2.8
	Rest of world	21.0	0.1	22.3
Oilseeds	India	1.8	0.2	1.9
	EU27	19.0	0.5	22.1
	Rest of world	5.7	0.5	7.3
Sugarcane/beet	India	1.0	13.0	14.1
	EU27	11.1	0.4	12.7
	Rest of world	9.0	0.7	10.9

Table 11.7. The percentage change in real land prices of biofuel feedstock crops by 2020 in different scenarios relative to the baseline scenario.

		Scenario		
Biofuel feedstock crops	Location	Non-India biofuels	India biofuels	Global biofuels
Wheat	India	4	5	10
	EU27	12	0	11
	Rest of world	5	1	6
Coarse grains	India	5	7	12
	EU27	8	0	7
	Rest of world	77	0	89
Oilseeds	India	13	9	24
	EU27	46	1	56
	Rest of world	37	1	45
Sugarcane/beet	India	6	156	174
	EU27	24	0	27
	Rest of world	36	2	47

Table 11.8. The percentage change in crop production by 2020 in different scenarios relative to the baseline scenario.

		Scenario		
Crop	Location	Non-India biofuels	India biofuels	Global biofuels
Wheat	India	−2.3	−3.2	−4.8
	EU27	−0.1	0.1	−0.2
	Rest of world	−2.6	0.2	−2.9
Coarse grains	India	−1.4	−1.2	−2.5
	EU27	−3.1	0.0	−3.0
	Rest of world	13.5	0.0	11.8
Oilseeds	India	4.3	−1.7	1.7
	EU27	23.7	0.1	21.6
	Rest of world	13.3	−0.3	12.1
Sugarcane/beet	India	−0.9	68.0	68.3
	EU27	8.6	0.3	8.0
	Rest of world	15.8	0.3	14.0

Table 11.9. The percentage change in input-use per hectare in India by 2020 in three scenarios relative to the baseline scenario.

		Scenario		
Crop	Input	Non-India biofuels	India biofuels	Global biofuels
Wheat	Labour	1	0	2
	Capital	1	0	2
	Fertilizer	5	4	9
Coarse grains	Labour	1	−1	0
	Capital	1	−1	0
	Fertilizer	5	6	11
Oilseeds	Labour	3	−1	1
	Capital	3	−1	1
	Fertilizer	11	7	20
Sugarcane/beet	Labour	1	19	20
	Capital	1	18	19
	Fertilizer	6	112	127

Table 11.10. The percentage change in input-use per hectare outside India by 2020 in three scenarios relative to the baseline scenario.

		Scenario		
Crop	Input	Non-India biofuels	India biofuels	Global biofuels
Wheat	Labour	2	0	3
	Capital	0	0	1
	Fertilizer	5	0	6
Coarse grains	Labour	11	0	12
	Capital	17	0	18
	Fertilizer	65	0	72
Oilseeds	Labour	9	1	11
	Capital	8	1	10
	Fertilizer	25	1	30
Sugarcane/beet	Labour	7	1	9
	Capital	11	1	14
	Fertilizer	28	1	34

outside India. It is obvious that the biofuel policy in India will have only a small effect on intensification in the rest of the world. Biofuel policies outside India could have a much larger impact on intensification of crop production. This is especially relevant in the case of coarse grains, and to a lesser extent for oilseeds and sugarcane/beet. The Non-India biofuels scenario would also generate a limited intensification in India, especially in the case of oilseed production due to the extra demand for crude vegetable oil.

Animal Feed

Biofuel production also has consequences for the animal feed sector. While molasses is a by-product of sugar production that is used for either animal feeding or ethanol production, the production of biodiesel and ethanol results in oilcakes and DDGs as by-products, respectively, which are both used as animal feed. Table 11.11 shows that the net effect of biofuel policies on the price of animal feed is that the price would increase, as the rise in crop prices would more

Table 11.11. Percentage change in use of production factors and livestock output per hectare in India and the EU by 2020 in three different scenarios compared to the baseline scenario.

	Scenario					
	India			EU27		
Factor	Non-India biofuels	India biofuels	Global biofuels	Non-India biofuels	India biofuels	Global biofuels
Land	5.7	2.8	7.5	6.1	0.4	3.8
Unskilled labour	4.3	3.1	6.3	11.0	0.7	12.1
Skilled labour	5.9	−1.3	2.9	10.9	0.8	12.0
Capital	4.8	2.2	5.9	10.4	0.7	11.6
Feed price	2.9	6.7	9.8	0.0	0.5	0.6
Production per ha	0.6	−1.1	−0.6	2.2	0.0	1.6

than compensate the increase in supply of by-products of biofuel production. Especially important is the increase in land prices, which would result in intensification, i.e. higher crop yields per hectare. The use of molasses for biofuel use in India could result in an increase in the price of molasses and as a result the price of animal feed would rise more than the price of land for livestock. Consequently, the farmers would reduce the use of crops for animal feed production and increase the use of pasture for grazing.

Table 11.12. Percentage change in production volume in India by 2020 in different scenarios compared to the baseline scenario.

	Scenario		
Sector	Non-India biofuels	India biofuels	Global biofuels
Crops	0.81	1.71	2.49
Livestock	−0.15	−0.76	−0.82
Agri-processed	0.96	−2.68	−1.91
Industry	0.61	−0.37	−0.01
Services	0.39	−0.20	0.11

Production and Welfare

In this section we evaluate the consequences of biofuel policies on social welfare and production in India. Table 11.12 shows that the effect of biofuel directives in other countries would be positive for all sectors of the economy in India, except for livestock. Crop production would expand as a result of higher biofuel production. The increase in import prices would be smaller than the increase in export prices in the Non-India biofuels scenario. As a result, the price of imported intermediate inputs would be reduced, which could increase the value added of the commodities produced. This positive 'terms of trade effect' would also be responsible for the increases in social welfare as discussed below. In the India biofuels scenario the production of crops will be much higher, but because of their use for (implicitly or explicitly) subsidized production of biofuels, this may not contribute to economic growth or social welfare.

Table 11.13 shows the decomposition of welfare effects by 2020 in India, the EU and the rest of the world for the Non-India and India biofuels scenarios. The Non-India biofuels scenario by 2020 depicts an increase in social welfare in India, whereby the change in terms of trade is responsible for the largest benefit; import prices will be reduced by 1.5% compared to export prices. For the EU, the Non-India biofuels scenario would generate a welfare benefit too, mainly as a consequence of terms of trade benefits, but also because distortions from production and consumption taxes in the economy will be reduced.

Table 11.14 shows the welfare effects on a selection of other regions. For the USA, the allocation effect of a biofuels policy would be highly negative as is the case for

Table 11.13. Decomposition of welfare effects by 2020 in India, the EU and the rest of the world, changes compared to the baseline scenario (in million constant 2007 US$).

	India		EU27		EU-Non-India	
Particulars	Non-India biofuels	India biofuels	Non-India biofuels	India biofuels	Non-India biofuels	India biofuels
Allocation	4,642	−11,349	11,107	3,727	−89,357	2,449
Endowment	1,306	2,662	103	14	5,614	1,896
Technology	1,297	642	1,625	138	15,221	735
Population	347	−7	84	13	−2,038	−197
Terms of trade	9,221	6,258	13,399	1,518	−21,921	−7,464
Investment-savings	116	150	2,407	407	−2,557	−565
Total	16,929	−1,644	28,725	5,817	−95,038	−3,146

Table 11.14. Welfare effects by 2020 of a global biofuels directive, changes compared to the baseline scenario (in million constant 2007 US$).

Particulars	USA	Brazil	South America, others	Africa	China	Asia, others
Allocation	−37,972	1,474	−764	−1,294	−49,955	266
Endowment	2,935	290	1,653	2,769	−3,327	3,380
Technology	2,430	510	1,108	3,090	5,731	1,358
Population	−234	158	48	−2,548	−106	97
Terms of trade	28,568	6,888	401	−88,352	33,425	21,582
Investment-savings	7,725	72	68	−55	−7,644	−2,793
Total	3,452	9,392	2,514	−86,390	−21,876	23,890

China. In both these regions, the cost of biofuels will be higher than the cost of fossil fuels by 2020. For the USA, this may be compensated by a positive terms of trade and investment-savings effect (consistent with the results of Gehlhar *et al.*, 2010), while the net effect on China will be negative. For Brazil, the distorting effect of biofuels production is absent, while they would have a positive terms of trade effect as a consequence of reduced fossil fuel imports.

The big loser of the global biofuel policies is Africa (including Middle East), which will have to pay more for its food without having much benefit of lower fossil fuel prices, and for the oil-exporting countries having less income from their oil exports.

The biofuel policies in India clearly will have a negative allocation effect in India by 2020, mainly as a consequence of the distorting effect of biofuel subsidies. This negative effect would partly be compensated by a positive terms of trade effect, because the costs of import of fossil fuel will decrease, whereby this benefit will obviously occur at the cost of the oil-producing regions. The welfare effects in other countries would differ, but for Africa (not having biofuel policies), the terms of trade effects would be highly negative, both for food and for export of fossil fuels.

Food Consumption

The introduction of biofuels also has consequences for the food consumption in India. Table 11.15 shows that food consumption in India will be reduced as a consequence of the higher agricultural prices in both the Non-India and India biofuels scenarios by 2020. This also suggests that food security effects will be negative, although the effects would be smaller than 1%. In the Non-India biofuel scenario, the reduction in food consumption could be accompanied by higher expenditures on industry and services, because total gross domestic product will increase. The effect of India biofuels will be a reduction in the private demand for industrial commodities.

Trade Balance

A look at the effect of biofuel policies on trade reveals that these effects will be relatively small. The implementation of biofuel policies outside India could lead to an increase in the exports of crops and processed food. The net export of industrial goods could increase, because India may pay about 6% less for its crude oil, which will reduce the oil import bill. The price of crude oil being less important in the case of services, the sector could lose a little bit of its comparative advantage, and therefore the net exports of services will be reduced by 2020.

An Indian biofuels policy implies again a reduction in crude oil price, but the main effect would be a reduction in the need to import crude oil, because oil imports would be replaced by domestically produced biofuels. However, because of the increased use of crops for the production of biofuels in India, less crops would be available for use as food. This implies that imports of primary and processed food commodities will have to be increased, which is indeed the case, as shown in Table 11.16 and Table 11.17.

Table 11.15. Percentage change in food consumption in India by 2020 relative to the baseline scenario.

	Scenario	
Sector	Non-India biofuels	India biofuels
Crops	−0.446	−0.7
Livestock	−0.712	−0.562
Agri-processed	−0.19	−0.613
Industry	0.933	−0.358
Services	0.912	0.033

Impact on Poverty

The production of biofuels also influences poverty in India. As an indication of poverty, we compared the change in the price of crops as an indicator of the cost of living of poor people, with the change in wages of unskilled labour. Table 11.18 shows that under the Non-India biofuels scenario by 2020, the wage rate in agriculture will rise at about the same rate as for the price of crops in India. This suggests that the effect of biofuel policies in the poor rural areas would probably be not very large. However, in industry, i.e. in urban areas, the wages are not likely to change much (+0.1%), while the price of crops would increase by 4.3%. This shows that the urban poor will be able to buy less food for their income. When looking at the Indian biofuel policy, the impact on poverty seems to be much more negative. The price of crops will rise by 11.5%, while the wage rate for unskilled labour in agriculture would increase by just 3.0% by 2020. These effects would be reinforced if biofuel policies in India and elsewhere are combined.

Table 11.16. Net export of India as fraction of Indian production by 2020.

	Scenario			
Sector	Baseline	Non-India biofuels	India biofuels	Global biofuels
Crops	0.25	0.65	−0.29	0.04
Livestock	−0.17	−0.14	−0.23	−0.19
Agri-processed	4.62	5.26	1.88	2.55
Industry	0.84	0.93	1.09	1.15
Services	0.54	0.42	0.51	0.39

Table 11.17. The change in net export of India by 2020 in different scenarios compared to the baseline scenario (in million constant 2007 US$).

	Scenario		
Sector	Non-India biofuels	India biofuels	Global biofuels
Crops	1369	–1826	–640
Livestock	22	–68	–39
Agri-processed	1315	–4328	–3048
Industry	2239	6416	8156
Services	–2856	–779	–3381
All commodities	2087	–585	1047

Table 11.18. Percentage change in crop prices and wages in India by 2020 in three scenarios, relative to the baseline scenario.

	Scenario		
Sector	Non-India biofuels	India biofuels	Global biofuels
Price of crops	4.3	11.5	16.5
Unskilled wage – agriculture	4.3	3.0	6.3
Unskilled wage – industry	0.1	–0.1	0.0

Table 11.19 shows more detailed results per crop type. In the India biofuels scenario, the price of sugarcane would increase by 77%, while the price effects for other crops would be 15% or less in all the scenarios. The increase in the price of rice and wheat, which are the two staple food crops in India, would be much less than the rise in average crop price. Nevertheless, the price increase of crops would be higher than the wage increase for unskilled labour in India and the consumption of crops could be reduced.

Conclusions

The current biofuel policies in India and other countries are based on the use of first-generation biofuels, such as ethanol made from conventional sugarcane and starch crops, and biodiesel produced from vegetable oils. The use

Table 11.19. Percentage change in crop prices in India by 2020 in three scenarios, relative to the baseline scenario.

	Scenario		
Crop	Non-India biofuels	India biofuels	Global biofuels
Rice	1.9	3.3	5.3
Wheat	1.4	3.2	4.7
Coarse grains	3.0	5.1	8.1
Oilseeds	7.6	6.8	15.2
Sugarcane	4.2	77.3	86.9
Horticulture	5.3	7.5	13.0
Plant-based fibres	4.7	6.1	10.9
Other crops	4.3	7.3	11.8

of these crops for biofuel production has depicted various effects on poverty, societal welfare, land use, trade, food security, etc. in India.

The biofuel policies outside India will have a negative net effect on poverty in India by 2020. The effect would be less on the rural poor in India, because they would benefit from the increased wages in agriculture, while the urban poor would only experience higher food prices. As a result, the consumption of crops and livestock in India is likely to decrease, although the societal welfare effects would be positive. These positive welfare effects will be caused by a positive 'terms of trade effect'. This effect is the result of import prices increasing less than export prices. The price of imported intermediate inputs will be reduced, which could increase the value added of the commodities produced and thus societal welfare.

The National Biofuel Policy in India will also have substantial effects. Global sugarcane production will increase by 18% and sugarcane prices by 27% in 2020. The societal welfare effects in India could be negative, because biofuel production (implicitly or explicitly) is subsidized. The increased use of resources for biofuel crop production cannot be used in other sectors, implying a reduction in production in these other sectors.

The results presented in this chapter are based on the MAGNET economic model and the calculations are extremely rough. Especially relevant is the question to what extent the urban and rural poor will benefit from the increased demand for labour as a consequence of biofuel policies by 2020. Our observations are consistent with the findings in the literature (Chakravorty *et al.*, 2012). However, further empirical validation and more refined analyses are needed as regional and longer-term effects from biofuel policies on agricultural productivity, rural development and technological change have only been partially considered.

References

Agoramoorthy, G., Hsu, M.J., Chaudhary, S. and Shieh, P. (2009) Can biofuel crops alleviate tribal poverty in India's drylands? *Applied Energy* 86(Suppl. 1), S118–S124.

Chakravorty, U., Hubert, M.-H. and Marchandi, B.U. (2012) Does the US biofuel mandate increase poverty in India? *Proceedings of the 2ndAnnual Summer Conference, Association of Environmental and Resource Economists*, 3–5 June, Asheville, North Carolina.

FAO and OECD (Food and Agriculture Organization of the United Nations and the Organization for Economic Co-operation and Development) (2012) *OECD-FAO Agricultural Outlook 2012–2021*. Paris, France.

Findlater, K.M. and Kandlikar, M. (2011) Land use and second-generation biofuel feedstocks: the unconsidered impacts of Jatropha biodiesel in Rajasthan, India. *Energy Policy* 39(6), 3404–3413.

Gehlhar, M., Winston, A.R. and Somwaru, A. (2010) *Effects of Increased Biofuels on the US Economy in 2022*. Economic Research Report. United States Department of Agriculture, Washington, DC.

GoI (Government of India) (2009) *National Policy on Biofuels*. Ministry of New & Renewable Energy, New Delhi.

Hertel, T.W. (ed.) (1997) *Global Trade Analysis: Modeling and Applications*. Cambridge University Press, Cambridge, UK.

IEA (International Energy Agency) (2011) *World Energy Outlook 2011*. Paris, France.

Mahapatra, A.K. and Mitchell, C.P. (1999) Biofuel consumption, deforestation, and farm level tree growing in rural India. *Biomass and Bioenergy* 17(4), 291–303.

Pohit, S., Biswas, P.K. and Ashra, S. (2011) Incentive structure of India's biofuel programs: Status, shortcomings and implications. *Biofuels* 2(4), 389–403.

Raju, S.S., Parappurathu, S., Chand, R., Joshi, P.K., Kumar, P. and Msangi, S. (2012) *Biofuels in India: Potential, Policy and Emerging Paradigms*. National Centre for Agricultural Economics and Policy Research (ICAR), New Delhi.

Ravindranath, N.H., Sita Lakshmi, C., Manuvie, R. and Balachandra, P. (2011) Biofuel production and implications for land use, food production and environment in India. *Energy Policy* 39(10), 5737–5745.

Sasmal, S., Goud, V.V. and Mohanty, K. (2012) Characterization of biomasses available in the region of North-East India for production of biofuels. *Biomass and Bioenergy* 45, 212–220.

Schaldach, R., Priess, J.A. and Alcamo, J. (2011) Simulating the impact of biofuel development on country-wide land-use change in India. *Biomass and Bioenergy* 35(6), 2401–2410.

Sorda, G., Banse, M. and Kemfert, C. (2010) An overview of biofuel policies across the world. *Energy Policy* 38(11), 6977–6988.

USDA (United States Department of Agriculture) (2012) *India Biofuels Annual 2012*. Foreign Agricultural Service, New Delhi.

Woltjer, G. and Kuiper, M. (2014) *The MAGNET Model, Module Description*. LEI, Wageningen UR, The Hague, The Netherlands.

12 Input Subsidy versus Farm Technology – Which is More Important for Agricultural Development?

Praduman Kumar[1]* and P.K. Joshi[2]

[1]*Division of Agricultural Economics, Indian Agricultural Research Institute, New Delhi;* [2]*International Food Policy Research Institute, New Delhi, India*

Introduction[1]

Input subsidy and technology are the two significant factors for the development of agriculture in India. Concerns are often expressed about a decrease or increase in input subsidy and inadequate investment in farm technology development. Policy planners often face questions such as what would happen to output supply, factor demand, agricultural prices and farmer income under alternative input subsidy and farm technology scenarios, and what would be the impact of input subsidy and technological innovation on the welfare of producer and consumer?

The rising costs of farm inputs discourage their use and lead to a reduction in agro-commodity supply and profitability of farmers. The decline in supply of these commodities raises their market prices, causing hardships to the consumers. On the other hand, a rise in crop prices is needed not only to counteract the rising input costs but also to provide sufficient margin to the farmers, which may be conducive to investment in agriculture. The situation can be managed by manipulating price and non-price factors through subsidy, investment in irrigation, capital inputs, technology development, market development, etc. This study has been undertaken to develop a food crops model with emphasis on two major crops – rice and wheat – which account for more than 70% of total food grains production and are the backbone of India's food security and household nutritional security. An attempt is also made to estimate the producer and consumer core systems for these two commodities. More specifically, models have been developed to evaluate the effects of price and non-price factors on factor demand, output supply, output demand, prices and farmers' income. These models have been simulated to suggest the adjustments needed in price and non-price factors to attain the specific goals and, finally, an attempt has been made to find which is more important between input subsidy and technology for agricultural development.

Analytical Approach

For the study, the partial models were designed to simulate the effects of macroeconomic developments and policies on the quantities and prices of agro-commodities produced and factors used in their production.

* Corresponding author, e-mail: pkumariari@gmail.com

The models were so designed as may be applied to either individual states or several states linked through national markets for crop output and factor demand. These markets are connected through supply–demand channels of producers and consumers. The producer behaviour core consists of factor demand and output supply equations in the product market. The consumer behaviour core consists of demand equations of agro-commodities.

For the study, a simplified version of the unified approach, described by Kumar *et al.* (1985), has been used. In the unified model, the system contains factor demand and output supply, output demand and crop net income equations. This model was used to analyse the impact of input subsidies, technology and demand shifters on prices, supply, demand and income for two major food crops, viz. rice and wheat. The policies and programmes considered were: input subsidies, irrigation investment, agricultural research and input–output policies. The empirical model had several blocks of equations. The first block was of the producer core system, consisting of factor demand and output supply equations (yield and acreage equations). The second block was of the consumer core system containing the consumer demand equation and indirect demand equations. The third block was for the TFP response to its sources. All the three blocks of equations were solved simultaneously to build price, supply and demand, and income models to undertake a simulation exercise to answer the policy questions such as what adjustments are needed in price and non-price factors to attain the specific goals of welfare of both producers and consumers.

Producer core system

The theory of profit function (Lau and Yotopoulos, 1972; Chand and Kumar, 1986) or cost function (Binswanger, 1974; Kumar *et al.*, 2010) provides a set of factor demand and output supply equations as:

$$\text{Factor demand: } X = X(P, p, Z, T) \tag{12.1}$$

$$\text{Output supply: } Q = Q(P, p, Z, T) \tag{12.2}$$

where X is a vector of *k* variable inputs, P is a vector of crop output prices, p is a vector of variable input prices, Z is a vector of fixed inputs, T is a vector of technology and Q is a vector of crop output supply. The output supply and factor demand functions are expressed in growth form as:

$$\dot{Q} = E_Q^P \dot{P} + E_Q^p \dot{p} + E_Q^Z \dot{Z} + E_Q^T \dot{T} \tag{12.3}$$

$$\dot{X} = E_F^P \dot{P} + E_F^p \dot{p} + E_F^Z \dot{Z} + E_F^T \dot{T} \tag{12.4}$$

where the dot on a variable indicates the rate of change; E parameters are the elasticities of output supply and factor demand; $Q = [Q_r, Q_w]$ is the output vector for rice and wheat; and $X = [N, B, M, F, O]$ is a vector of human labour (*N*), bullock labour (*B*), machine labour (*M*), fertilizer (*F*) and other inputs (*O*), respectively; $P = [Pr, Pw]$ is a vector of rice and wheat prices; and $p = [w, b, m, f, i]$ is the vector of input prices corresponding to human labour, bullock labour, machine labour, fertilizer and other inputs.

Acreage response model

The crop-area response elasticity for rice and wheat was estimated using the Nerlove dynamic model (Nerlove, 1958) based on the concept of adaptive expectations. The reduced estimable equation was specified as follows:

$$A_t = \mathrm{f}(A_{t-1}, P_{t-1}, TFP_{t-1}, \text{State dummy}, \text{Period dummy}) \tag{12.5}$$

where A_t is the crop area under cultivation at time *t*, A_{t-1} is the 1 year lagged area, P_{t-1} is the 1 year lagged crop price and TFP_{t-1} is the 1 year lagged TFP index.

TFP response model

The issue of sustainability of crop productivity is emerging fast. The productivity attained during the green revolution period has not been sustained in the post-green revolution period, posing a challenge to shift the production function by improving the technology index. It is possible through technology movers, judicious use of natural resources and harnessing of biodiversity. During the green revolution era, large investments were made in agricultural research, extension and irrigation development. The total factor productivity (TFP) can be induced by such factors as research, extension, human capital development (literacy), irrigation and climate. The data were analysed with state dummy fixed effects as such specifications would knock out the climate effects. As an input to public investment decisions, it is useful to understand the relative importance of these productivity-enhancing factors in determining productivity growth. Multiple regression analysis was carried out to assess the determinants of TFP. The TFP index was regressed on the following variables:

TFP = f (Research stock, Extension stock, Literacy, Irrigation, State dummy, Period dummy)

where:

Research stock = Total research stock per hectare of crop area;
Extension stock = Total extension stock per hectare of crop area;
Literacy = Rural literacy rate in per cent;
Irrigation = Irrigated area to total crop area in per cent;
State dummy = Name of individual state; and
Period dummy = 1971–1980, 1981–1990 and 1991–2000.

Following Evenson *et al.* (1999), the research stock variable was constructed by summing up research investment of 5 years by assigning weights as 0.2 in year t–2, 0.4 in year t–3, 0.6 in year t–4, 0.8 in year t–5 and 1.0 in year t–6. The extension stock variable was constructed by summing up 3 years' extension investment by assigning weights as 1.0 in year t–1, 0.8 in year t–2 and 0.4 in year t–3.

Supply growth model

Crop area (AREA), total factor productivity (TFP), supply elasticity and input–output price environment are the major sources of supply growth. The supply growth equation for commodity can be expressed as:

$$\dot{S} = E_S^P \dot{P} + \textstyle\sum_{i=1}^{k} E_S^{p_i} \dot{p}_i + C\dot{R}AREA + T\dot{F}P \quad (12.6)$$

where:

$\dot{S}$ = Supply growth for the commodity;
E_S^P = Yield response elasticity with respect to the product price;
$\dot{P}$ = Output price growth;
E_S^{pi} = Elasticity of factor demand for the ith input;
$\dot{p}_i$ = Input price growth of the ith input;
$C\dot{R}AREA$ = Acreage growth of the commodity;
$T\dot{F}P$ = TFP growth of the commodity; and
k = Number of inputs.

Income model

The net income (I) from a crop is given by Eqn 12.7:

$$I = P^*Q(P, p, Z, T) - p^*F(P, p, Z, T) \quad (12.7)$$

where:

P = Price of a commodity;
p = Vector of input price [w, b, m, f, i];
F = Vector of input-use [N, B, M, F, O];
T = Vector of technology;
w = Wages;
b = Animal labour price;
m = Machine labour price,
f = Fertilizer price,
i = Price of other inputs (irrigation, plant protection, etc.);
N = Human labour use;
B = Animal labour use;
M = Machine labour use;
F = Fertilizer use;
O = Use of other inputs (irrigation, plant protection, etc.); and
Z = Acreage.

The growth in net income in terms of elasticity can be measured as in Eqn 12.8:

$$\dot{I} = E_I^P \dot{P} = E_I^w \dot{w} + E_I^b \dot{b} + E_I^m \dot{m} + E_I^r \dot{r} + E_I^i \dot{i} + E_I^z \dot{z} + T_I^T \dot{T} \quad (12.8)$$

Using the formulae developed by de Janvry and Kumar (1981), different income elasticities were computed using Eqns 12.9–12.16 given in Box 12.1.

Box 12.1. Formulae developed by de Janvry and Kumar (1981) to compute income elasticities.

$$
\begin{aligned}
E_I^P = {} & \frac{PQ}{I}\left(1 - E_Q^{w/P} - E_Q^{b/P} - E_Q^{m/P} - E_Q^{f/P} - E_Q^{i/P}\right) + \frac{wN}{I}\left(E_N^{w/P} + E_N^{b/P} + E_N^{m/P} + E_N^{f/P} + E_N^{i/P}\right) \\
& + \frac{bB}{I}\left(E_B^{w/P} + E_B^{b/P} + E_B^{m/P} + E_B^{f/P} + E_B^{i/P}\right) + \frac{mM}{I}\left(E_M^{w/P} + E_M^{b/P} + E_M^{m/P} + E_M^{f/P} + E_M^{i/P}\right) \\
& + \frac{rF}{I}\left(E_F^{w/P} + E_F^{b/P} + E_F^{m/P} + E_F^{f/P} + E_F^{i/P}\right) + \frac{iO}{I}\left(E_O^{w/P} + E_O^{b/P} + E_O^{m/P} + E_O^{f/P} + E_O^{i/P}\right)
\end{aligned} \tag{12.9}
$$

$$E_I^w = \frac{PQ}{I}E_Q^{w/P} - \frac{wN}{I}\left(1 + E_N^{w/P}\right) - \frac{bB}{I}E_B^{w/P} - \frac{mM}{I}E_M^{w/P} - \frac{fF}{I}E_F^{w/P} - \frac{iO}{I}E_O^{w/P} \tag{12.10}$$

$$E_I^b = \frac{PQ}{I}E_Q^{b/P} - \frac{bB}{I}\left(1 + E_B^{b/P}\right) - \frac{wN}{I}E_N^{b/P} - \frac{mM}{I}E_M^{b/P} - \frac{fF}{I}E_F^{b/P} - \frac{iO}{I}E_O^{b/P} \tag{12.11}$$

$$E_I^m = \frac{PQ}{I}E_Q^{m/P} - \frac{mM}{I}\left(1 + E_M^{m/P}\right) - \frac{wN}{I}E_N^{m/P} - \frac{bB}{I}E_B^{m/P} - \frac{fF}{I}E_F^{m/P} - \frac{iO}{I}E_O^{m/P} \tag{12.12}$$

$$E_I^f = \frac{PQ}{I}E_Q^{f/P} - \frac{rF}{I}\left(1 + E_F^{f/P}\right) - \frac{wN}{I}E_N^{f/P} - \frac{bB}{I}E_B^{f/P} - \frac{mM}{I}E_M^{f/P} - \frac{iO}{I}E_O^{f/P} \tag{12.13}$$

$$E_I^i = \frac{PQ}{I}E_Q^{i/P} - \frac{iO}{I}\left(1 + E_O^{i/P}\right) - \frac{wN}{I}E_N^{i/P} - \frac{bP}{I}E_B^{i/P} - \frac{mM}{I}E_M^{i/P} - \frac{rF}{I}E_F^{i/P} \tag{12.14}$$

$$E_I^T = \frac{PQ}{I}E_Q^T - \frac{wN}{I}E_N^T - \frac{bB}{I}E_B^T - \frac{mM}{I}E_M^T - \frac{fF}{I}E_F^T - \frac{iO}{I}E_O^T \tag{12.15}$$

$$E_I^Z = \frac{PQ}{I}E_Q^Z - \frac{wN}{I}E_N^Z - \frac{bB}{I}E_B^Z - \frac{mM}{I}E_M^Z - \frac{fF}{I}E_F^Z - \frac{iO}{I}E_O^Z \tag{12.16}$$

Consumer core system

Following the consumer demand theory, the commodity demand equations in growth form can be expressed as:

Per capita Consumer Demand

$$\dot{d} = E_d^P \dot{P} + E_d^{I_c} \dot{I}_c \tag{12.17}$$

$$P = [P, Ps, Pc]$$

$$\dot{d} = E_d^P \dot{P} + E_d^{P_s} \dot{P}_S + E_d^{P_c} \dot{P}_C + E_d^I \dot{I} \tag{12.18}$$

Indirect Demand

$$ID = \left(SEED * CRAREA + OU\right) / POP \tag{12.19}$$

$$OU = \left(FEED + WAST + INDUSE\right) / POP \tag{12.20}$$

$$\dot{ID} = s_1\,\dot{SEED} + s_1\,\dot{CRAREA} + s_2\,\dot{OU} \tag{12.21}$$

Total Demand

$$\dot{D} = S^*\,\dot{ID} + (1 - S)\dot{d} + \dot{POP} \tag{12.22}$$

where:

P = Price of the *i*th output;
I_c = Per capita consumer income;
SEED = Seed rate;
CRAREA = Area under crop;
OU = Other uses;
FEED = Feed demand;

WAST = Wastages;
POP = Population;
ID = Indirect demand per capita for seed, feed, wastages and industrial uses;
S = Share of indirect demand in total demand;
s_1 = Share of seed demand in total indirect demand; and
s_2 = Share of other uses in total indirect demand.

The unified model

Each supply and demand relationship in every crop market has both endogenous and exogenous variables. In the market, price and quantity of a commodity are determined by its demand and supply. The exogenous shifters, viz. technology movers, population and income growth, and indirect demand within domestic and international markets are not determined within the markets *per se*. The equilibrium product prices are determined by equating output supply to its demand for each crop ($S_i = D_i$; $i = 1,..., n$). These equations are solved for prices and substituted into supply, demand and income equations. The endogenous variables are expressed as:

$P = P(p, Lr, Lw, LIT, RES, EXT, IRR, ID, Ic, POP)$

$S = S(p, Lr, Lw, LIT, RES, EXT, IRR, ID, Ic, POP)$

$D = D(p, Lr, Lw, LIT, RES, EXT, IRR, ID, Ic, POP)$

$I = I(p, Lr, Lw, LIT, RES, EXT, IRR, ID, Ic, POP)$

where *p* terms measure the input prices effect, *Lr* and *Lw* are the area under rice and wheat, respectively, the terms *LIT*, *RES*, *EXT* and *IRR* are the supply shifters, and *ID*, *Ic*, *POP* are the demand shifters and measure the shifters' influence on policy variables, viz. product price (*P*), supply (*S*), demand (*D*) and farmers income (*I*). The exogenous shifters play a critical role in the policy model.

These models can be cast in growth rates of the endogenous variables, exogenous variables (shifters) and elasticity of demand and supply curves in all markets. The model measures the impact on the growth rate of endogenous variables with a small change in the growth rate of shifter variables. The model allows asking questions such as what would happen to output supply, factor demand, agricultural prices and incomes under alternative technological change scenario, or what would be the effect of input subsidy and labour-saving technological change on producer and consumer welfare. The empirical unified models for rice and wheat have been developed and analysed for policy concern in the study.

Estimates of Producer Core System

The econometric application of new production theory based on the duality relationship between production function and variable profit/cost function was a major development in generating supply response estimates. Following Binswanger (1974), translog cost function was used to derive a system of factor demand equations and to estimate factor demand and output supply elasticities (for details see Kumar *et al.*, 2010). Information on yield and input-use and their prices was obtained from the records of the 'Comprehensive Scheme for the Study on Cost of Cultivation of Principal Crops' of the Directorate of Economics and Statistics, Government of India, New Delhi, for the period 1972–2009 (DES, 2009).

Input demand elasticity

The restricted estimates of the parameters of translog cost function model were obtained by jointly estimating the four factor share equations, viz. human labour, animal labour, machine labour and fertilizers for rice and wheat crops. Most of the restricted estimates were highly significant for all the factor share equations for both the crops. The parameters of share equations, though having little economic meaning, were used to compute the elasticity for factor demand. The input demand elasticities were estimated with respect to own and cross prices for human labour, animal labour, machine labour, fertilizers and other inputs (irrigation, plant protection and others). The factor demand elasticities for rice and wheat crops are given in Table 12.1. As expected, all own input price elasticities of demand had statistically significant negative signs. The magnitude of factor demand elasticity differed significantly across crops and inputs, depending on the level of agricultural development and technology used.

For rice, the own price elasticity of input demand was estimated to be highest for machine labour (–0.61) followed by irrigation and plant protection (–0.54), animal labour (–0.41), human labour (–0.25) and fertilizer (–0.21). The estimates indicate that demand for machine labour is sensitive to its price. On the policy front, a reduction in machinery prices through subsidy is expected to expand mechanization in rice farming and enhance rice productivity. The demand for animal labour, machine labour and fertilizers is influenced significantly by the change in wages. With a raise in wages, the demand for animal labour and machine labour is likely to increase and for fertilizer use is likely to decline.

For wheat crop, the animal labour demand is sensitive to animal labour wages. The demand elasticity was estimated to be as high as –0.62 for animal labour, followed by –0.34 each for machine labour and fertilizers, –0.31 for human labour and was highly inelastic for irrigation (–0.06). Cross-price elasticities across machine labour and human labour were positive and significant and indicated the substitutive relationship between human labour and machine labour for rice. For wheat, a substitutive relationship was observed for animal labour and machine labour; this is because mechanized operations are easier for wheat than for rice and are more farmer-friendly.

A rise in high animal labour charges will induce a higher use of machinery, as it results in substitution of animal labour with machine labour. In wheat production, the technology being used was such that animal labour could be easily substituted by machine labour. Wheat is an irrigated crop and, therefore, irrigation demand is not sensitive to the rise in irrigation price.

Table 12.1. Estimates of partial elasticities of factor demand for rice and wheat crops, India (authors' calculations).

	Input prices				
Crop inputs	w/P	b/P	m/P	f/P	i/P
Rice					
Human labour	–0.2484	0.1847	0.0472	–0.0582	0.0747
t-Value	–11.6	17.7	4.0	–6.0	6.0
Animal labour	0.6402	–0.4072	–0.0617	–0.0785	–0.0928
t-Value	17.7	–10.9	–2.6	–3.3	–3.6
Machine labour	0.2648	–0.1000	–0.6142	0.1713	0.2781
t-Value	4.0	–2.6	–8.7	3.7	5.5
Fertilizer	–0.2420	–0.0941	0.1268	–0.2137	0.4230
t-Value	–6.0	–3.3	3.7	–5.4	13.1
Other inputs	0.2007	–0.0719	0.1331	0.2734	–0.5353
Wheat					
Human labour	–0.3055	0.1528	0.0087	0.1625	–0.0186
t-Value	–8.59	8.69	0.28	6.50	–0.57
Animal labour	0.3715	–0.6200	0.1920	–0.2125	0.2691
t-Value	8.69	–13.09	4.06	–4.60	5.43
Machine labour	0.0164	0.1492	–0.3370	0.2216	–0.0502
t-Value	0.28	4.06	–3.46	4.19	–0.48
Fertilizer	0.3093	–0.1664	0.2232	–0.3368	–0.0294
t-Value	6.50	–4.60	4.19	–5.19	–0.44
Other inputs	–0.0233	0.1387	–0.0333	–0.0193	–0.0628

b, cost of animal labour (Rs/h); f, cost of fertilizer (NPK) (Rs/kg); i, cost of irrigation and other inputs (Rs/ha); m, cost of machine labour (Rs/h); P, price of commodity (Rs/100 kg); w, wage (Rs/h)

The subsidy on farm machinery and fertilizers would induce a higher use of modern inputs and improve farming efficiency and productivity.

Yield response elasticities

The yield response elasticities for rice and wheat crops were derived from the factor demand elasticity matrix and the results are presented in Table 12.2. The yield response elasticities have shown the response of commodity price and input price on supply of rice and wheat. The yield response elasticity with respect to crop output price was estimated as 0.22 for rice and 0.27 for wheat. The input responses were highly inelastic, nearly zero. The crop price had the dominating influence on the supply of commodities and, therefore, a positive price policy will enhance food supply.

Acreage response elasticities

Following Nerlove's adjustment model (Nerlove, 1958), the acreage elasticity with respect to lagged acreage was estimated to be 0.85 for rice and 0.86 for wheat (Table 12.3). The price of rice has a significant positive impact on its acreage. However, wheat price has not shown a significant effect on its acreage. Technology (TFP) has depicted a positive and significant effect on area under the crop. With the development of technology, the cropping pattern will shift in favour of major food crops.

Table 12.2. Yield response elasticity for rice and wheat, India (authors' calculations).

Prices	Rice	Wheat
Commodity price (P)	0.2249	0.2667
Input price		
Human labour (w/P)	–0.0786	–0.0615
Animal labour (b/P)	–0.0369	–0.1271
Machine labour (m/P)	–0.0335	–0.0336
Fertilizer (f/P)	–0.0155	–0.0347
Other inputs (i/P)	–0.0603	–0.0099

b, cost of animal labour (Rs/h); f, cost of fertilizer (NPK) (Rs/kg); i, cost of irrigation and other inputs (Rs/ha); m, cost of machine labour (Rs/h); P, price of commodity (Rs/100 kg); w, wage (Rs/h)

TFP elasticities

To study the factors influencing TFP for rice and wheat crops, multiple regression analysis was carried out. The TFP index was regressed on rural literacy, research stock, extension stock and irrigation. The state dummy was also included in the estimation to control fixed effects. A term for interaction between research and extension was also included in the estimation. The elasticity of TFP with respect to various sources at mean level was computed and is given in Table 12.4.

The results revealed that irrigation, agricultural research and extension services delivery are the important sources of TFP. These TFP elasticities were used to build the supply model described in the subsequent section. Additional investments on irrigation and rural literacy have been found to be highly productive and rewarding and would go a long way in stepping up TFP in India and will shift the food supply functions upwards.

Supply elasticities

The supply elasticities derived using yield, acreage and TFP elasticities with respect to their exogenous variables are given in Table 12.5. The supply elasticities with respect

Table 12.3. Acreage response elasticities for rice and wheat, India.

Variable	Rice	Wheat
Lagged crop area	0.8530**	0.8610**
Commodity price	0.0494**	–0.0167
TFP	0.0618**	0.0798**

**Significant at 1% level

Table 12.4. Elasticity of TFP with respect to sources for rice and wheat, India.

Sources of TFP	Rice	Wheat
Rural literacy	0.1221	0.7959
Research stock	0.0443	0.0464
Extension stock	0.0873	–0.0915^ns^
Irrigated area (%)	0.5842	0.7354

ns, non-significant coefficient

Table 12.5. Supply response elasticities for rice and wheat, India (authors' calculations).

Sources	Yield response		Acreage response		TFP response		Supply response	
	Rice	Wheat	Rice	Wheat	Rice	Wheat	Rice	Wheat
Commodity price	0.2249	0.2667	0.0494	−0.0915			0.2742	0.2667
Input price								
Human labour	−0.0786	−0.0615					−0.0786	−0.0615
Animal labour	−0.0369	−0.1271					−0.0369	−0.1271
Machine labour	−0.0335	−0.0336					−0.0335	−0.0336
Fertilizers	−0.0155	−0.0347					−0.0155	−0.0347
Other inputs	−0.0603	−0.0099					−0.0603	−0.0099
TFP sources								
Rural literacy					0.1221	0.7959	0.1296	0.8594
Research stock					0.0443	0.0464	0.0470	0.0501
Extension stock					0.0873		0.0927	
Irrigated area					0.5842	0.7354	0.6203	0.7941
Supply shifters								
Acreage			0.8530	0.8610			0.8530	0.8610
TFP	1.00	1.00	0.0618	0.0798				

to input prices are very low and less than one, the lowest being for fertilizer prices and the highest for wages. The commodity supply is not sensitive to fertilizer price. The supply is highly responsive to commodity price. This has obvious implications on the determination of the level of government price support for agricultural output. A raise of 10% in the commodity price will induce additional supply of 2.7% for both rice and wheat.

Among non-price factors, acreage, irrigation, literacy and research are the powerful instruments that need to be simulated to attain supply growth at the desired levels. Irrigation, literacy and research are the major sources of TFP growth. A 10% growth in irrigation will increase output supply by 6.2% for rice and 7.9% for wheat. Rural education also enhances farming efficiency. A 10% raise in education level would induce substitution of traditional labour with machine labour and would result in 1.3% increase in rice production and 8.6% rise in wheat production.

Using commodity supply elasticities with respect to its sources and growth rate of each source, the contribution of each source to supply growth was computed for both rice and wheat crops and the results are presented in Table 12.6. The output price has revealed the supply growth of 45% for rice and of 31% for wheat. The input prices have depicted a negative supply growth for both wheat and rice. The net price effect on supply growth was estimated to be 0.2% for rice and −9.6% for wheat. The acreage could contribute 43.6% to rice supply growth and 27.7% to wheat supply growth. The TFP sources accounted for about half of the rice supply growth and two-thirds of the wheat supply growth. Among the TFP sources in rice supply, contribution of irrigation was highest (37.2%), followed by research and extension (11.2%) and literacy (7.8%). Irrigation and literacy were found to be the most important sources for wheat supply growth. Higher investments on irrigation, rural education and agricultural research will induce a substantial growth in supply of both these food crops.

Crop income elasticities

To compute crop net income elasticities with respect to input and output prices, data provided in Appendix Table A12.1 are required along with the elasticities of factor demand and output supply. These elasticities for rice and wheat crops are presented in Table 12.7. The income elasticity with respect to output

Table 12.6. Sources of supply for rice and wheat in India (authors' calculations).

	Annual growth (%)		Supply elasticities		Sources of supply (%)	
Sources	Rice	Wheat	Rice	Wheat	Rice	Wheat
Commodity price	6.98	7.09	0.2742	0.2667	45.3	31.4
Input price						
Human labour	10.20	10.08	−0.0786	−0.0615	−19.0	−10.3
Animal labour	10.02	10.65	−0.0369	−0.1271	−8.8	−22.4
Machine labour	8.01	8.01	−0.0335	−0.0336	−6.3	−4.5
Fertilizers	4.40	4.54	−0.0155	−0.0347	−1.6	−2.6
Other inputs	6.58	6.99	−0.0603	−0.0099	−9.4	−1.2
TFP sources						
Rural literacy	2.53	2.88	0.1296	0.8594	7.8	41.0
Research stock	5.24	3.51	0.0470	0.0501	5.8	2.9
Extension stock	2.48	2.62	0.0927		5.4	0.0
Irrigated area	2.53	2.88	0.6203	0.7941	37.2	37.9
Supply shifters						
Acreage	2.16	1.94	0.8530	0.8610	43.6	27.7

Table 12.7. Income elasticities with respect to input and output price for rice and wheat, India (authors' calculations).

	Income elasticity	
Price	Rice	Wheat
Output price		
Commodity price	3.4826	3.4172
Input price		
Human labour	−1.0310	−0.6692
Animal labour	−0.6165	−0.5583
Machine labour	−0.2515	−0.3692
Fertilizers	−0.3487	−0.4069
Other inputs (irrigation, etc.)	−0.3539	−0.4136

price was found to be quite high (highly elastic) for both the crops and was estimated to be 3.4 for wheat and 3.5 for rice. The income elasticities with respect to input price, such as for human labour, animal labour, machine labour, fertilizers and irrigation, were negative for both the crops. The negative income elasticities were the highest with respect to wages, followed by input prices for animal labour, irrigation, fertilizers and machine labour. Since paddy is a labour-intensive crop, a raise in human labour wages and animal labour price will have a bigger negative impact on income from paddy than from a wheat crop. With a rise in modern input prices, the decline in income will be slightly more in wheat than in a rice crop.

From these elasticities, the impact of pure price inflation on crop income can be measured assuming $dp/p = dw/w = db/b = dm/m = dr/r = di/i$. Since all the relative prices remain constant, the elasticities of output and derived factor demand with respect to relative prices are equal to zero. The elasticity of income with respect to pure price inflation was estimated to be 0.88 for rice and 1.0 for wheat. With a 10% pure price inflation, the income of the producer will increase by 8.8% from rice and 10.0% from wheat. Thus, even though pure price inflation is neutral on output level and factor use, it has a strong positive effect on crop income. If there is a 10% inflation in factor price, then to sustain the producer income from crop there would be a need to increase commodity price by 7.5% for rice and 7.1% for wheat.

Consumer demand elasticities

To estimate demand elasticities for cereals, a multi-stage (three-stage) budgeting framework was used (see for methodology Dey, 2000; Dey *et al.*, 2008; Kumar *et al.*, 2011) and the results are presented in Table 12.8. The income elasticities were less than one for all the cereals (highly inelastic), with magnitude of −0.03, −0.05, −0.06 and −0.04 for rice, wheat, maize and other coarse cereals, respectively. The own price elasticities were found negative for all the cereals, as expected. For rice and wheat, the own price elasticities were more than one, −1.30 and −1.81, respectively. For maize and other coarse cereals, these elasticities were less than unity: −0.45 for maize and −0.77 for other cereals. The own price elasticities were much higher than expenditure elasticities. The own price elasticities for demand were negative and cross price elasticities were positive, indicating substitution across cereal types. The implication is that food prices need to be kept low for achieving food security.

Aggregate Demand for Rice and Wheat

The aggregate demand for commodities is influenced by not only price and income factors included in the per capita consumer demand equations, but also by non-price factors (shifters) such as population, seed rate, acreage, feed, other industrial uses and trading of commodities. By using respective shares of consumer demand, seed and other uses (given in Appendix Table A12.2), the aggregate demand elasticities for commodities were computed and are presented in Table 12.9.

The own price elasticity of aggregate demand has been estimated to be −1.23 for rice and −1.61 for wheat. The cross-price elasticity of the substitute commodities was positive and less than one. The rice demand elasticity with respect to wheat price has been estimated to be 0.43 and the wheat demand elasticity with respect to rice price

Table 12.8. Consumer demand elasticities for cereals in India (authors' calculations).

Elasticity	Rice	Wheat	Maize	Other coarse cereals
Income	−0.03	−0.05	−0.06	−0.04
Price				
Rice	−1.30	0.45	0.04	0.00
Wheat	0.57	−1.81	−0.09	−0.03
Maize	2.22	−3.85	−0.45	0.26
Other coarse cereals	−0.09	−0.20	0.07	−0.77

Table 12.9. Source-wise demand elasticities for rice and wheat, India (authors' calculations).

	Consumer demand		Total demand	
Demand sources	Rice	Wheat	Rice	Wheat
Price				
Rice	−1.3003	0.5687	−1.2343	0.5062
Wheat	0.4505	−1.8059	0.4276	−1.6073
Maize	0.0356	−0.0943	0.0338	−0.0839
Other coarse cereals	0.0012	−0.0330	0.0012	−0.0293
Income	−0.0305	−0.0512	−0.0289	−0.0455
Indirect sources				
Seed			0.0130	0.0214
Acreage			0.0130	0.0214
Other uses[a]			0.0378	0.0885
Population	1.00	1.00	0.9492	0.8901

[a]Feed + wastage + industrial uses

has been estimated to be 0.51. With an increase of 1% in own price, the decline in demand would be 1.23% for rice and 1.61% for wheat. After adjusting the substitution effect, the net negative impact on demand would be 0.81% for rice and 1.1% for wheat. Income has revealed an inelastic negative effect on the aggregate demand for rice and wheat. It explains the observed phenomenon that the per capita cereals consumption is declining over time. Rice and wheat are strong substitutes for each other in the consumer food basket, whereas maize and other coarse cereals are the weak substitutes for rice and wheat.

Population and indirect demand (seed, feed, wastages and industrial uses) for food grains are the strong demand shifters. The aggregate demand elasticity with respect to population was estimated to be 0.95 for rice and 0.89 for wheat (Table 12.9), which are both less than one. Thus, their demand will grow at a rate lower than of the population growth, indicating a declining trend in the cereals consumption, as has been observed in the consumer expenditure data collected by various rounds of NSSO. This decline is basically attributable to the structural shift in the dietary pattern and increasing availability of a wide variety of other food commodities in the market.

Simulated Results of Unified Model

The exogenous variables in the unified model have been classified as price and non-price factors. The price factors included factor price and acreage. The non-price factors included growth in productivity through technology, population, consumer income, trade and other uses. The technology is influenced by the investment in research, extension, literacy, irrigation, infrastructure, etc. The estimated supply and demand models provide the elasticity of price and non-price factors, indicating the direct partial effects of each one of them on factor demand, output supply and demand, and crop net income. At equilibrium, the rice demand growth is equal to the rice supply growth ($D{\cdot}r = S{\cdot}r$) and the wheat demand growth is equal to the wheat supply growth ($D{\cdot}w = S{\cdot}w$). By solving these equations simultaneously, the equilibrium price determination equations were derived as per Equations (23) and (24):

$$\begin{aligned}\dot{P}r =\ & 0.0086P\dot{m} - 0.0040P\dot{o}c + 0.0622\dot{w} \\ & + 0.0453\dot{b} + 0.0277\dot{m} + 0.0160\dot{r} + 0.0416\dot{o} \\ & - 0.5569\dot{L}R - 0.1375\dot{L}w - 0.2267L\dot{I}T \\ & - 0.0394R\dot{E}S - 0.0615E\dot{X}T - 0.5413\dot{R}R \\ & + 0.0395\dot{I}D - 0.2266\dot{I}c + 0.7751P\dot{O}P\end{aligned} \tag{12.23}$$

$$\begin{aligned}\dot{P}w =\ & -0.0382\dot{P}m - 0.0154\dot{P}oc \\ & + 0.0480\dot{w} + 0.0750\dot{b} + 0.0244\dot{m} \\ & + 0.0215\dot{r} + 0.0170\dot{o} - 0.1629\dot{L}r \\ & - 0.4479\dot{L}w - 0.4837L\dot{I}T - 0.0358R\dot{E}S \\ & - 0.0180E\dot{X}T - 0.5440I\dot{R}R + 0.0545\dot{I}D \\ & - 0.0299\dot{I}c + 0.6590P\dot{O}P\end{aligned} \tag{12.24}$$

where *Pr*, *Pw*, *Pm*, and *Poc* are the commodity prices of rice, wheat, maize and other cereals, respectively.

By substituting the equilibrium price of rice and wheat in the farm input demand equation (Table 12.1), commodity demand (Table 12.9), commodity supply (Table 12.5) and net crop income (Table 12.7), the net effects of price and non-price sources on input demand (Table 12.10) and on supply, demand and net income (Table 12.11) for rice and wheat were computed. These equilibrium models are useful for commodity market planning as they provide instant results to understand the role of price factors (input prices, subsidy, etc.) and non-price factors, i.e. supply shifters (investment on research, extension, literacy, irrigation, etc.) and demand shifters (population, indirect demand, consumer income, etc.) on food security and welfare of producers (farmers) and consumers. Using the elasticities given in Table 12.10 and Table 12.11 and the growth observed in exogenous variables, the sources of growth for factor demand, output price, supply demand and income were computed for both rice and wheat crops and are shown in Table 12.12 and Table 12.13, respectively.

Table 12.10. Factor demand elasticities for rice and wheat, India (authors' calculations).

	Human labour		Animal labour		Machine labour		Fertilizers		Other inputs	
Sources	Rice	Wheat	Rice	Wheat	Rice	Wheat	Rice	Wheat	Rice	Wheat
Price of substitute of rice and wheat										
Maize	0.0021	0.0026	0.0035	0.0054	0.0053	0.0029	0.0018	0.0029	0.0046	0.0005
Other coarse cereals	–0.0010	–0.0012	–0.0016	–0.0025	–0.0025	–0.0014	–0.0009	–0.0014	–0.0022	–0.0003
Input price										
Wages	–0.2329	–0.2865	0.6655	0.4101	0.3030	0.0373	–0.2287	0.3303	0.2340	–0.0193
Animal labour	0.1959	0.1667	–0.3888	–0.5919	–0.0722	0.1645	–0.0844	–0.1512	–0.0476	0.1415
Machine labour	0.0540	0.0171	–0.0504	0.2091	–0.5972	–0.3276	0.1327	0.2325	0.1479	–0.0316
Fertilizers	–0.0543	0.1674	–0.0720	–0.2027	0.1811	0.2269	–0.2103	–0.3314	0.2819	–0.0183
Other inputs	0.0851	–0.0059	–0.0758	0.2949	0.3037	–0.0362	0.4319	–0.0153	–0.5130	–0.0602
Acreage										
Rice	0.8617	–0.1701	0.7732	–0.3453	0.6580	–0.1876	0.8810	–0.1875	0.7019	–0.0350
Wheat	–0.0342	0.9580	–0.0560	0.9147	–0.0845	0.9537	–0.0294	0.9537	–0.0736	0.9914
Supply shifters (TFP sources)										
Literacy	–0.0563	–0.0693	–0.0923	–0.1406	–0.1392	–0.0764	–0.0484	–0.0763	–0.1214	–0.0142
Research stock	–0.0098	–0.0120	–0.0160	–0.0244	–0.0242	–0.0133	–0.0084	–0.0133	–0.0211	–0.0025
Extension stock	–0.0153	–0.0188	–0.0250	–0.0381	–0.0378	–0.0207	–0.0131	–0.0207	–0.0329	–0.0039
Irrigated area	–0.1344	–0.1653	–0.2204	–0.3356	–0.3324	–0.1824	–0.1157	–0.1823	–0.2897	–0.0340
Demand shifters										
Indirect demand	0.0098	0.0121	0.0161	0.0245	0.0243	0.0133	0.0084	0.0133	0.0212	0.0025
Consumer income	–0.0066	–0.0081	–0.0109	–0.0165	–0.0164	–0.0090	–0.0057	–0.0090	–0.0143	–0.0017
Population	0.1925	0.2368	0.3156	0.4805	0.4761	0.2612	0.1656	0.2610	0.4149	0.0487
All sources										
Inputs price	–0.2329	–0.2865	–0.3888	–0.5919	–0.5972	–0.3276	–0.2103	–0.3314	–0.5130	–0.0602
Acreage	0.8275	0.7879	0.7172	0.5695	0.5735	0.7660	0.8516	0.7662	0.6283	0.9564
Supply shifters	–0.2158	–0.2654	–0.3538	–0.5387	–0.5336	–0.2928	–0.1856	–0.2926	–0.4651	–0.0546
Demand shifters	0.1957	0.2407	0.3209	0.4885	0.4840	0.2655	0.1684	0.2653	0.4218	0.0495

Table 12.11. Price, supply, demand and income elasticities with respect to sources for rice and wheat, India (authors' calculations).

	Price		Supply		Demand		Income	
Sources	Rice	Wheat	Rice	Wheat	Rice	Wheat	Rice	Wheat
Price of substitute of rice and wheat								
Maize	0.0086	−0.0382	0.0024	−0.0102	0.0068	−0.0181	0.0301	−0.1306
Other coarse cereals	−0.0040	−0.0154	−0.0011	−0.0041	−0.0005	−0.0066	−0.0140	−0.0527
Price of inputs								
Wages	0.0622	0.0480	−0.0616	−0.0748	−0.0562	−0.0457	−0.8144	−0.5050
Animal labour	0.0453	0.0750	−0.0245	−0.1071	−0.0238	−0.0976	−0.4587	−0.3021
Machine labour	0.0277	0.0244	−0.0259	−0.0271	−0.0238	−0.0252	−0.1550	−0.2858
Fertilizers	0.0160	0.0215	−0.0111	−0.0289	−0.0105	−0.0265	−0.2931	−0.3334
Other inputs	0.0416	0.0170	−0.0489	−0.0054	−0.0441	−0.0062	−0.2090	−0.3556
Acreage								
Rice	−0.5569	−0.1629	0.7003	−0.0434	0.6306	−0.0201	−0.9393	−0.5566
Wheat	−0.1375	−0.4479	−0.0377	0.7414	−0.0218	0.6718	−0.4789	−0.5307
Supply shifters (TFP sources)								
Literacy	−0.2267	−0.4837	0.0675	0.7304	0.0730	0.6627	−0.7895	−1.6529
Research stock	−0.0394	−0.0358	0.0362	0.0405	0.0333	0.0377	−0.1371	−0.1224
Extension stock	−0.0615	−0.0180	0.0759	−0.0048	0.0682	−0.0022	−0.2141	−0.0614
Irrigated area	−0.5413	−0.5440	0.4718	0.6490	0.4354	0.6004	−1.8850	−1.8590
Demand shifters								
Indirect demand	0.0395	0.0545	0.0108	0.0145	0.0123	0.0208	0.1377	0.1864
Consumer income	−0.0266	−0.0299	−0.0073	−0.0080	−0.0088	−0.0109	−0.0928	−0.1022
Population	0.7751	0.6590	0.2126	0.1758	0.2744	0.2232	2.6992	2.2519
All sources								
Input price	0.1928	0.1859	−0.1720	−0.2171	−0.1584	−0.2013	−1.9303	−1.7818
Acreage	−0.6944	−0.6108	0.6626	0.6980	0.6088	0.6518	−1.4182	−1.0873
TFP	−0.8688	−1.0815	0.6514	1.4151	0.6099	1.2986	−3.0258	−3.6958
Demand shifters	0.7880	0.6836	0.2161	0.1823	0.2779	0.2330	2.7441	2.3360

Sources of input demand

The own input price has a negative effect and the acreage a positive effect on factor demand (Table 12.10). The technology movers improve the input-use efficiency and cut the input-use in producing the same output. Thus, the factor demand elasticities with respect to technology are negative for all the inputs. The elasticities of factor demand with respect to demand shifters are positive and are dominating for the population. Among shifters, the acreage and population induce higher use of inputs.

All the sources of input demand at observed growth rates, given in Table 12.12, revealed that the demand for human labour will increase at the rate of 1.85% for rice and 0.82% for wheat per annum. The animal labour demand will increase at the growth rate of 2.77% for rice and 0.85% for wheat. The demand for machine labour will be higher for wheat than for rice. The fertilizer demand will grow at a high rate of 2.97% for wheat and 1.4% for rice. The demand for irrigation and plant protection chemicals would be higher for wheat than for rice. The demand for human labour and animal labour will grow faster for rice as compared to wheat.

Sources of commodity price

The input prices have an inflationary effect on the market price of both rice and wheat (Table 12.13). With increase in the price of input, its use decreases and, consequently, commodity supply decreases and commodity price increases. Across farm inputs, the input price effect on commodity prices is highest on wages, followed by animal labour

Table 12.12. Growth by different sources in input use for rice and wheat, India (authors' calculations).

	Human labour		Animal labour		Machine labour		Fertilizers		Other inputs	
Sources	Rice	Wheat	Rice	Wheat	Rice	Wheat	Rice	Wheat	Rice	Wheat
Price of substitutes of rice and wheat										
Maize	0.015	0.018	0.024	0.038	0.037	0.020	0.013	0.020	0.032	0.003
Other coarse cereals	−0.007	−0.008	−0.011	−0.017	−0.017	−0.010	−0.006	−0.010	−0.015	−0.002
Input price										
Wages	−2.376	−2.889	6.790	4.136	3.092	0.376	−2.333	3.331	2.388	−0.195
Animal labour	1.963	1.775	−3.896	−6.301	−0.723	1.751	−0.846	−1.610	−0.477	1.506
Machine labour	0.432	0.137	−0.404	1.674	−4.782	−2.623	1.063	1.862	1.184	−0.253
Fertilizers	−0.239	0.760	−0.317	−0.920	0.797	1.030	−0.926	−1.504	1.241	−0.083
Other inputs	0.560	−0.041	−0.499	2.061	1.998	−0.253	2.841	−0.107	−3.375	−0.421
Acreage										
Rice	1.860	−0.367	1.669	−0.745	1.421	−0.405	1.902	−0.405	1.515	−0.076
Wheat	−0.066	1.860	−0.109	1.776	−0.164	1.851	−0.057	1.851	−0.143	1.925
Supply shifters (TFP sources)										
Literacy	−0.143	−0.200	−0.234	−0.405	−0.353	−0.220	−0.123	−0.220	−0.308	−0.041
Research stock	−0.051	−0.042	−0.084	−0.086	−0.127	−0.047	−0.044	−0.047	−0.111	−0.009
Extension stock	−0.038	−0.049	−0.062	−0.100	−0.094	−0.054	−0.032	−0.054	−0.081	−0.010
Irrigated area	−0.341	−0.476	−0.559	−0.966	−0.843	−0.525	−0.293	−0.525	−0.734	−0.098
Demand shifters										
Indirect demand	0.014	0.017	0.023	0.034	0.034	0.019	0.012	0.019	0.030	0.004
Consumer income	0.000	0.000	−0.001	−0.001	−0.001	0.000	0.000	0.000	−0.001	0.000
Population	0.270	0.332	0.442	0.673	0.667	0.366	0.232	0.365	0.581	0.068
All sources										
Price of rice and wheat substitutes	0.008	0.010	0.013	0.020	0.020	0.010	0.006	0.010	0.017	0.001
Input price	0.340	−0.259	1.675	0.650	0.381	0.281	−0.201	1.972	0.961	0.555
Acreage	1.794	1.493	1.561	1.030	1.257	1.446	1.845	1.447	1.372	1.849
Supply shifters	−0.573	−0.767	−0.938	−1.556	−1.416	−0.846	−0.492	−0.845	−1.234	−0.158
Demand shifters	0.283	0.348	0.464	0.706	0.700	0.384	0.243	0.384	0.610	0.072
All sources	1.852	0.824	2.774	0.850	0.941	1.276	1.401	2.967	1.726	2.319

Table 12.13. Growth by different sources in commodity price, supply, demand and income for rice and wheat in India (authors' calculations).

	Price		Supply		Demand		Income	
Sources	Rice	Wheat	Rice	Wheat	Rice	Wheat	Rice	Wheat
Price of cereals								
Maize	0.060	−0.267	0.017	−0.071	0.047	−0.126	0.210	−0.911
Other coarse cereals	−0.028	−0.107	−0.008	−0.029	−0.003	−0.046	−0.098	−0.368
Price of inputs								
Wages	0.635	0.484	−0.629	−0.754	−0.573	−0.461	−8.309	−5.093
Animal labour	0.454	0.798	−0.246	−1.140	−0.238	−1.039	−4.596	−3.216
Machine labour	0.222	0.195	−0.207	−0.217	−0.191	−0.202	−1.241	−2.288
Fertilizers	0.070	0.098	−0.049	−0.131	−0.046	−0.120	−1.290	−1.513
Other inputs	0.274	0.119	−0.322	−0.038	−0.290	−0.043	−1.375	−2.485
Area under crop								
Rice	−1.202	−0.352	1.512	−0.094	1.361	−0.043	−2.028	−1.202
Wheat	−0.267	−0.369	−0.073	1.439	−0.042	1.304	−0.930	−1.030
Supply shifters (TFP sources)								
Literacy	−0.575	−1.393	0.171	2.103	0.185	1.908	−2.001	−4.759
Research stock	−0.206	−0.126	0.190	0.142	0.175	0.132	−0.718	−0.429
Extension stock	−0.152	−0.047	0.188	−0.013	0.169	−0.006	−0.530	−0.161
Irrigated area	−1.372	−1.566	1.196	1.869	1.104	1.729	−4.778	−5.352
Demand shifters								
Indirect demand	0.055	0.076	0.015	0.020	0.017	0.029	0.193	0.261
Consumer income	−0.001	−0.001	0.000	0.000	0.000	−0.001	−0.005	−0.005
Population	1.085	0.923	0.298	0.246	0.384	0.312	3.779	3.153
Sources of growth								
Coarse cereals	0.032	−0.374	0.009	−0.100	0.044	−0.172	0.112	−1.279
Input price	1.654	1.694	−1.452	−2.280	−1.339	−1.865	−16.812	−14.595
Cropping pattern	−1.469	−1.221	1.439	1.346	1.319	1.261	−2.958	−2.232
Technology movers (TFP)	−2.306	−3.132	1.745	4.101	1.632	3.763	−8.028	−10.702
Demand shifters	1.139	0.997	0.312	0.266	0.401	0.341	3.967	3.409
All sources	−0.949	−2.035	2.053	3.333	2.057	3.327	−23.718	−25.399

and machine labour and is the least on fertilizer prices. The inflationary pressure on input prices will increase the prices of both rice and wheat at the rate of 1.7% per year. Rice and wheat being the major staple cereals, accounting for more than half of the food expenditure of poor consumers, the rise in price of these cereals has a negative impact on welfare of the poor. The increase in area under crop or its substitutes will have a negative effect on crop output price. An expansion of 1% in acreage will lead to a decrease in commodity price by 1.47% for rice and by 1.22% for wheat. Supply shifters or technology movers (literacy, research, extension and irrigation) have a negative effect on the market price of a crop. At the observed past growth of technological development, the commodity prices are expected to decline at the rate of 2.3% for rice and 3.1% for wheat.

Among different inputs, irrigation has emerged as the most important source contributing to food security, due to inducing a considerable decline in commodity prices (1.37% for rice and 1.57% for wheat annually). Irrigation, literacy and research investment contribute to a higher input efficiency and supply and lower the unit cost and market price of cereals, benefiting both producers and consumers. Among the demand shifters, population plays a dominating role in generating demand and raising prices by 1.09% for rice and 0.92% for wheat annually. Consumer income has the minimum effect on cereal prices. It seems that the positive effects of input prices and demand shifters on market prices of rice and wheat have nullified the negative effect of acreage and TFP sources. The growth in product price by all sources will decline at the rate of 0.95% for rice and 2.03% for wheat. The income elasticity with respect to commodity price is highly elastic (Table 12.7); therefore, in the absence of minimum support price[2] (MSP), the producer income will decline substantially for both rice and wheat.

Sources of supply, demand and farm income

As seen in Table 12.11, the net effects of price factors on supply, demand and farm income are negative, as expected. The capital inputs (machinery) have mild and negative effects on output supply and farm income. A subsidy of 10% on fertilizer price will increase fertilizer use by 2.1% for rice and 3.3% for wheat (Table 12.10) and would lead to a mild increase in the commodity supply (0.11–0.29%), commodity demand (0.11–0.26%) and farmers' income (2.93–3.33%). The input subsidy providing 10% reduction in input prices will help in a decline in commodity prices by 1.9% and would raise commodity supply by 1.7% for rice and 2.2% for wheat and will raise profitability by 19.3% from rice and 17.8% from wheat cultivation. However, the input subsidy will not be feasible in the long-run. A viable solution can be found by appropriate adjustments in technology movers, mainly irrigation led by literacy, research and extension investment.

Unlike price factors, the technology movers that influence the TFP can have a stronger effect on factor demand and output supply, but have a negative impact on commodity price. Thus, crop income to farmers would decline substantially because it is highly elastic with respect to output price. With 1% increase in technology, the supply will increase by 0.65% for rice and by 1.41% for wheat and the price will decline by 0.87% for rice and by 1.08% for wheat. Due to the decline in price, the consumer demand will increase by 0.61% for rice and 1.30% for wheat and net income will decline by as high as 3.0% from rice and 3.7% from wheat. The area under crop will increase the supply, reduce the price, increase the demand and reduce the income. However, the demand shifters, viz. indirect demand (seed, industrial use and trade) and population have positive effects on commodity prices, their supply and demand, and income from crops.

Under the assumption that the factor price inflation will continue at the same rate as observed in the past – no change in acreage, technology movers and demand shifters – the price of rice will increase by 1.65% and its supply, demand and income will decline by 1.45%, 1.34% and 16.8%, respectively (Table 12.13). Similarly, for the wheat crop, the output price will increase by 1.69% and wheat supply, demand and income will decline by 2.01%, 1.86% and 14.6%, respectively. The acreage would have a negative impact on output price with annual growth rate of 1.47% for rice and 1.22% for wheat. The supply will increase at the annual growth rate of 1.44% for rice and 1.35% for wheat. The demand will grow at the annual rate of 1.32% for rice and 1.26% for wheat. However, income will decline at an annual rate of 2.96% from rice and 2.23% from wheat.

The technology movers will shift the supply function upward and supply will increase at an annual growth rate of 1.74% for rice and 4.1% for wheat. The price of the commodity will decline by 2.3% for rice and by 3.13% for wheat. The decline in price will have a positive impact on demand with the growth rate of 1.63% for rice and 3.76% for wheat. The profitability will decline at an annual rate of 8.03% in rice and 10.7% in wheat. The demand shifters, as a result of indirect demand and population growth, will increase the price at the rate of 1.14% for rice and 1.0% for wheat. The higher output price has a positive response on supply, which will grow at the annual rate by 0.31% for rice and 0.27% for wheat and profitability will increase at the rate of 3.97% from rice and 3.40% from wheat. Taking all price and non-price factors together, the supply will increase at the rate of 2.05% for rice and 3.33% for wheat annually. The price will decline at the rate of 0.95% in rice and 2.03% in wheat. This will have an adverse influence on profitability. The crop net income is likely to decline by 23.7% in rice and by 25.4% in wheat. To safeguard farmers' interests, government intervention becomes essential for price, production and income stabilization.

Fertilizer Subsidy and Crop Price Stabilization

The price growth model provides the output price elasticities with respect to input price, acreage, technology movers and demand shifters. If we withdraw the fertilizer subsidy and depend exclusively on technology to ensure complete product price stability ($dp/p = 0$), then the required adjustment in technology was computed and is given in Table 12.14. The withdrawal of fertilizer subsidy will have a negative impact on the supply of rice and wheat and their prices will increase. The technological changes induce output supply. The positive and negative impacts can be neutralized exclusively by adjusting the technology sources at the

Table 12.14. Technology versus fertilizer subsidy withdrawal: required growth in TFP sources for rice and wheat, India (authors' calculations).

Particulars	Output price elasticity with respect to fertilizer price and TFP sources		Elasticity of TFP sources with respect to fertilizer price		Required change in TFP sources (%) to counter withdrawal of 10% subsidy on fertilizers	
	Rice	Wheat	Rice	Wheat	Rice	Wheat
Fertilizer price	0.0160	0.0215				
TFP sources						
Literacy rate	−0.2267	−0.4837	0.0704	0.0444	0.704	0.444
Research stock	−0.0394	−0.0358	0.4051	0.6000	4.051	6.000
Extension stock	−0.0615	−0.0180	0.2595	ns	2.595	ns
Research and extension	−0.1009	−0.0538	0.6646	0.6000	6.646	6.000
Irrigated area	−0.5413	−0.5440	0.0295	0.0395	0.295	0.395
All sources	−0.8688	−1.0815	0.0184	0.0199	0.184	0.199

ns, non-significant

desired levels to compensate withdrawal of 10% fertilizer subsidy. The results are presented in the last column of Table 12.14. For this, the literacy rate will have to be increased by 0.44–0.70%, investment on research and extension needs to be increased at the growth rate of 6.0–6.6% annually, and irrigation must be increased at the growth rate of 0.30–0.40% per annum. The required growth in technology (TFP) is estimated to be 0.18–0.20% annually to compensate for the 10% fertilizer subsidy burden and ensure food price stability. The model can also be used to estimate the required adjustments in TFP and its sources to attain price stabilization under population pressure.

Conclusions

Technology, acreage and population are the most powerful instruments that need to be manipulated not only to neutralize factor price inflation but also to safeguard the interest of producers and consumers, while the input-price subsidy is likely to have a weak effect on commodity supply. It is the technology that has a substantial impact on food supply. Public policies such as investments in irrigation, rural literacy, research and extension are crucial to increase commodity supply at a higher growth rate. The input subsidy has a positive effect on input use, crop supply and farm income, but technology shifters have a positive and strong influence on commodity supply and a substantial negative effect on farmer income because of the decline in market price in the absence of MSP policy. Also, the input subsidy to farmers and price subsidy to consumers will not be feasible in the long-run as they involve a substantial share of public resources. A viable solution can only be found with appropriate adjustments in the non-price factors. An effective MSP programme is essential to protect the welfare of farmers. To compensate for the impact of a 10% reduction in fertilizer subsidy, the TFP growth will have to be increased from the present level by 0.18% for rice and 0.20% for wheat by adopting appropriate measures.

Notes

[1] The authors thank the Agricultural Economics Research Association (India), New Delhi, India for permission to reproduce their article on which this chapter is based: Kumar, P. and Joshi, P.K. (2014) Input subsidy vs farm technology – which is more important for agricultural development? *Agricultural Economics Research Review* 27(1), 1–18.

[2] This is the price announced by the Government of India every year for procurement of a food commodity.

References

Binswanger, H.P. (1974) A cost function approach to the measurement of factor demand elasticities of substitution. *American Journal of Agricultural Economics* 56(20), 377–386.

Chand, R. and Kumar, P. (1986) Supply response functions for major crops of Punjab in green revolution period – A profit function approach. *Indian Journal of Economics* 47(1), 91–109.

de Janvry, A. and Kumar, P. (1981) The transmission of cost inflation in agriculture with subsistence production: A case study in northern India. *Indian Journal of Agricultural Economics* 36(3), 1–14.

Dey, M.M. (2000) Analysis of demand for fish in Bangladesh. *Aquaculture Economics & Management* 4, 63–79.

Dey, M.M., Garcia, Y.T., Kumar, P., Piumsombun, S., Haque, M.S., Li, L., Radam, A., Senaratne, A., Nguyen, T.K. and Koeshendrajana, S. (2008) Demand for fish in Asia: A cross-country analysis. *The Australian Journal of Agricultural and Resource Economics* 52, 321–338.

Evenson, R.E., Pray, C.E. and Resegrant, M.W. (1999) *Agricultural Research and Productivity Growth in India.* Research Report No. 109. International Food Policy Research Institute, Washington, DC.

Joshi, P.K. and Kumar, P. (2011) Food demand and supply projections for India: 2010–2030. In: *Trade, Agricultural Policies and Structural Changes in India's Agrifood System; Implications for National and Global Markets*. TAPSIM Project No. KBBE212617. International Food Policy Research Institute, New Delhi.

Kumar, P. and Joshi, P.K. (2014) Input subsidy vs farm technology – which is more important for agricultural development? *Agricultural Economics Research Review* 27(1), 1–18.

Kumar, P., Mruthyunjaya and Chand, R. (1985) Stabilisation of prices, supply and income for paddy and wheat in Punjab: A unified approach. *Indian Journal of Agricultural Economics* 40(3), 382–389.

Kumar, P., Shinoj, P., Raju, S.S., Kumar, A., Rich, K.M. and Msangi, S. (2010) Factor demand, output supply elasticities and supply projections of major crops of India. *Agricultural Economics Research Review* 23(1), 1–14.

Kumar, P., Kumar, A., Shinoj, P. and Raju, S.S. (2011) Estimation of demand elasticities for food commodities in India. *Agricultural Economics Research Review* 24(1), 1–14.

Lau, L.J. and Yotopoulos, P.A. (1972) Profit, supply and factor demand functions. *American Journal of Agricultural Economics* 54(1), 11–18.

Nerlove, M. (1958) *Dynamics of Supply: Estimation of Farmers' Response to Price*. John Hopkins University Press, Baltimore, Maryland.

Appendix

Table A12.1. Cost and return structure of paddy and wheat in India.

Item	Paddy	Wheat
Yield (t/ha)	3.69	3.36
Price (Rs/t)	7,650	9,930
Human labour (Rs/ha)	33,658	48,456
Animal labour (Rs/ha)	10,571	8,557
Machine labour (Rs/ha)	3,014	4,245
Fertilizers (Rs/ha)	3,407	5,729
Other inputs (Rs/ha)	1,819	5,128
Net income (Rs/ha)	11,837	17,962

Table A12.2. Demand sources for rice and wheat in India, 2009 (Joshi and Kumar, 2011).

	Demand (thousand t)		Share in total demand	
Demand sources	Rice	Wheat	Rice	Wheat
Human demand				
Households	90,673	63,138	0.9194	0.7606
Home away	2,947	10,750	0.0299	0.1295
Indirect demand				
Seed	1,280	1,780	0.0130	0.0214
Feed	791	3,895	0.0080	0.0469
Wastages	539	1,227	0.0055	0.0148
Industrial uses	2,396	2,225	0.0243	0.0268
Total	98,626	83,015	1.0000	1.0000

13 High-value Production and Poverty: The Case of Dairy in India

Anneleen Vandeplas,[1,2]* Mara P. Squicciarini[1] and Johan F.M. Swinnen[1]
[1]*LICOS, Center for Institutions and Economic Performance KU Leuven, Belgium;* [2]*European Commission, Brussels, Belgium*

Introduction[1]

Rising incomes and expanding urbanization are rapidly changing dietary patterns in India and other Asian countries. People are not only consuming more food, but also diversifying increasingly towards high-value commodities, such as fruits, vegetables, dairy products, meat, fish, eggs, etc. (Kumar and Birthal, 2007; Pingali, 2007). In contrast to many other livestock-based livelihoods in other countries or regions of the world, because of specific cultural norms, it is not very common to eat cow (or even buffalo) meat in most parts of India. As a result, amongst all livestock products, milk and milk products are consumed most. With progressive urbanization, Indian households face wider ranges of product choice, and increasingly rely on markets to buy their daily cup of milk rather than rearing their own cows or buffaloes. These factors have contributed to a rapid surge in the market demand for milk and milk products in India over the past decade.[2]

To many, the rising demand for high-value products represents a great opportunity for the development of domestic high-value agricultural production, and of the dairy sector in particular. India is the largest milk-producing country in the world and livestock is traditionally considered as a sector that has strong potential for pro-poor growth in the country (Das, 2006; Goswami, 2007). Dairy animals have traditionally been serving as the main source of draught power in the fields, as well as for subsistence milk production. The Indian dairy sector is dominated by small production units (hence, with a small number of animals). This means that many rural households are engaged in dairying, and potentially depend on it as a source of income. In 2002, it was estimated that more than 70 million rural households in India derived direct income or employment from the dairy sector (Sharma *et al.*, 2002).

Livestock is widely considered as an important income-generating resource for landless households, with livestock being more equitably distributed than land (Ahuja, 2004). In addition, dairy production may have important beneficial gender implications, as it offers an opportunity for income generation to women who usually stay at home – which may increase their intra-household bargaining power. Finally, milk has been promoted worldwide as an important instrument in the fight against undernutrition, which can be due to lack of access to food or due to an inadequately balanced food intake and is a pervasive problem in India, especially among

* Corresponding author, e-mail: anneleen.vandeplas@kuleuven.be

young children. The poor households who rear their own milk animals are often expected to consume more milk themselves. For all these reasons, the Indian government has set up a number of support programmes (at the regional as well as the central level) to promote dairying among poor rural households. It also maintains high import tariffs for dairy products in international trade in an attempt to protect its dairy producers, even if the continued strife in India for self-sufficiency in major agricultural commodities may be a contributing factor as well.

While demand and supply projections by major policy institutions tend to disagree on whether India will be a net importer or a net exporter of milk powder by 2020 (see EC, 2011; Kumar and Joshi, Chapter 4, this volume), industry experts expect that domestic demand growth for milk and milk products will outpace supply growth over the coming years (VGBO, 2009) and that India will need to increase its dairy imports. Now that a strong growth in demand for milk is being observed, the dilemma between protecting producers (by keeping tariffs high) and protecting consumers (by allowing for cheap imports) becomes more important, but not less complex. Already, Mishra and Roy (2011) have argued that milk was the major contributor to Indian food price inflation in 2010. Milk price inflation may also increase incentives for adulteration of milk, increasing the likelihood of food safety crises, as experienced by China in 2008 (Gale and Hu, 2009).

The policy debate addressing this dilemma is often based on ad hoc claims and assertions, with little hard micro-level evidence to support these. To date, the lack of accurate official data has complicated meaningful analysis as well as the formulation of appropriate policy recommendations.

In this chapter, we have used a unique household-level dataset on 1000 rural households in Andhra Pradesh (India) to shed some light on the ongoing micro-level dynamics in dairy production systems in India, so as to inform the debate on how supply is responding to high demand growth. Second, we have studied the extent to which dairy production has potential to act as a motor of pro-poor growth in India. In particular, we have studied the profile of a typical milk producer in India to find out to what extent the poorest rural households are effectively involved in dairy production. This is expected to shed some light on our understanding of the poverty implications of growth in high-value production, notably in the dairy sector.

Data Collection and Sampling

Our analysis is based on micro-level data from a survey of 1000 rural households, which was conducted between April and June 2010 in Andhra Pradesh (AP), a state in the south of India. Andhra Pradesh is the fifth largest state in India by population. Although it does not belong to the top milk-producing states in India in terms of milk availability per capita, milk availability per capita in AP exceeds the all-India average by 30%. As shown below, AP strongly outperforms the all-India average in terms of dairy sector growth, which is why it presents an interesting case for the study of dynamics in the Indian dairy sector. It should be clear, nevertheless, that trends in the AP dairy sector are not necessarily representative for India as a whole.

The survey for data collection was set up to be representative for the rural population of the southern half of AP, covering the Rayalaseema region (in particular, the districts of Kurnool, Cuddapah, Ananthapur and Chittoor) and the southern part of coastal AP (more specifically, the districts of Nellore, Prakasam, Guntur and Krishna; see Fig. 7.1, this volume).

First, the study region was divided into four zones based on milk production per (rural) capita, and on whether dairy production systems were predominantly buffalo- or cow-based. This may correlate with weather and relative humidity conditions.

Within each region, one district was sampled at random. From the selected districts, 50 rural villages were randomly selected (of which 7 were in Chittoor, 12 in Cuddapah, 16 in Kurnool and 15 in Guntur) from the district-level list of villages obtained from the Government of Andhra Pradesh (GoAP, 2009).

In each village, a census was organized to record the number of female adult dairy animals that each household owned. Based on this number, the households were classified into four categories or strata.[3] A fixed number of households were selected from each strata, so as to oversample households with larger herd sizes and obtain a set of 20 households per village.

The selected 1000 households were interviewed using a questionnaire that contained detailed modules on dairy production, input markets and output markets, investments, hygienic practices, services offered by buyers, future intentions regarding dairy production, but also all required modules to estimate total household income and total household consumption.

Apart from the household interviews, we surveyed village-level milk traders in each of the survey villages. As such, we obtained a sample of 91 milk traders, spread over the 50 sample villages. Finally, a village-level questionnaire was administered to a group of village elders and key informants in the village, to get at important village-level variables such as local prices, general access to markets and infrastructure, and geographic variables.

Descriptive Statistics

Dairy farming in AP, as in most other parts of India, is dominated by small production units. This means that, while a relatively large proportion of the population owns some dairy animals, the number of dairy animals usually remains very low. For example, in our study region, 51% of the rural households had at least one female adult dairy animal at the time of the survey (see Table 13.1); however, only 3% had more than five dairy animals. As we will see below, for those who own at least 1 female adult dairy animal, the average number of dairy animals per household is only 2.5. This small scale of the typical livestock holding has important implications for the organization of dairy supply chains in AP and elsewhere in India; it is also a crucial determinant in the feasibility of implementing relevant food safety and quality standards in the dairy sector.

Table 13.1. Distribution of dairy farm size in Andhra Pradesh[a] (authors' survey).

Dairy farm size	Share in sample (%)	Share in population (%)
No dairy animals	20.0	49.0
1–2 dairy animals	37.0	33.6
3–5 dairy animals	35.6	14.4
>5 dairy animals	7.4	3.0

[a]The population under study is the rural population of the southern part of Andhra Pradesh, as defined in the text.

Table 13.2 shows that the average household in the population under study is headed by a male adult of 46 years, with 3.3 years of education. His wife is 40 years old and has 1.8 years of education. About 27.7% of the households belong to a scheduled caste (SC) or a scheduled tribe (ST), castes which have historically experienced social discrimination in India.[4] While the majority of the households adhere to Hindu religion, 11.6% and 7.5% classify themselves as Christians and Muslims, respectively.

A little more than half of the surveyed households owned dairy animals in 2010, and almost all of them produced milk. The average number of female adult dairy animals in 2010 stood at 1.3 when we take all households into consideration, and 2.5 when we only consider households with at least one dairy animal. Milk productivity was dramatically low, standing at 3.3 l per animal per day in 2010. Only 30% of dairy animals belonged to an improved breed – which means that they had resulted from crossbreeding of local dairy animals (cows or buffaloes) with (usually exotic) high-yielding breeds. The remaining animals were predominantly local breeds.

Households owned 1.1 ha of land on average, and cultivated 1.2 ha of land. This suggests that they rented in some land from other households that had migrated to the cities. More than half of the households under study were landless, and almost 34% depended on agricultural labour as their main source of income.

In most cases, agricultural wage labour is seasonal and for casual work. The literature suggests that, in general, agricultural wages

Table 13.2. Descriptive statistics of population under study, 2010 (authors' survey).

Household characteristics	Average	Standard deviation
Age of household-head (years)	46.3	11.3
Age of household-head's spouse (years)	40.2	11.2
Education level of household-head (years in schooling)	3.3	4.8
Education level household-head's spouse (years in schooling)	1.8	3.4
Scheduled caste/Scheduled tribe households (%)	27.7	
Christian households (%)	11.6	
Muslim households (%)	7.5	
Households having dairy animals (%)	51.0	
Households having dairy animals in milk (%)	49.9	
No. of female adult dairy animals per household	1.3	1.9
No. of female adult dairy animals per household (out of households with at least one dairy animal)	2.5	2.0
Productivity per animal per day (yearly average) (l)	3.3	2.7
Owned land (ha)	1.1	2.0
Cultivated land (ha)	1.2	2.0
Landless households (%)	54.0	
Households with agricultural labour as main source of income (%)	33.8	
Village characteristics		
Distance to closer dairy plant (km)	18.4	13.9
Village population size (no.)	699.0	607.3
No. of milk buyers in village	2.9	1.5

in India have not increased in line with wage growth in other sectors, such as the rural non-farm sector or the urban labour sector (Binswanger, 2012). Hence, agricultural wage labourers would benefit a lot from additional income-generating opportunities. They constitute a major target group of social policies for poverty reduction such as the Mahatma Gandhi National Rural Employment Guarantee Act (MGNREGA), which guarantees a minimum of 100 person-days of unskilled manual wage work to each rural household in a year (Zepeda *et al.*, 2013).

The village population in the study region was around 700 on average. Villages were on average 18.4 km away from the closest dairy processing plant; and there were approximately three milk buyers active per village in the region.

Dynamics in Dairy Production and Demand

This section explores how dairy production systems are evolving over time in AP. This is important, as it provides some insights in how milk supply is likely to respond to the observed strong growth in its demand.

Starting from roughly the same level in 2003 (230 g per capita; see DAHD, 2006), the per capita availability of milk in AP exceeded the all-India average by 30% at the time of the study (342 and 263 g per capita per day, respectively, in 2010; see DAHD, 2010).

Official statistics estimate that annual growth of milk production over the period 2005–2010 amounted to 7.5% in AP as compared to 4% for India as a whole (DAHD, 2010). This implies that milk production growth has been faster in AP than in the rest of India. The number of dairy animals in AP is reported to have increased at the rate of 5% per year according to official statistics, indicating that most of the increase in milk production must have come from an increased stock (hence, increased herd size per dairy farm, or a growing number of dairy farms) rather than from increased animal productivity.

Our data suggest that official growth rates are somewhat over-estimated. Over the period 2005–2010, herd size growth has averaged 3.2% per year in AP, while milk production growth has averaged 5%.

The trends in dairy characteristics presented in Table 13.3 reveal that the number of households involved in dairying (defined as owning at least one female adult dairy animal) has substantially increased over time, from 38% in 2005 to 51% in 2010. The average number of dairy animals per household has increased only slightly, from 1.1 to 1.3 between 2005 and 2010. If we consider only those households who own at least one dairy animal, we even see a decrease in the average herd size, from 2.9 to 2.5 dairy animals per household. If we focus on median herd size, rather than average herd size – to avoid that our results are driven by a few outliers – we see that it has remained constant, at two dairy animals per household. In addition, if we look at the share of households that are engaged in dairying and have more than five dairy animals, this share has remained constant at 6%. Hence, dairying has largely remained a small-scale activity, and commercial dairy farming is still in its infancy in the region under study.

In general, average wealth increased across the rural households in the population under study over the period 2005–2010: average land ownership increased slightly from 1.0 ha to 1.1 ha per household; and assets ownership increased as well. Finally, the number of milk buyers in the village has increased from two milk buyers per village in 2005 to almost three in 2010, which is a statistically significant change.[5]

For a better understanding of the dairy dynamics in the 5 years preceding our survey, Table 13.4 shows the growth in herd size, milk production and number of milk buyers, as reported by the households in our sample. Almost two-thirds of the households claimed to have kept their dairy herd at par over the 5-year period preceding our survey. One-fourth reported to have increased their dairy herd and one-eighth reported to have reduced their herd size.[6] Interestingly, not many large expansions had been taking place. Most herd size changes reflect the addition (or removal) of one dairy animal – often through natural growth (or, conversely, decline). When we look at milk production growth, almost 40% of the households reported an increase in milk production between 2005 and 2010. One-third of the households had kept milk production constant and about one-fourth reported a decrease. Interestingly, the number of milk buyers per village grew more strongly. On average, the number of milk buyers increased in almost 85% of the villages. This provides at least some suggestive empirical support to the claim by staff from major milk companies in AP that milk buyers are increasingly 'scrambling' for milk, especially during the summer (Squicciarini and Vandeplas, 2010).[7]

Hence, it seems that, in spite of strong reported demand growth, dairy supply in the region under study had not moved forward in a commensurate way by the time of the survey, and it was too early to speak of agricultural transformation in the sector, at least in this region of India.

Table 13.3. Changes in household characteristics, 2005–2010 (authors' survey).

	2005		2010	
Characteristics	Average	Standard deviation	Average	Standard deviation
Households involved in dairy farming (%)	38.0		51.0	
No. of female adult dairy animals per household	1.1	2.1	1.3	1.9
No. of female adult dairy animals per household (household with at least one dairy animal)	2.9	2.5	2.5	2.0
Median herd size (households with dairy animals) (no.)	2.0		2.0	
% of dairy household with >5 dairy animals	6.0		6.0	
Owned land (ha)	1.0	1.9	1.1	2.0
Owned land (households with dairy animals) (ha)	1.3	2.5	1.5	2.5
Asset index	–0.7	0.8	–0.1	0.9
Average no. of milk buyers per village	2.0	1.0	2.9	1.5

Table 13.4. Reported trends in dairy production, 2005–2010[a] (authors' survey).

Indicator	Growth	At par	Decline
Herd size (%)	25.1	62.4	12.5
Milk production (%)	39.3	33.4	27.3
Number of milk buyers per village (%)	84.3	5.0	10.7

[a]Milk production growth was calculated as increase in milk production per day per household for the period October–December between 2005 and 2010. The increase in number of milk buyers was calculated as the difference in the average number of buyers in a village, as reported by individual households, in 2010 and 2005, respectively.

Dairy Production and Income

This section considers a set of poverty and wealth indicators in order to unravel the potential linkages between these indicators on the one hand, and being involved in dairy farming on the other hand. If dairy farming is an activity in which many poor households are involved, development of the dairy sector could have important benefits for poverty reduction and pro-poor growth.

As Table 13.5 shows, livestock holdings correlate with diverse socio-economic characteristics: land owned, asset ownership, total income and income from different sources. Rural households with more than five dairy animals have on average twice as much land as rural households with one or two dairy animals, who, in turn, have on average twice as much land as rural households without a dairy animal. This could mean two things. Either dairy production increases household resources to the extent that it enables households to accumulate land and other assets faster than other households, or causality runs in the other direction, meaning that owning land facilitates dairy activities. If the latter is true, this would mean that rather than being a substitute for land in terms of productive assets, and thus an alternative source of income generation for landless households, livestock rearing may rather be complementary to growing crops.

Having access to land certainly facilitates access to fodder: on the one hand, dairy animals feed on crop residues that are left in the field after the harvest; on the other hand, achieving high yields requires sufficient green fodder (mostly grasses). Markets for green fodder do not seem to work well: key informants in the AP dairy sector testified that green fodder is fairly expensive to buy, while it is much cheaper for farmers to grow it themselves. If green fodder markets were working properly, farmers should be indifferent between producing green fodder crops on the one hand and producing food crops and buying green fodder from the proceeds on the other hand. This is apparently not the case in AP. Earlier literature has pinpointed the unavailability of fodder as a major constraint to dairy production as well (Birthal and Jha, 2005; Hall *et al.*, 2007; Singh *et al.*, 2012).

Average income for households with three or more dairy animals is almost twice the income for households with less than three (or no) dairy animals. This is not only driven by higher levels of income from dairying, but also – and to a major extent – by higher levels of income from land and cropping. Households without dairy animals do have higher incomes from business and/or wages, but this reflects a high dependency on agricultural labour as their principal source of income. We find a high and statistically significant correlation (of more than 30%) between three indicators that are often considered as determinants of rural poverty: belonging to a SC or ST household, being landless and having to rely on agricultural labour as a major source of income.

Amongst non-dairy producers, a higher incidence of poverty (as reflected by these three indicators) is seen. The incidence of households with SC/ST status amongst non-dairy producers is 31.5%, as compared to 27.7% in the general population. The difference is more pronounced for the other indicators. The incidence of households with agricultural labour as their main source of income amongst non-dairy producers is 45.3%, as compared to 33.8% in the general

Table 13.5. Producer profile according to livestock holdings (authors' survey).

Indicators	No dairy animals	1–2 dairy animals	3–5 dairy animals	>5 dairy animals	Total population
Share of population (%)	49.0	33.6	14.4	3.0	100.0
Land owned (ha)	0.6	1.3	1.8	2.4	1.1
Land cultivated (ha)	0.7	1.4	2.1	2.5	1.2
Total income (Rs/year)	91,950	116,565	252,215	280,387	128,930
Including dairy (Rs/year)	130	23,224	45,489	85,304	16,977
Including crops/land (Rs/year)	39,194	64,053	173,664	146,825	70,114
Including business/ wages (Rs/year)	51,576	34,590	41,054	63,560	44,716
Income per capita (Rs/year)	24,854	25,840	48,674	44,492	29,194
Asset index	−0.2	−0.1	0.2	0.5	−0.1
SC/ST status (%)	31.5	27.9	17.4	12.4	27.7
Agricultural labour as main source of income (%)	45.3	27.4	14.9	10.3	33.8
Landless (%)	51.5	31.7	24.7	15.1	39.9

population. The incidence of landless households amongst non-dairy producers is 51.5%, as compared to 39.9% in the general population. The larger the dairy herd size, the lower is the frequency of SC/ST, agricultural labour-dependent and landless households. In terms of wealth indicators, such as land ownership, income and the asset index, we find similar results: the larger the dairy herd size, the higher these wealth indicators are on average.[8]

Of course, these results are only descriptive, and they should not be interpreted as establishing causality. On the one hand, they may indicate that richer households are more likely to be engaged in dairy production; on the other hand, they may reflect that households which are engaged in dairy production increase their incomes and are able to accumulate more wealth.

Using the same dataset, Vandeplas *et al.* (2013) have shown that – after controlling for endogeneity of the decision to engage in dairy production – producing milk has a positive and significant impact on income, as well as on land accumulation. A negative (but not always significant) impact is found on the accumulation of non-productive assets (such as motorbikes, fridges and mobile phones). A potential explanation could be that households that are engaged in dairy production are, in general, households with a positive attitude towards agriculture; and that in such cases, investing in land may be wiser than investing in other assets – especially in view of projected increases in food prices.

Vandeplas *et al.* (2013) have also found that, while asset accumulation is not driven by dairy production, it does have a positive impact on the likelihood of being involved in dairy production. This does suggest that wealthier households are more likely to engage in dairy production than poorer households.

This is in line with earlier literature arguing that, although the rural poor may benefit strongly from livestock ownership, they are less likely to own livestock. If they own livestock at all, they may be more likely to own 'small' animals such as poultry and sheep or goats, which require less capital and are generally less risky than 'large' animals such as cows and buffaloes (Dolberg, 2001; IFAD, 2001). Based on a review of major donors' experiences, Ashley *et al.* (1999) have concluded that, in general, dairy development programmes aiming at increasing milk supply have contributed little to poor rural livelihoods. One of the reasons is that it is usually the wealthier households who self-select into such programmes.

Of course, 'wealth' is a relative concept. The income level per capita for the average household in our dataset amounts to Rs29,194 or less than US$650 per year. Hence, even sustaining the livelihoods of the 'wealthier' amongst these households may have favourable implications for development.

Conclusions

A look at the current micro-level trends in dairy production in India reveals that, although India has witnessed strong growth in incomes as well as demand for high-value food commodities over the past decade, agricultural transformation had not reached the dairy sector in AP at the time of our survey in 2010. Dairy production has increased slightly, but more on the extensive margin (more households engaging in dairy) rather than on the intensive margin (households engaged in dairy scaling up their operations).

Before being able to speak of dairy development, however, it is especially important for households engaged in dairy to increase productivity, which is still low compared to other major dairy-producing countries. The literature suggests that Indian dairy farmers face important constraints in accessing input markets, such as markets for fodder and land (e.g. Singh *et al.*, 2012). Such constraints will need to be overcome before we can be confident that domestic supply growth will be able to keep up with domestic demand growth in India. While macro-level projections have shown that China is very likely to increase its net imports of milk powder over the coming decades, no conclusive results have been reached at this level for India so far.

Looking at the potential implications of dairy development for pro-poor growth, we have found positive correlations between participation in dairy production, incomes per capita, land accumulation and other assets (such as motorbikes, fridges and mobile phones). However, we find that the rural poor are less likely to be dairy producers than wealthier households. A reason could be constrained access to land.

One may argue that there could be important indirect benefits of dairy development, which we have not measured in this chapter, for example through job creation for unskilled manual workers (Carter *et al.*, 1996; Humphrey *et al.*, 2004; Maertens *et al.*, 2011). However, our data have suggested that this channel for poverty reduction may not be very promising either: in our dataset of 1000 rural households, we have found only four households that derive their income mainly from working as a dairy labourer, out of which three cases were concerned with casual labour, rather than labour which is hired on a more permanent basis. This compares to 242 rural households that derive their income mainly from working as a crop labourer (out of which 218 work as casual labour).

To sum up, while dairy participation correlates positively to major livelihood indicators, there may still be important constraints that weaken the dairy sector's current potential for pro-poor development. While further research is needed to corroborate our findings, we do expect that they can contribute importantly to the policy debate on dairy development in India.

Notes

[1] Mara P. Squicciarini is a Doctoral Fellow of the Research Foundation Flanders (FWO). Research for this study was conducted before Anneleen Vandeplas joined the European Commission, while she was a Postdoctoral Fellow of the Research Foundation Flanders (FWO). Any opinions expressed in this book are those of the authors and do not necessarily reflect the views of their respective institutions.

[2] Kumar and Birthal (2007) have reported that per capita consumption of milk in India rose from 40.4 kg/year in 1980 to 66.2 kg/year in 2000 and projected that this will continue to increase to a level between 90.6 and 107.0 kg/year by 2025. More recent National Sample Survey Organization (NSSO) consumption data show

that the average expenditure on dairy products by urban and rural Indian households was Rs82 and Rs50, respectively, in 2006, up from Rs74 and Rs43 in 2000.

[3] Category 1: no female adult dairy animal; Category 2: 1–2 female adult dairy animals; Category 3: 3–5 female adult dairy animals; and Category 4: more than 5 female adult dairy animals. We only counted female adult dairy animals (having had at least one calf) as dairy animals. Hence, we have not taken into account male or immature dairy animals in our size classification.

[4] Scheduled castes (SC) and scheduled tribes (ST) represent 16.2% and 8.2% of the total Indian population, respectively (Census of India, 2011). We found slightly higher shares of SC/ST households in our study, probably due to a higher representation of SC/ST amongst the rural population.

[5] These include different types of milk buyers: the government cooperatives, the Mutually Aided Cooperative Societies (MACSs), the private formal sector and finally the informal sector, including direct marketing (from farmers to neighbours, hotels and sweetshops). More details on these different categories can be found in Squicciarini and Vandeplas (2010).

[6] Note that this included households which did not have any dairy animals in 2005 nor in 2010: these were considered to have kept their herd size at par as well. If these were dropped from the analysis, the figures were roughly 25%, 50% and 25% for households which kept their herd size at par, households who increased and households who reduced their dairy herd, respectively.

[7] Strictly speaking, as milk buyers can visit different villages in one day, an increased reported number of milk buyers may as well reflect the fact that the same milk buyer is visiting a larger number of villages. However, this does also lead to a higher level of competition at the village-level, so we have treated both cases in an identical way.

[8] Total income was calculated as the sum of net incomes from dairy, cropping, business and wages, and from other sources (such as pensions or remittances). Production costs were subtracted. In case milk or crops were used for subsistence rather than for sales, we used imputed sales value. Income per capita was calculated as total income per adult equivalent, in which we counted male adults as 1 adult equivalent, women as 0.8 and children below 15 as 0.5. Our asset index is a variable which has been created through principal component analysis based on ownership of a list of 'large' assets such as diesel engines, pickups, or cars as well as 'small' assets such as fridges, bicycles and mobile phones. It is centred around 0 and has a variance of 1 and should be interpreted as a relative measure of asset ownership, rather than as an absolute measure.

References

Ahuja, V. (2004) *Livestock and Livelihoods: Challenges and Opportunities for Asia in the Emerging Market Environment.* National Dairy Development Board of India, Anand, Gujarat; and Food and Agriculture Organization of the United Nations, Rome.

Ashley, S.D., Holden, S.J. and Bazeley, P.B.S. (1999) *Livestock in Poverty-focused Development.* Livestock in Development, Somerset, UK.

Binswanger, H. (2012) India 1960–2010: Structural change, the rural non-farm sector, and the prospects for agriculture. Working Paper for the Stanford Symposium Series on *Global Food Policy and Food Security in the 21st Century.* Center on Food Security and the Environment, Stanford University, California.

Birthal, P.S. and Jha, A.K. (2005) Economic losses due to various constraints in dairy production in India. *Indian Journal of Animal Sciences* 75(12), 1470–1475.

Carter, M.R., Barham, B.L. and Mesbah, D. (1996) Agricultural export booms and the rural poor in Chile, Guatemala and Paraguay. *Latin American Research Review* 31(1), 33–65.

Census of India (2011) *Census of India* 2011. Available at: http://www.censusindia.gov.in (accessed 13 February 2013).

DAHD (Department of Animal Husbandry Dairying & Fisheries) (2006) *Basic Animal Husbandry Statistics 2006.* Government of India, New Delhi. Available at: http://dahd.nic.in/dahd/statistics/animal-husbandry-statistics.aspx# (accessed 13 February 2013).

DAHD (Department of Animal Husbandry Dairying & Fisheries) (2010) *Basic Animal Husbandry Statistics 2010.* Government of India, New Delhi. Available at: http://dahd.nic.in/dahd/upload/BAHS_2010.pdf (accessed 13 February 2013).

Das, A. (2006) WTO negotiations and India's stand: Agriculture, NAMA and services. *InfoChange News & Features,* December. Available at: http://infochangeindia.org/trade-a-development/backgrounder/wto-negotiations-and-indias-stand-agriculture-nama-and services.html (accessed 13 February 2013).

Dolberg, F. (2001) A livestock development approach that contributes to poverty alleviation and widespread improvement of nutrition among the poor. *Livestock Research for Rural Development* 13(5).

EC (European Commission) (2011) *Agricultural Commodity Markets Outlook 2011–2020: A Comparative Analysis of Projections*. OECD-FAO, FAPRI and USDA. European Commission DG Agriculture and Rural Development: Agricultural Trade Policy Analysis Unit, Brussels, Belgium.

Gale, F. and Hu, D. (2009) Supply chain issues in China's milk adulteration incident. Paper prepared for presentation at *the IAAE 2009 Conference*, Beijing, China, 16–22 August.

Goswami, B. (2007) Can Indian dairy cooperatives survive in the new economic order? Paper presented at the *WTO Public Forum How Can the WTO Help Harness Globalization*, Geneva, Switzerland.

GoAP (Government of Andhra Pradesh) (2009) *Andhra Pradesh Land Record Information*. Andhra Pradesh State Unit, National Informatics Centre, Hyderabad. Available at: http://apland.ap.nic.in (accessed 13 February 2013).

Hall, A., Sulaiman, R.V., Dhamankar, M., Bezkorwajnyj, P. and Prasad, L. (2007) *Reframing Technical Change: Livestock Fodder Scarcity Revisited as Innovation Capacity Scarcity, Part 1. A Review of Historical and Recent Experiences*. UNUMERIT Working Paper 2008-002. UNUMERIT, The Netherlands.

Humphrey, J., McCulloch, N. and Ota, M. (2004) The impact of European market changes on employment in the Kenyan horticulture sector. *Journal of International Development* 16, 63–80.

IFAD (International Fund for Agricultural Development) (2001) *Rural Poverty Report 2001 - The Challenge of Ending Rural Poverty*. IFAD, Rome.

Kumar, P. and Birthal, P.S. (2007) Changing consumption pattern in South Asia. In: Joshi, P.K., Gulati, A. and Cummings, R.J. (eds) *Agricultural Diversification and Smallholders in South Asia*. Academic Foundation, New Delhi, pp. 151–187.

Maertens, M., Colen, L. and Swinnen, J. (2011) Globalization and poverty in Senegal: A worst case scenario? *European Review of Agricultural Economics* 38(1), 31–54.

Mishra, R. and Roy, D. (2011) Explaining inflation in India: The role of food prices. *Brookings NCAER India Policy Forum* 8, 2011–2012.

Pingali, P. (2007) Westernization of Asian diets and the transformation of food systems: Implications for research and policy. *Food Policy* 32(3), 281–298.

Sharma, V.P., Singh, R.V., Staal, S. and Delgado, C. (2002) Annex I: Critical issues for poor people in the Indian dairy sector on the threshold of a new era. In: Delgado, C. and Narrod, C. (eds) *Impact of Changing Market Forces on Structural Change in the Livestock Industries of Selected Fast-growing Developing Countries. Final research report of phase I – Project on Livestock Industrialization, Trade and Social-Health-Environment Impact in Developing Countries*. Agriculture and Consumer Protection Department, FAO, Rome.

Singh, K.M., Singh, R.K.P., Jha, A.K., Singh, D.K. and Singh, V.K. (2012) A Study of Fodder Value Chain in Bihar. Available at: http://dx.doi.org/10.2139/ssrn.2030610 (accessed 13 February 2013).

Squicciarini, M. and Vandeplas, A. (2010) *The Dairy Sector of Andhra Pradesh: Findings from a Company Survey*. LICOS, KU Leuven, Belgium.

Vandeplas, A., Squicciarini, M. and Swinnen, J. (2013) *Agricultural Transformation and Poverty: The Case of Dairy in India*. TAPSIM Deliverable D5.5. LICOS, KU Leuven, Belgium.

VGBO (Victorian Government Business Office–India) (2009) *Market Opportunities in Indian Dairy Value Chain*. Department of Business and Innovation, Australia.

Zepeda, E., McDonald, S., Panda, M., Kumar, G. and Sapotka, C. (2013) *Employing India: Guaranteeing Jobs for the Rural Poor*. Carnegie Endowment for International Peace, Washington, DC.

14 Changing Structure of Retail in India: Looking Beyond Price Competition

Devesh Roy,[1]* Shwetima Joshi,[2] P.K. Joshi[1] and Bhushana Karandikar[3]
[1]International Food Policy Research Institute, New Delhi, India; [2]University of Texas, El Paso, USA; [3] Indian School of Political Economy, Pune, India

Introduction

The food and grocery segment in organized retail has been growing at a fast pace in India. The compound annual growth rate of food and grocery in organized retail has been estimated to be 16% during 2004–2007 (Joseph *et al.*, 2008). The India Retail Report (GoI, 2007) put the growth rate of the organized food and grocery sector at 42% in 2006 over 2005. The report by Planet Retail (2004) projected the sales of the top five grocery retailers in India to grow from US$1 billion in 2007 to US$15 billion in 2012, a 15-fold increase in 5 years. It is expected that these changes comprising the growth of the organized sector vis-à-vis unorganized retail would be more pronounced with rising incomes, rapid urbanization and unfolding globalization.

The concept of shopping in India has indeed changed in recent years in terms of formats available to consumers to buy from.[1] A young population (median age of 28 years) and a rising middle class are the key drivers for the fast emergence of consolidated formats such as hypermarkets and supermarkets. In this chapter, based on a survey of over 1850 consumers in three of India's cities, viz. Mumbai, Mysore and Pune, we argue that the demand for those features that can lead to back-end development (to deliver for example food safety) are generally missing among supermarket customers. We find there are marked differences across regions with customers in Mysore being more discerning of product features such as food safety, product origins and customization.

The limited demand for these product features has significant implications for the supply chains. Non-price competition among the retailers based on items such as product certification, traceability, hygiene or quality standards in procurement, handling or storage of products if they exist, could have a significant bearing on the incentives for supply chain coordination. In order to provide for these attributes retailers, domestic or foreign, would have to invest in back-end development. Based on our survey, the product features other than price that are demanded by customers seem to be 'convenience' and 'quality' (in terms of visual traits) that do not necessitate modernizing the back-end (for example, in transport, storage). With price and superficial product attributes (such as quality based on looks) being the leading factors, customers continue to mix across types of retailers.

One change that seems to have happened is diversification in the set of products being

* Corresponding author, e-mail: d.roy@cgiar.org

demanded, including in their provenance. There is a rising demand for some niche products sometimes uniquely sold by the supermarkets that are sourced from imports. Our survey results show that 'imported' could be becoming the new niche.

Further, imports become an outlet for customers demanding attributes such as food safety. In this sense, imports bring out the latent demand for customized, safe products for various food items.

With this background, this chapter addresses the following research questions:

- What is the role of non-price factors relative to price in the purchases made in the supermarkets? What are the reasons for the observed purchase decisions across different formats of retail: is it: (i) food safety; (ii) food quality; (iii) guarantee of an assured supply; (iv) relative prices between modern and traditional retail; or (v) peer effect?
- What are the characteristics of the customers who buy imported/niche products?
- What are the factors that determine the distribution of purchases by customers between modern and traditional retail? How are these factors related to ones determining the choice between domestic and imported products?

The issue of customers' demand for characteristics in food is germane to the current policy debate in India following the recent announcement to liberalize foreign direct investment (FDI) in retail.[2] In the food sector, proponents of FDI project it as a tool to improve the supply chain by increasing back-end investment. In this regard, the government ordinance is reliant on command and control approaches to improve supply chain coordination and develop the back-end. The legislation, for example, outlines the minimum investment size (US$100 million), 50% of which has to be used in the back-end and 30% procurement has to occur from small manufacturing enterprises (SMEs) within India (units with total plant and machinery investment not exceeding around Rs12.5 million at initiation).[3]

However, a system to raise investments in the back-end by order could fail if penalties for not doing it or rewards for doing it are expected to be small. Instead, customer preferences over products and/or their attributes could foster important changes in the supply chain by creating demand pull for them.

In this chapter, based on a survey of over 1850 consumers in 14 supermarkets (of different retail chains) in three Indian cities, we find that most customers mix across retail formats while making their food purchases. Further, the attributes that customers demand are not particularly demanding in terms of supply-chain coordination.

Regression results also show that customers meet their demand for attributes such as food safety or better customization by relying on imported products. Conditional on being aware about imported products, several customers opt for them for their ability to provide non-price attributes. A conditional (on awareness about their presence) mixed process model establishes this aspect of customers' choice.

Survey Design and Data Collection

Data for this study were collected through on-site customer surveys in outlets of supermarkets in Mumbai, Mysore and Pune. In each city, six to eight outlets of different supermarket chains were covered in the survey. The cities covered are: tier 1, Mumbai – high population and high income, a metro city; tier 2, Pune – smaller population and income lower than tier 1 cities; and tier 3, Mysore. Regionally, Mumbai and Pune are in western India and Mysore is in southern India. Even though Mysore is a relatively small city, it has a high density of supermarkets and also has specialized outlets dealing in organic produce (for example, 'Prakriti', 'Nesara Organic' outlet, 'Green Market' among several others), i.e. perceived by customers as safer food.

Customers were randomly selected for on-site interview using a structured questionnaire that contained three modules. Module 1 gathered information on the identity of the respondent, while module 2 collected information on the individual and household characteristics. These characteristics for example, included age, education, family composition and expenditure characteristics such as share of food in total and in monthly expenditure on groceries. The final module 3 obtained detailed information on the features of purchases of goods and the allocation between supermarkets and traditional retail.

Data and Descriptive Statistics

Profile of customers in the sample

Table 14.1 presents summary statistics about the profile of customers in the sample. Even though there does not exist a sampling frame from which the set of customers going to supermarkets could be chosen, we expect the sample to be representative of middle-income class customers.

In the three cities, the average customer in our sample is middle aged (less than 40 years). From census data in 2011, about 68% of Mumbai's population is in the age group 15–59 years, while the share of this group is 61% and 62% in Pune and Mysore, respectively. The average age of the customers in supermarkets is along the expected lines since younger customers have a greater propensity to shop in these stores.

There seems to be a definite pecking order in the choice of shopping outlets in terms of socio-economic status. Based on 2011 census data, in both the Mumbai and Pune samples, the proportion of customers with graduate or above degree is much higher than the figures in the 2011 census data (less than 12% in both cities). The same is the case in Mysore. Since education levels are positively correlated with incomes, the set of customers does seem to belong to above-average income class in the three locations.[4]

Other covariates also point to segmentation by socio-economic class. A disproportionately large share of customers in supermarkets in these cities belongs to the high caste. According to the census data, the share of the three largest scheduled caste and scheduled tribe groups in the population is 5.87%, 14.15% and 27% in Mumbai, Pune and Mysore, respectively.[5] Also, among the customers who are employed, a larger share works in the private sector.

The profile of the customers in the supermarkets is thus significantly different from the average of the population in these cities. Belonging to a better-educated and higher-income class, *a priori* we would expect a greater valuation of non-price attributes relative to the average of the population in the three cities.

Further, even though the range of expenses on food in total expenditure averages between 30 and 35% in the three cities, in two cities, nearly half of the expenditure in supermarkets comprises spending on food (Table 14.1).

Choice between Modern and Traditional Retail in Shopping: The Role of Non-price Attributes

The survey asked questions on the choices to be made in situations where supermarkets and traditional outlets are at price parity. For a product that is available at the same or a comparable price in wet (traditional) markets as well as in supermarkets, only 7% customers would never buy it from wet markets, 22% of customers would buy from wet markets rarely, while the majority, i.e. 39%, would buy sometimes. Thus, there is a large proportion of consumers (nearly 26%) who would often buy from the wet markets. An almost identical percentage of customers would never buy from the wet markets as much as who would always buy from the wet markets (i.e. 7%) (Table 14.2). The thick middle in this distribution points to the duality in the retail sector.

In analysing the choice between supermarket and traditional retail, data show no clear demarcation between formats and substantial mixing across the two types. Price emerges as the dominant factor in customers' choice with only 8% of customers in favour of supermarkets over traditional retail, irrespective of the price differential.

To understand the drivers of the choice of supermarkets over traditional retail (with

Table 14.1. Demographic profile of supermarket customers in Mumbai and Pune sample.

Characteristics	All (n =1848)	Mumbai (n =598)	Pune (n = 602)	Mysore (n = 648)
Basic demographic profile of sample customers				
Age	37.41	37.8	39.2	35.25
	(11.8)	(13.24)	(10.1)	(11.64)
Percentage of customers with				
Less than matriculation	1.83	1.37	2.54	5.5
Only matriculate degree	5.87	3.90	5.92	7.8
Only intermediate degree	10.62	5.77	11.51	14.6
With graduate degree	47.29	53.65	46.53	41.7
With post-graduate degree	29.7	35.31	33.50	20.3
Percentage of customers belonging to				
Scheduled caste/tribe	10.11	3.57	8.74	18.02
Backward caste	19.73	5.61	12.62	39.29
Minorities	14.54	8.16	23.3	12.17
Upper castes	52.85	82.65	55.74	20.18
Household size and composition (no. of members)				
of <2 years	1.05	1.03	1.09	1.1
	(0.31)	(0.33)	(0.29)	(0.30)
of 2–5 years	1.15	1.10	1.20	1.28
	(0.45)	(0.34)	(0.53)	(0.46)
of >18 years	3.24	2.90	3.50	3.18
	(1.74)	(2.07)	(1.25)	(2.17)
Household size	4.22	3.66	4.79	4.5
	(3.54)	(2.26)	(4.40)	(3.20)
Average monthly food expenditure (Rs)	7842	7342	8631	5001
	(5331.30)	(5330.57)	(5253.51)	(3341.76)
Average share of food in expenditure in the supermarket (%)	47.2	48.2	45.5	34.18
	(22.9)	(20.2)	(25.65)	(15.88)[a]
Average distance of home from supermarket (km)	5.69	7.5	3.68	2.66
	(7.92)	(8.19)	(8.08)	(18.68)

The figures within the parentheses denote standard deviation

[a]In Mysore, customers actually answered for the share of food in budget, while in Mumbai and Pune, it was the share of food in total expenditure in the supermarket.

Table 14.2. Choice between supermarkets and traditional market in unbranded equally priced food product, showing percentage of customers with the choice of wet market over supermarket for an equally priced product.

Choice	All	Mumbai	Mysore	Pune
Never	6.21	2.55	5.10	10.83
Rarely	20.55	15.11	18.37	28.26
Sometimes	42.74	46.01	50.00	32.15
Often	24.11	28.01	21.26	23.18
Always	6.38	8.32	5.37	5.58

differing levels of engagement on a Likert scale), we asked the following question: For a product that is available in both retail formats, what usually are the possible reasons for choosing supermarket over traditional retail? Multiple options in answer to this question were possible.

Figure 14.1 summarizes the responses of customers for the combined samples of Pune and Mumbai. Figure 14.2 presents the

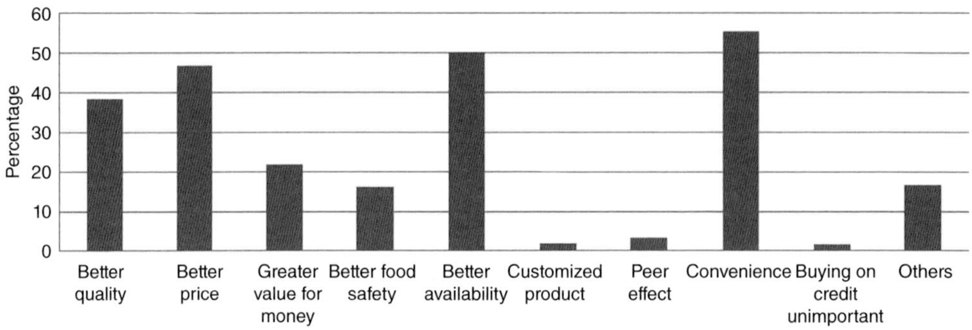

Fig. 14.1. Choice factors in favour of supermarkets when traditional retail is an available option (Mumbai and Pune).

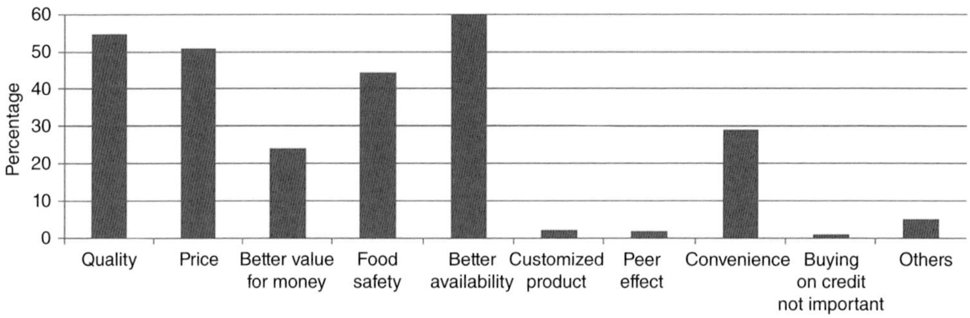

Fig. 14.2. Choice factors in favour of supermarkets when traditional retail is an available option (Mysore).

case of Mysore separately since the factors determining choice turn out to be quite different. On average, the most commonly cited reason for choice of supermarkets over traditional retail was shopping convenience (expressed mostly in terms of geographical proximity to home or workplace and availability of many products at one place) in Mumbai and Pune, while it was better availability of products in Mysore.

Among customers in supermarkets in Mumbai and Pune, better availability was the second most important factor while in Mysore it was third. Apart from geographical proximity and availability of multiple products under one roof, the attribute 'convenience' included factors such as presence of clearly marked shelves, no need to bargain on prices and the availability of sales persons for providing information.

A large number of customers choosing supermarkets for better price and only a few opting for it for reasons such as food safety is quite informative (more so in the case of Mumbai and Pune). This indicates near parity (in terms of perception) between traditional retail and supermarkets about providing non-price attributes.

Some customers also stated other reasons for choosing supermarkets over traditional retail. Most of the reasons suggested in 'others' options are however correlated with the ones in Fig. 14.3. The three principal reasons in the set of 'others' comprised discounts, greater variety of goods available and ambience (Fig. 14.3). Discount is an alternative framing of the price attribute. Ambience could be taken as a correlate of convenience. Among the fewer customers who cited hygiene and food safety (with responses such as organic, local food with short duration of storage, etc.) in a follow-up question after choosing others, most were customers from Mysore in the sample.

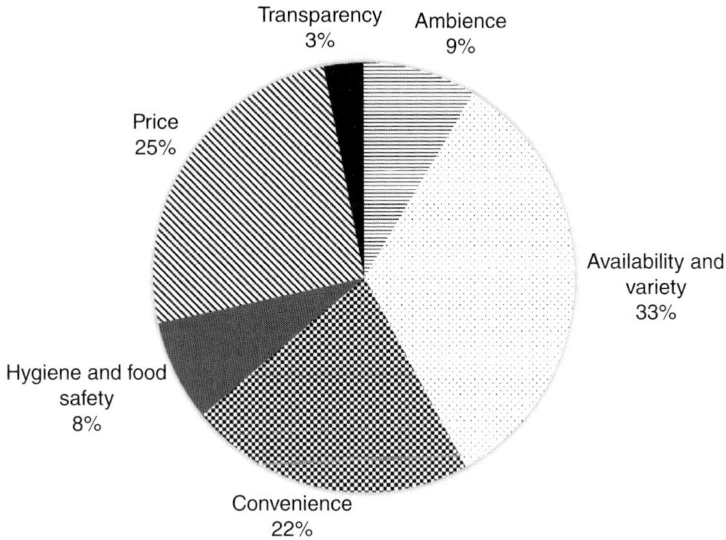

Fig. 14.3. Breakdown of other reasons for choosing supermarkets (Pune, Mumbai and Mysore combined).

Suppressing price as the factor, the respondents were then asked: for a food product that is equally priced in supermarkets and traditional retail, will you buy from wet or traditional markets? Given the mixing across formats, the response to this question was not binary as factors other than prices could also play a significant role in customers' choices. Thus, customers were asked to respond on a scale with five options for choosing equally priced and unbranded fresh food products. The possible choices in favour of wet markets (traditional retail) were: 1. Never, 2. Rarely, 3. Sometimes, 4. Often, and 5. Always. Table 14.2 presents the responses of the customers across the three cities.

Summary statistics in Table 14.2 reveal that there is a significant mass in the two categories of 'sometimes' and 'often' in all the three cities. This is clear evidence of significant mixing across retail formats in the current situation.

Purchase of niche and imported products

With demand for non-price attributes affecting the choice of supermarkets over traditional retail only in a limited way, another factor that could lead to back-end development in the supply chain can be diversification in the set of demanded products. If a new set of products is demanded that has traditionally not been sold, supply chains for such products would need to be developed. However, if these products are imported, this could weaken the force towards domestic supply-chain investment.

The products that were earlier thought to be niche in India, such as broccoli, are now more generally demanded and supplied locally. A new set of niche products has developed that is to a large extent unique to supermarkets (vis-à-vis traditional retail) in India and many are imported. Based on several focus group discussions with customers, retailers and farmers prior to the survey, we created a master list of products that are commonly perceived to be 'niche'. The 11 niche groups created based on stocking patterns in different supermarkets are listed in Table 14.3.

With the set of niche products in Table 14.3, what is striking is that more than one-third of the surveyed customers purchase olive oil, i.e. almost completely imported and is generally highly priced and used to be demanded by only a few.[6] With a large number of buyers, several consumers rationalized their purchase as this being the

Table 14.3. Niche products with the shares of customers buying them.

Niche products	Number of customers	Percent-age
Exotic fruits like kiwi, dragon fruit	447	23.91
Basil, celery	220	13.48
Red cabbage, broccoli, colour capsicum	618	33.24
Sweetcorn, baby corn, cherry tomatoes	703	37.81
Prunes	108	5.80
Olive oil	628	33.7
Exotic cheese	291	15.65
Exotic fruit juices and beverages	287	15.43
Chocolates in different flavours, meat delicacies	381	20.49
Ham	173	9.30
Flavoured tea and coffee	220	11.83

healthier oil, i.e. a non-price factor.[7] Also, note the purchases of some animal source products such as ham and meat delicacies. These products are almost exclusively sold in the supermarkets and are mostly imported like olive oil (Reardon and Minten, 2011).

Strikingly, in 8 out of the 11 cases, the customers who choose supermarkets over traditional retail expecting better food safety in the former are likely to demand niche products, many of which are imported.

For several niche products, the preference for supermarkets is also guided by another non-price attribute, viz. the availability of customized products or year-round availability. The requirement of uninterrupted availability could be one factor for relying on imports. Specifically, better availability as a factor in supermarket purchases has a bearing on buying sweetcorn, baby corn, cherry tomatoes, prunes, exotic fruit juices and beverages, chocolates in different flavours and flavoured tea and coffee. From the point of role of customization of products in the supermarkets, it has a bearing on the purchases of exotic fruits and chocolate products as well.

Choice between imported and domestic products

On the choice between imported product and one from domestic sources, the survey question at first related specifically to staple food. Notably, many customers were not aware about all the imported staple products available in the supermarkets. Conditional on awareness about them, the revealed choices are ordered with 24% customers having the highest preference for imported products and 58% customers with the weakest preference for it. This question was asked for the hypothetical situation of price parity between a domestic and imported variant of the product.

The distribution of customers showing lack of preference for imported products is largely driven by the nature of the product, i.e. non-perishable staples (for example, pulses) and exporting country characteristics, mainly reputation. Because of domestic supply shortfall, pulses currently face one of the lowest trade barriers and are commonly mixed with Indian pulses. However, a sizable number of consumers do not know staples like pulses are imported. In Mumbai and Pune, 64% and 63% customers, respectively, know that staples in the supermarkets are among imported products while in Mysore only 48% customers are actually aware of this.

In pulses where imports are significant (about 4 Mt in 2011–2012, sourced mainly from Australia, Canada and Myanmar) and retailers generally mix domestic and imported variants to save on costs, there is probably a strategic withholding of information by the retailers. In pulses, customers' preferences tend to be rigid both in favour of localness as well as no mixing, and there is a lower weightage of attributes such as food safety. Hence, pulses from Australia or Canada would not be preferred even though exports from there in general have a reputation for delivering quality.

Focusing on customers who show a clear preference for an imported product over its domestic variant, the main reason cited for this choice by the highest number of customers was greater value for money. From the follow-up questions, value for money meant several things. It meant better quality and higher food safety, while for some other customers this meant that the same quantity implied more volume (for example, one imported chocolate having more volume than the domestic counterpart). The last description of value for money is largely an indirect formulation of the price attribute, i.e. the actual price for some imported products to be lower than the listed price.

Hence, among several buyers, unlike domestic products, the imported ones carry a reputation or expectation of delivering on food safety.[8] Similarly, customers might demand a better customization (another non-price attribute) of the product and continuous availability to meet specific needs. Overall, this is an important stratification in the market where imports become a medium for expression of customers' latent demand for non-price attributes.

However, as expected, the response in favour of imported products is product and country-of-origin specific, more pronounced in cases where food safety is a greater concern (like animal source foods). The reputation of the exporter is also quite important (with products from countries like China being discounted where competition based on price becomes the centre piece).

Most importantly, relative to domestic products, imports are associated with higher expectation of delivering features such as food safety. Figure 14.4 shows that nearly one-third consumers have food safety as one of the determining factors in favour of imported products over its domestic variant under price-parity.

Since products are sold in a dual marketing system where customers buy both from supermarkets as well as from traditional retail, it is not directly possible to link imports of products directly with supermarkets.[9] However, there are clearly products that can be taken as *supermarket leaning*, especially if one looks at their year-round availability. Specifically, some fruits and vegetables and animal-source foods' imports such as exotic cheese have a greater likelihood of having supermarkets as their destination.[10]

Overall, the imports of fresh vegetables that are likely to be sold in the supermarkets are quite small. Figure 14.5 presents some examples of imports of fresh vegetables for illustration. The characteristics of the low level of imports (that could be sold in supermarkets) has been generic over the past decade.

In comparison, imported frozen vegetables that are largely sold by supermarkets have more sizable imports. The sizable imports of frozen processed vegetables have taken place mainly from China and Thailand. In 2008, the sales of frozen processed food via supermarkets/hypermarkets expanded by 13% in current value terms over the previous year (Euromonitor, 2008).

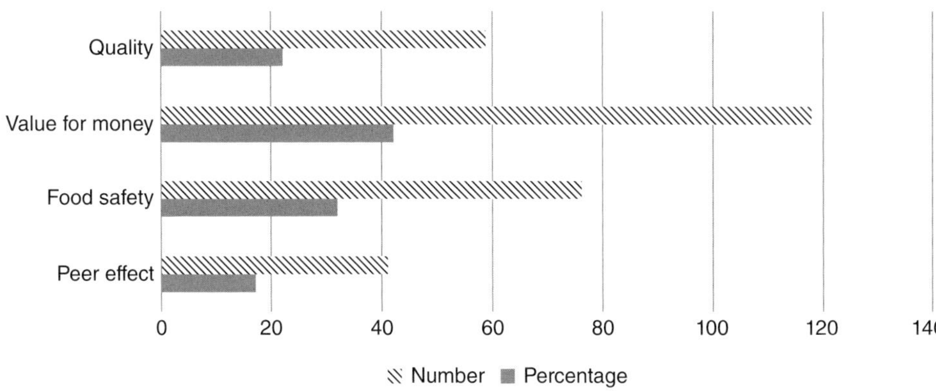

Fig. 14.4. Factors driving the choice of imported over domestic product.

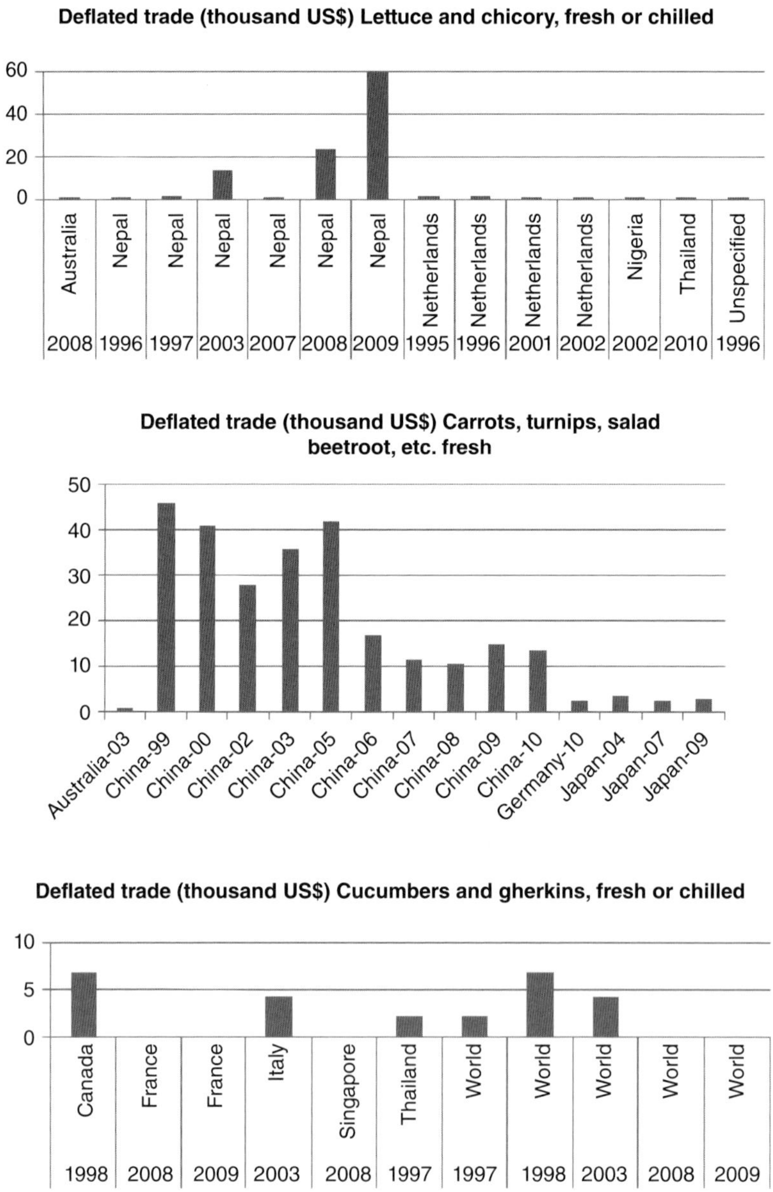

Fig. 14.5. Imports of selected vegetables.

Moreover, the patterns in imports of fruits (relative to vegetables) are quite different. Imported fruits have made successful inroads into the Indian supermarkets where imports in 2007–2008 equalled nearly US$203 million. A large proportion of imported fruits including the common ones like apple, are sold in supermarkets even though there is a customs duty of over 50%. Apples are thus the top-most imports with a large share of supermarkets followed by oranges, grapes and pears. The USA, China and Chile are the major exporters.

Figure 14.6 presents the evolution and distribution of imports of fruits (apples and citrus fruits) by exporters. Even among the

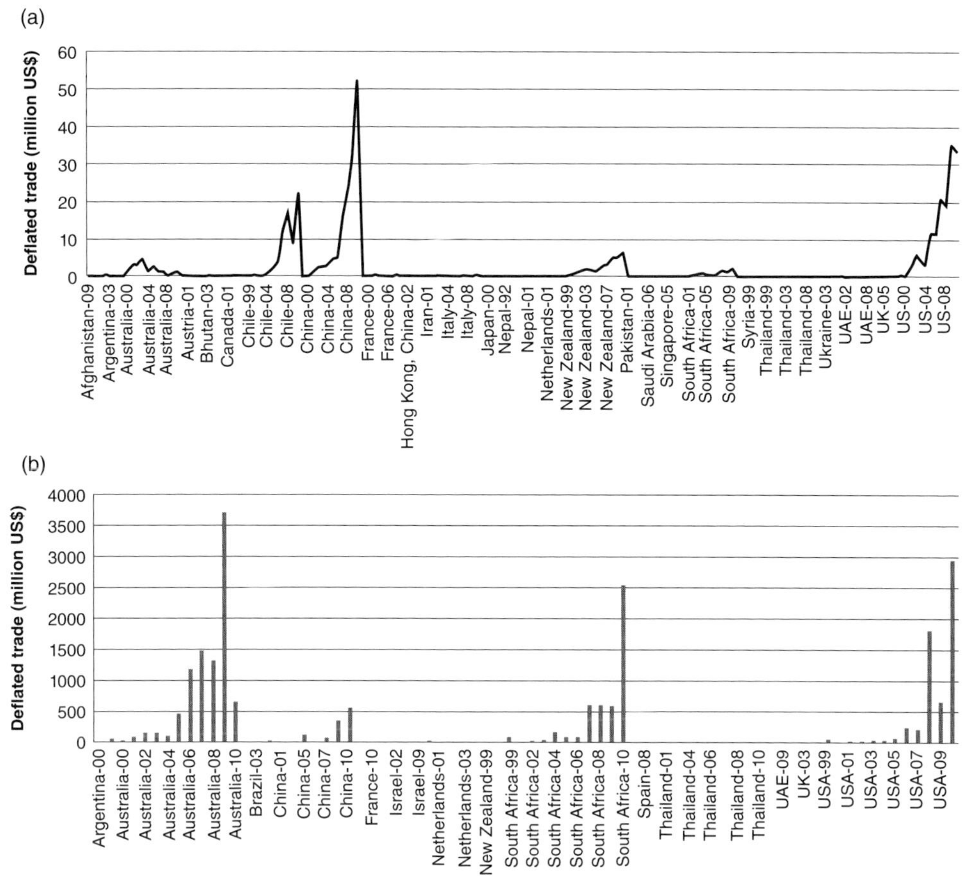

Fig. 14.6. Supermarket-leaning imports of important fruits into India: (a) fresh apples, peaches and quince and (b) dried citrus fruits (FAOSTAT, 2013).

high and middle income countries, the imports of fresh products are mostly from non-European countries.

In products such as cherries, peaches, plums and apricots, the imported variants of which are commonly stocked in supermarkets, countries like the USA, Chile, Australia and China have greater shares. Similarly in melons and papaya, apart from neighbouring countries the only country that has had exports to India has been the USA. There has also been a continued increase in the share of China in several fruit products. In comparison, the exports from Europe in these supermarket-leaning products have been negligible.

Among the animal-source foods, though there has been growth in imports in meat products since early 2000, it has been sporadic and countries hardly have a sustained trade relationship. The imports of meat products that would go either to supermarkets or the hotels, restaurant and institutions (HRI) sector have been small.

There are several possible reasons for this, for example high import duties, high cost of refrigerated transportation and consumer preference for fresh meat over frozen. The most important reason apart from consumer preferences for low-value of imports of meat could be infrastructure constraints. In an extensive survey of supermarkets by the International Finance Corporation (IFC), lack of consistent supply of electricity emerged as the main constraint affecting the supermarkets in India.

It is also possible that supermarkets have been experimenting with different products at small volumes to gauge their demand. Indeed, the HRI sector is the biggest absorber of meat imports though some imported brands have started getting a foothold in supermarkets in India. Zwanburg (The Netherlands), for example, is the best-selling meat brand in supermarkets in India. Imported cheese, which many supermarkets have on their shelves, has a 5% market share. This market is still dominated by a local brand – Amul. Happy Cow (Austria), Laughing Cow (France) and Kraft (Australia) are the prominent imported brands selling in Indian supermarkets.

The import of cheese reached US$4.79 million in 2007–2008. Denmark, The Netherlands, France and Italy are the major sources of imported cheese. Kraft, Happy Cow and Laughing Cow are some of the popular brands. Whey, which is a by-product of cheese production, is another major product of import which is sourced primarily from the USA, France and Denmark.

The main points that emerge from this summary analysis of imports of supermarket-leaning products are:

1. In products that have relatively lower trade barriers such as pulses, there is a preference for localness. In other products that have significant trade barriers in India, such as fruits and vegetables and animal-source foods, the import volumes are still quite small. Specific products, such as apples, have experienced reasonable growth in imports in spite of high trade barriers. The case of fruit imports as supermarket-leaning products highlights the problems in the domestic chains where import potential remains high as demand for high-value food items has been expanding over time.
2. There has been a trend shift in imports of the majority of supermarket-leaning high-value items since 2003–2004. Across exporters also there has been a gradual shift with increased share of China and Thailand in several products as well as a rising share of the USA. Europe seems to be the continent left out in this expansion. However, in animal source foods, particularly cheese, European countries have been increasing their footprint.
3. Some of the niche products, for example olive oil and exotic juices and animal-source products, are available only as imports and largely concentrated in supermarkets. Some new processed products are also becoming common in Indian supermarkets; for example, there is an increasing demand for another niche product, viz. pasta, by Indian consumers, and the supermarkets increasingly stock Italian brands, e.g. San Remo, Barilla and Agnesi.

Even though there is an increasing presence of imports in the Indian supermarkets, there is no direct way of linking supermarkets with imports. At a summary level, it is natural that the imports of supermarket-leaning products should be higher in areas where there is a greater presence of supermarkets. It follows that since supermarkets are more widespread in south India and near the Mumbai–Pune corridor in western India, the imports of such products could be comparatively high in relevant ports (Fig. 14.7).

Figure 14.7 shows the imports of apples, oranges and pears by ports in India drawn from the customs data at the 8-digit disaggregated level. The ports at which imports of fruits arrive are Mumbai, Jawahar Lal Nehru Port (JNPT), Calcutta, Madras, Cochin and Tuticorin. The first two ports are in Mumbai while the last three are in south India.

Since the prime areas for the presence of supermarkets in India are western India and south India, the imports into these ports on a much larger scale than into the eastern port of Calcutta is indicative of the role of supermarkets in these imports.

Econometric Analysis of Customers' Choice of Imports

There are several imported products that have geographical indicators or are commonly recognized as being from a certain country or region. Examples of such products in the supermarkets include Belgian

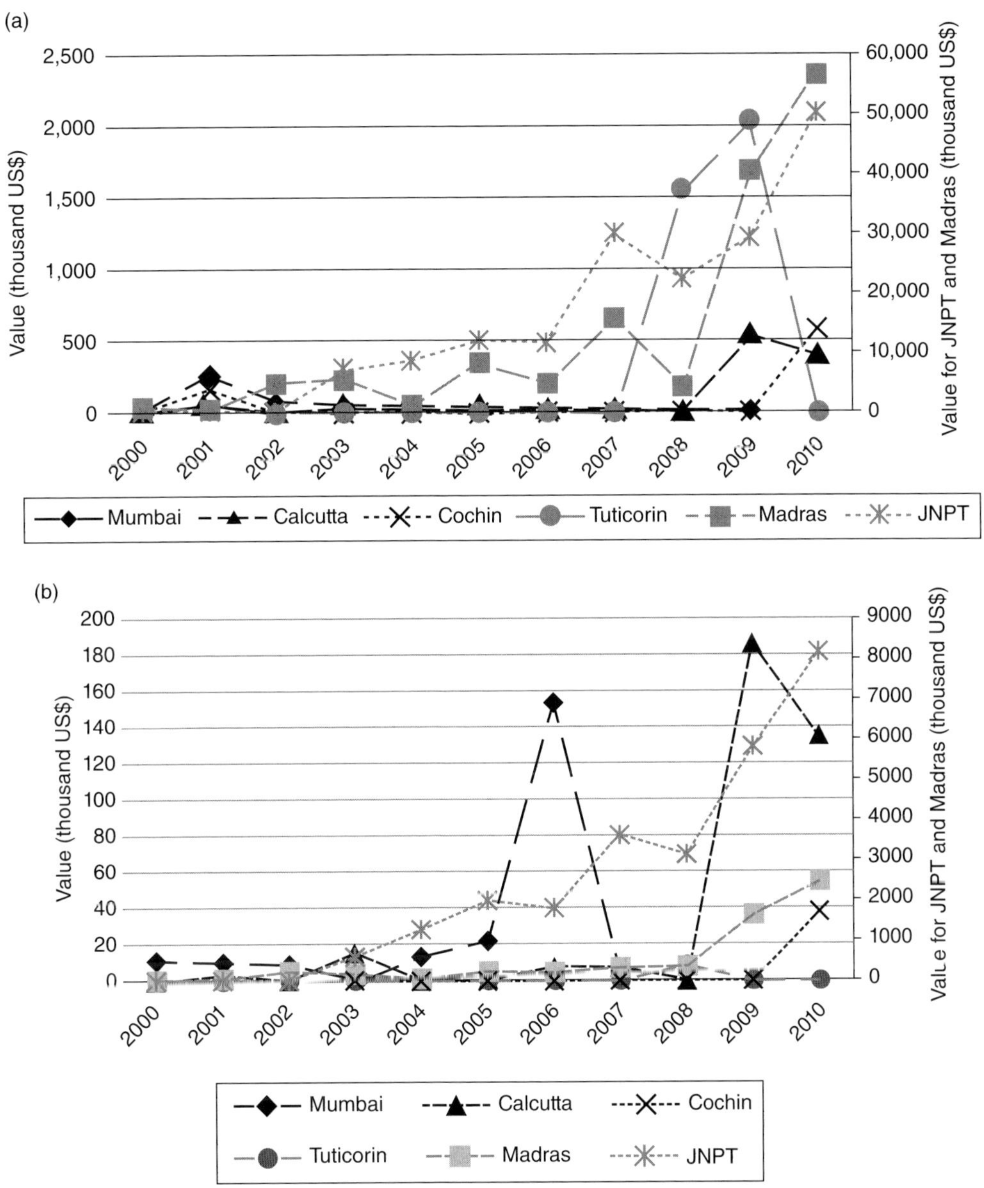

Fig. 14.7. Imports of (a) apples, (b) pears and (c) fresh oranges by port in India (JNPT, Jawaharlal Nehru Port).

chocolates or Swiss cheese. At the same time, the list of imported products and source countries for products sold in supermarkets has been quite dynamic. New sets of food products (for example, garlic, pulses, processed vegetables, pasta) keep being added to the shelves of supermarkets in India as imported products. Currently, several fruits and to some extent vegetables and also staple food items, such as pulses, are being imported and are sold by the supermarkets.

However, surprisingly in our data (considering the highly educated profile of the respondents), more than 40% of the surveyed customers did not know that staple foods and fruits and vegetable items were imported and stocked in the supermarkets.[11]

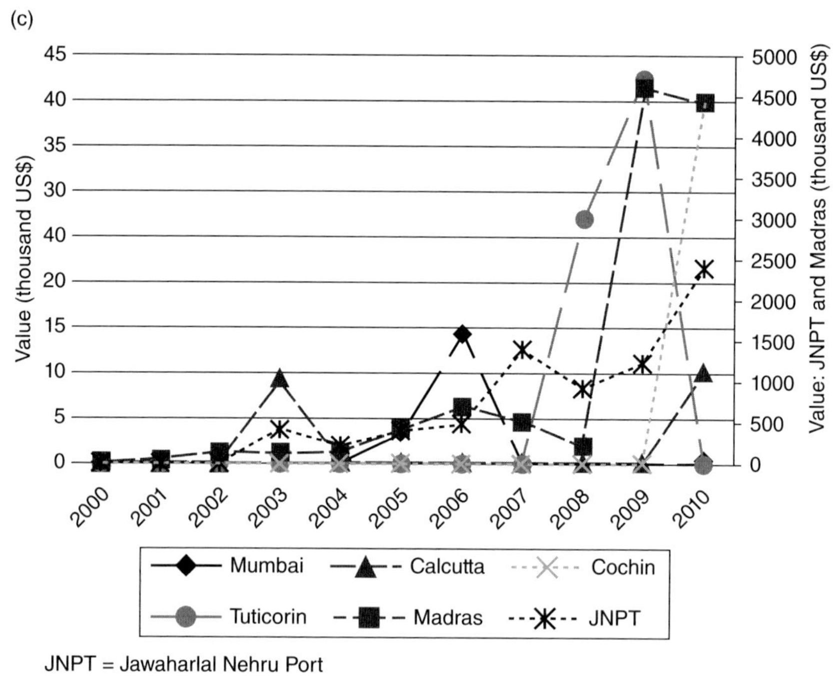

Fig. 14.7. Continued.

We have generated an awareness indicator where customers are taken as relatively more aware about imports if they know that *both* staples and fruits and vegetables are imported and sold at supermarkets. To account for differences in customers' awareness about imported products with a choice problem between an imported and a domestic product, we adopted an empirical framework where valuation is contingent on awareness that products are imported (i.e. staples like pulses or vegetables like garlic).

We thus modelled this choice problem in a framework of conditional mixed processes where in a sequential setting the customers' relative awareness about the existence of an imported variant of a product was assessed and upon that knowledge, the choice between an imported and a domestic product was analysed.

Assuming normal standard errors and linear functions, econometric models can be combined. Since the normal distribution has a natural multi-dimensional generalization, models can be combined into multi-equation systems in which the errors share a multivariate normal distribution (Roodman, 2011). The estimation is done using the cmp routine in stata, where a truly recursive data-generating process is posited and fully modelled.

The first stage of the model has the binary dependent variable, i.e. whether or not the customer is aware of products like staples and fruits/vegetables being imported and sold in the supermarkets (the one where they shop). In the recursive process, the second stage involves the three-level ordered probit models, where the first level is indifference between domestic and imported variant, the second level is preference for domestic variant and the third level is preference for the imported one.

Since preference between imported and domestic products could be affected by the experience of consuming imported products, we control for whether or not the customer has demanded niche products that could be imported. Results from estimation of a conditional mixed process model are presented in Table 14.4.

In the second stage of the conditional process, our variables of interest are related to reasons for purchasing at supermarkets

Table 14.4. Results of conditional mixed process models (probit and ordered probit) in choice of imported products.

Reason for buying at supermarkets	Awareness of imported attribute not commonly known and non-geographically marked products	Ordered choice of imports over domestic products (1=indifferent between the two, 2=would prefer domestic products, 3=would prefer imported products)	cut_1	cut_2
Better quality	0.109* (0.0644)	−0.0455 (0.0616)		
Value for money	0.0944 (0.0718)	−0.134* (0.0692)		
Better price	−0.144** (0.0627)	0.0108 (0.06)		
Higher food safety	−0.312*** (0.0741)	0.271*** (0.0721)		
Year-round availability	0.125** (0.0626)	−0.0428 (0.0602)		
Supply of customized product	0.0553 (0.235)	−0.0591 (0.169)		
Peer effect	0.184 (0.174)	0.293* (0.162)		
Exotic fruits like kiwi, dragon fruit	0.153** (0.0719)	−0.00902 (0.0665)		
Basil, celery	−0.00644 (0.0847)	0.0243 (0.0806)		
Red cabbage, broccoli, colour capsicum	−0.0526 (0.0724)	−0.0525 (0.0683)		
Sweetcorn, baby corn, cherry tomatoes	0.269*** (0.0659)	−0.154** (0.0626)		
Prunes	0.0988 (0.129)	−0.101 (0.119)		
Olive oil of all types	0.128** (0.0651)	−0.166*** (0.0627)		
Cheese such mozzarella, cottage, parmesan	0.288*** (0.0944)	−0.211**(0.0854)		
Exotic fruit juices and beverages	−0.156* (0.0908)	0.0875 (0.0821)		
Chocolates in different flavours	0.157* (0.0842)	0.0306 (0.0784)		
Meat and fish delicacles such as Parma ham	−0.0319 (0.107)	0.0136 (0.0951)		
Flavoured coffee, tea	−0.145 (0.102)	0.217** (0.104)		
Grocery expenditure in rupees	6.99E−06 (6.71E−06)	2.13E−06 (6.62E−06)		
Gender (1=female, 0 otherwise)	−0.077 (0.0623)	0.0146 (0.06)		
Age (in years)	2.33E−05 (0.00254)	−0.00047 (0.00247)		
Higher education (graduate and above=1, 0 otherwise)	0.431*** (0.0788)	−0.400*** (0.0803)		
Pune dummy	0.138 (0.102)	−0.182* (0.1)		
Mumbai dummy	0.265*** (0.0901)	−0.175** (0.0851)		
Upper caste dummy (1=yes, 0 otherwise)	0.216** (0.0875)	−0.127 (0.0843)		
Managerial position in the private sector (1=yes, 0=no)	0.303** (0.136)	−0.116 (0.113)		
Constant	−0.543*** (0.145)		−1.04*** (0.1)	−0.6*** (0.1)
Observations (no.)	1771	1771	1771	1771

Robust standard errors in parentheses
***p < 0.01, **p < 0.05, *p < 0.1

over traditional retail. As expected, more educated and higher-income consumers (such as managers in the private sector) are relatively more aware about imported products in supermarkets. Customers who choose supermarkets for better price are less likely to know about imports of staples and fruits and vegetables. Surprisingly, customers who choose supermarkets expecting better food safety are on average less likely to know about these imports.

The conditional mixed process results, however, show that conditional on being more aware about imports, the customers who look for food safety while choosing supermarkets are more inclined towards imported products. The coefficients show the lower probability of these customers being indifferent between domestic and imported products and the greater probability of preferring imported over domestic product. Imports thus serve as an outlet for meeting the demand of non-price attributes.

The results from the conditional mixed process model further show that within the class of supermarket customers, both awareness and propensity to buy imported products vis-à-vis its domestic counterpart, are higher across the upper segment of the society (in terms of caste status, consumption expenditure as well as education levels).

Moreover, customers in big cities like Mumbai are on average more aware of different imported products (such as fruits and vegetables and pulses). Women are relatively less aware of the imported status of some specific products which they were asked about. Just as in the case of choice of niche products, customers who prefer supermarkets for year-round availability and value the customization of products have a greater preference for imported products.

Overall, the customers who have a comparatively high valuation of non-price attributes (food safety, customization, year-round availability) are more likely to value imported products. This characteristic also underlies the choices of niche products. In summary, there is a significant overlap between the product categories, i.e. imported and domestic.

Conclusions and Policy Recommendations

In this chapter, a study on different aspects of the emerging modern retail sector in India is presented from the point of view of customers' choices. The research questions addressed are related to choices over products and product attributes in organized retail vis-à-vis the traditional marketing outlets. In particular, the valuation of non-price attribute is assessed in comparison with price attribute or its correlated traits. At the level of choice of products, the study looks at the interrelated choice of niche and imported products.

The findings suggest that subject to regional differences in degree, the attributes that are dominant have in general limited pull for supply chain coordination even though the class of customers shopping in supermarkets comprises the upper strata of the population (in terms of socio-economic profile). Non-price attributes, specifically food safety, are not of high importance in customers' choices. The advent of supermarkets has introduced an element of convenience in shopping and, for a significant majority of customers, it has been an attractive feature of supermarkets.

Supermarkets seem to have played a role in diversifying the consumption portfolio of its clients, particularly with imported niche products. Results show that the customers who purchase niche and/or imported products in supermarkets tend to have similar characteristics valuing non-price attributes such as food safety, customization of the product and year-round availability. Further, there is a definite move in the niche space towards imported products. The erstwhile niche products for example, broccoli or baby corn, are readily available domestically including in the traditional markets and are generally not imported. The niche products sold almost exclusively by supermarkets such as olive oils and exotic cheese are beginning to fill in the niche space in supermarkets.

Several imports carry a reputation for delivering on quality and safety or provide value for money, yet survey results show that in some products localness carries a premium. This has happened in the light of a diversifying set of products (staples, fruits and vegetables, etc.) and exporters (for example, more exports from lower income countries). Incidentally, pulses are the biggest food imports after edible oils with a diversifying set of suppliers including several from developed countries such as Australia and Canada. The products where imports get discounted are importantly the ones where issues like food safety, customization of product, year-round availability and product differentiation are less of a concern.

Several policy implications follow from the analysis here. The high valuation of price or superficial non-price attributes on domestically produced goods in the supermarkets imply that there are no significant demand pull pressures for improving supply-chain coordination. Such a demand rises if customers demand attributes such as food safety, customization of product or hygiene. Part of the reason for lack of demand is that there is no credible system of certification that would differentiate a safe product from an unsafe product. Instituting such a system would create demand pull for improving supply-chain coordination.

The rise in demand for products that were earlier considered niche implies that it is a good opportunity to engender production diversification as density of supermarkets rises. Data indicate that the scope for diversification in India to be driven by supermarkets would mostly be in products that are currently being imported. Therefore there will be a need to minimize the barriers to direct firm–farm linkages. As the niche space is being increasingly taken up by imports, policies to diversify the set of processed items domestically might be worthwhile.

Similar to the case in fresh vegetables, supermarkets have much less of a direct role in animal-source foods. Meat and dairy products are sold by supermarkets only in a limited way. Improving infrastructure conditions such as power supply would provide a good link between livestock farmers and organized retail. The survey shows that there is little uptake of animal-source foods from supermarkets. Improving hygiene through cleanliness, storage and packaging could differentiate animal-source foods in supermarkets from others.

Notes

[1] Based on different reports from retail consulting firms, there are the following different formats defined largely in terms of the retail space. Large stores where floor spaces range from 20,000 to 50,000 square feet. Large outlets like supermarkets contribute up to 30% of all food and grocery organized retail sales. Supermarkets can be classified into mini-supermarkets, typically 1000 square feet to 2000 square feet, and large supermarkets ranging from 3500 square feet to 5000 square feet having a strong focus on food and grocery and personal sales. Convenience stores are relatively small stores of 400–2000 square feet located near the residential areas (Kearney, 2006).

[2] Technically the policy refers to multi-brand retail. Relaxation of foreign investment norms in wholesale and single brand retail had happened before in 2006.

[3] The back-end investments are defined in the legislation to include processing, manufacturing, distribution, design improvement, quality control, cold chain, warehouses and packaging.

[4] Note that those pursuing a post-graduate degree also described themselves as being in the category of graduate degree and above.

[5] Scheduled caste and tribes constitute the lowest strata in terms of social identity.

[6] Note that the clubbing of disparate products into one category in the table is because of stocking patterns across different supermarkets.

[7] Earlier olive oil treated as a luxury item was used selectively as a massage oil for infants.

[8] Still the awareness about food safety is generally low. Since certification and labelling are largely absent in Indian markets, to gauge the food safety consciousness of the consumers we asked a simple question

regarding their propensity to check for expiration date on products. Only 38% customers check expiry date on all products, 49% on some products and about 11% of customers never check for expiration date.
[9] Irrespective to product differentiation, it is difficult to have a price parity between imported and domestic products. Indian import tariffs and non-tariff barriers are among the highest in the world. These apply to food products such as meat, fruits and vegetables. Import tariffs on most food products range from 30% to over 50% (Euromonitor, 2008).
[10] The same products could also be demanded by the HRI sector, though one would expect their intake to be on a smaller scale. Examples of such products could be frozen vegetables, packaged fish and smoked meat products among several others.
[11] In reality, in the supermarkets that customers were interviewed, these imported products were in stock.

References

Euromonitor (2008) WTM Global Trends Report. Euromonitor International. London. Available at: http://www.wtmlondon.com/RXUK/RXUK_WTMLondon (accessed December 2015).

FAOSTAT (2013) FAOSTAT database, Food and Agriculture Organization of the United Nations, Rome. Available at http://faostat.fao.org/site/362 (accessed November 2015).

GoI (Government of India) (2007) *India Retail Report 2007*. Ministry of Commerce and Industry, New Delhi.

Joseph, M., Soundararajan, N., Gupta, M. and Sahu, S. (2008) *Impact of Organized Retailing on the Unorganized Sector*. ICRIER Working Paper No. 222. New Delhi.

Kearney, A.T. (2006) A Confederation of Indian Industry – A.T. Kearney Report, Retail in India: Getting organized to drive growth. Available at: https://www.atkearney.com/documents/10192/553806/retail_india_app.pdf (accessed December 2015).

Planet Retail (2004) Consumers Forsake Wet Markets in Asia. *Planet Retail, Daily News* 7 June.

Reardon, T. and Minten, B. (2011) Surprised by supermarkets: diffusion of modern food retail in India. *Journal of Agribusiness in Developing and Emerging Economies* 1(2), 134–161.

Roodman, D. (2011) Fitting fully observed recursive mixed-process models with cmp. *Stata Journal* 11(2), 159–206.

15 Conclusions and Way Forward

P.K. Joshi[1]* and Floor Brouwer[2]
[1]*International Food Policy Research Institute, New Delhi, India;*
[2]*LEI Wageningen, UR, The Hague, The Netherlands*

Introduction

India has registered impressive economic growth of 7–8% following economic liberalization in 1991. Consequently, there has generally been a reduction in poverty levels in both rural and urban areas, and the impact is clearly visible during the post-reform period. Agriculture also performed better during the previous decade with an average annual growth rate of more than 3.5%. The agricultural sector witnessed a structural transformation which was not just restricted to the farm sector. Unfolding patterns include integration of agriculture with other non-farm activities (e.g. processing and retailing), farmers linkages with markets, and opening-up of the economy for many agricultural commodities for trade. The paradigm shift was a result of various policy initiatives and reforms in agricultural and non-agricultural sectors. The policy changes and reform measures have implications for agricultural production, food consumption, food prices and trade. These changes have significantly contributed to raising incomes and reducing poverty. A question, however, remains about the nature and pattern of food supply, demand, trade, poverty and income inequality under different economic growth scenarios. This book is intended to answer that question by projecting implications of policies and trade on farm and non-farm incomes, poverty and inequity. This chapter summarizes the key findings of various studies and provides action plans for strengthening the agriculture sector and strategies for promoting agricultural trade for increasing farm incomes and reducing poverty in the country.

Evolution of Agricultural and Trade Policies

India has undergone significant macroeconomic modifications through economic liberalization to realize the growth potential and step towards greater economic integration, both domestically and globally. Chapter 2 documents useful information on the key areas of concern that pose challenges and opportunities to overcome the barriers to higher growth and more income for farmers. This chapter also covers near stagnating public investments, increasing pressure on subsidies, limited access to formal credit, slowing of irrigation expansion, and the limited role of the private sector in agriculture. For increasing agricultural productivity, the government intervention through mega-programmes like

* Corresponding author, e-mail: p.joshi@cgiar.org

Rashtriya Krishi Vikas Yojna (RKVY) (National Agricultural Development Programme), National Food Security Mission (NFSM) and the new initiatives for a second green revolution in eastern India are laudable. A similar initiative, National Horticultural Mission (NHM), has been operational successfully and there is a need for such programmes in other high-value sectors such as livestock and fisheries.

Technology has played an important role in Indian agriculture beginning with the success of high-yielding varieties since 1960s. The public sector played a key role in developing and disseminating improved technologies that transformed Indian agriculture. The private sector is gradually finding its niches in some areas, especially in hybrids. For example, despite all the controversies surrounding the success of Bt cotton in India, it benefited farmers by raising their incomes. At the national level, India became a net exporter of cotton from being a net importer. The public and private sector-driven technologies will significantly contribute to increasing agricultural production and improving livelihoods of the poor, especially in the backward areas.

Investment in agriculture is the key to improving agriculture in a sustainable manner, be it in irrigation, infrastructure (roads, markets), agricultural R&D, extension services, etc. Investment to enhance quality and delivery of public goods will have to come from the public sector. The subsidies have outlived their significance, and it is time to rationalize them to boost investments. Agricultural input subsidies have increased manifold and, in certain cases, have even proved to be environmentally damaging, as observed in the case of fertilizer and power. These subsidies often do not reach the targeted beneficiaries, particularly the marginal and small farmers who dominate the agricultural sector.

Agricultural trade policy in India will remain subservient to food-security concerns. Despite large reserves of foreign exchange, agricultural trade policies are driven by food-security concerns. Liberalization of agricultural trade had aroused apprehensions among the policy makers that the domestic market would be flooded by imports, and may adversely affect the farmers. Contrary to the perception, production of high-value commodities, such as horticulture, livestock and marine products, increased significantly, and contributed a boost to their exports. Yet India is still a very small player in the global market in the high-value commodities segment. However, there is enormous scope to expand further export of these commodities but this is constrained due to effective compliance with sanitary and phytosanitary (SPS) issues. The other factor is the lack of world class physical infrastructure (such as processing, ports, cold chains, etc.). There is a need for promoting best practices for compliance of SPS issues and to attract the private sector for making investment in key areas in partnership with the public sector to further boost exports.

Demand–supply Projections for Food Commodities

Driven by rising population and a growing economy, the consumption and consequently demand for food commodities is continuously increasing in India. On the other hand, the country is facing the problems of plateauing productivity, decreasing farm sizes and diminishing natural resources. Under such a scenario, Chapter 4 attempts to find the answer to a significant question, 'Will India be able to produce enough to meet its growing food demand or will it be open for imports of food commodities by 2030?' The chapter not only makes projections for the demand and supply of key food commodities in India by 2020 and 2030, but also presents their trade potential. The study revealed that India will continue to meet the demand for wheat through domestic production even by 2030. In the case of rice, another major staple crop, the country is not likely to remain in surplus and may even become deficit to the extent of 3–5 Mt by the year 2030, if appropriate measure are not taken. Among other food commodities, the domestic supply of pulses and oilseeds will continue to remain short of demand in future, with implications for their strategic trade policy.

In the non-food grains sector, the chapter projects a substantial supply–demand gap in vegetables and fruits by 2030 unless effective measures are taken to minimize postharvest losses. According to the projections, India will be able to meet the domestic demand for eggs, with a marginal surplus, but will remain deficit in total meat production in the years to come.

A significant message from the study is that policies that can help in maintaining the total factor productivity (TFP) growth in the long-run will be able to keep a balance between domestic production and demand for different food commodities. This emphasizes the need for strengthening the efforts at increasing production potential through public investments in irrigation, infrastructure development, agricultural research and efficient use of natural resources.

Economic Growth, Trade and their Implications on Poverty

It is debated often that high economic growth leads to wider income disparities. In order to understand better this dynamic, an analysis of income distribution across rural and urban India under different GDP growth scenarios has been done using the computable general equilibrium (CGE) model (Chapter 6). It is found that a high economic growth scenario of 8% and above does not favour rural households due to the decline in agricultural growth. With GDP growth moving up from 8% to 10%, the rural share in real income comes down from 61% to 56%. The 10% GDP growth seems to benefit the industrial and services sectors and not the agricultural sector.

The analysis in terms of income groups shows that 10% GDP growth will largely benefit the urban households of the top 30% income percentile. The study confirms that growth is not trickling down and real income of rural households of the lowest 30% and middle 30–70% categories will come down on moving up from the 8% to 10% GDP growth scenario during 2010–2020. The study also reveals that on moving from the 8% to 10% growth regime, incomes become redistributed from rural to urban regions, particularly for the lower and middle income groups.

A major policy implication drawn from the study is 'how to improve agriculture sector in the higher GDP growth scenario?' as agricultural growth is going to be one important factor for reducing poverty in the rural areas of India. Supplementary measures are needed to tackle the dipping rural poor income with increasing GDP growth.

Trade and implications on poverty

India is currently inadequately integrated with world trade, but merchandise trade is growing fast and therefore trade agreements will assume considerable significance in the time to come. This aspect under different scenarios is covered in Chapter 5. It shows that by 2030, India will be growing relatively faster compared to the rest of the world, mostly due to higher technological progress in the manufacturing sector. In terms of production and exports, Indian industry, including agri-processing, will become more important and India would become a net exporter of industrial commodities and surprisingly a net importer of services. With respect to agriculture, the restrictions on land and water availability, in combination with a rise in demand for foods, imply that the net imports of crops would increase much more in the times to come. The rise in imports is, however, not likely to be enough to meet the rising demands for food, so the pressure on land would increase, resulting in fast rising land and crop (and livestock) prices, much more than in the rest of the world. This will lead to intensification of agriculture in terms of higher fertilizer use, better access to capital, adoption of labour-saving approaches and higher agricultural investments.

As mentioned above, the chapter investigates the impacts of alternative growth and trade policies on the Indian economy and the rest of the world, notably the European Union (EU), using a global trade model, MAGNET (Modular Applied GeNeral Equilibrium Tool). With a rise in annual economic growth from 6% to 10%, the chapter presents the economic scenario under two trade agreements: a bilateral Free Trade Agreement (FTA) with the EU and a multilateral trade agreement in the context of the World Trade Organization (WTO). The study reveals that the WTO agreement gives a much smaller reduction in tariffs for Indian imports from the EU than the FTA gives. The study shows that total imports by India of products from the EU would change only slightly in the presence of a WTO agreement (2.7%), and would increase considerably (by 52%) with a FTA in 2015. In particular, the imports of processed agricultural products could increase a lot (close to eight-fold increase), while the import of industrial commodities would almost double. In relative terms, crop imports will also show a large (three-fold) increase.

The impact of high Indian growth on production in the rest of the world would be relatively small. The largest impact would be felt in industry, with industrial production being 2.3% lower in the EU and 1.7% lower in the rest of the world by 2030. The EU, as also Africa, will benefit relatively more from the rising agricultural imports by India than other regions of the world.

An important inference emerging from the study is that trade agreements in general are good for food security with further improvements over time. A FTA with the EU will not influence food prices to a great extent, but will increase income of the poor as the wages for unskilled labour would rise. On the other hand, a WTO agreement has depicted a greater reduction in agriculture prices in India.

Implications of Social Programmes and Price Policies

Four policies are analysed to assess their impact on food consumption, agricultural production and income. These include: (i) social protection programme, namely the Mahatma Gandhi National Employment Guarantee Act (MGNREGA); (ii) minimum support policy programme; (iii) input subsidies in agriculture; and (iv) biofuel policy.

Impact of MGNREGA on food consumption

As stated earlier, the Government of India has launched various programmes from time to time to alleviate poverty. One of the latest programmes overarching all the earlier schemes was implemented through the legislation entitled 'National Rural Employment Guarantee Act' (NREGA), now renamed the Mahatma Gandhi National Rural Employment Guarantee Act (MGNREGA). It is the largest employment-providing programme ever started by a country in the world. This cross-cutting scheme stipulates a legal guarantee of providing wage employment for 100 days in a financial year to adult members of any rural household willing to do unskilled manual work at the statutory minimum wages. Chapter 3 in the book examines the changes in food consumption and nutritional security of rural poor households and assesses the impact of MGNREGA by comparing the food consumption patterns of MGNREGA-beneficiary and non-beneficiary households.

The study reveals that implementation of MGNREGA is a direct way of providing employment and as a consequence enhances income of the rural poor. At the national level, the benefits of MGNREGA have reached 22.5% of the rural households: 30% BPL (below poverty line) households, 37% agricultural labourers, 27% sub-marginal farmers (holding <0.5 ha land) and 21% landless households. MGNREGA has been successful in reducing the poverty level by 4% by providing wage employment, on average, for 43 days per household in a year. Additionally, MGNREGA provides equal employment benefits to all the categories of households and contributes substantially to women empowerment by not only providing them employment but paying wages equal to men. The study on state-wise socio-economic dynamics of rural households reveals that economically weaker states of the country have benefited more and have implemented MGNREGA more vigorously.

The chapter reveals that the higher income received by MNREGA beneficiaries is leading to an increase in food consumption level – of both cereals and non-cereals. Another impact of MGNREGA is seen in diet diversification towards non-cereal high-value commodities, such as fruits, meat, eggs and fish, etc. As a result, a substantial increase is seen in calorie intake as well as protein intake of different categories of households, leading to a decrease in the number of undernourished and nutrition-deficient households by 8–9%.

In a nutshell, the chapter concludes that the impact of MGNREGA has been positive and it has contributed to increasing household food consumption, changing dietary patterns and providing food and nutritional security to the poor rural households of India.

Long-term impact of MGNREGA

The impact of MGNREGA has also been assessed on GDP, industrial sector, household income and labour supply to agriculture using the generalized equilibrium model (Chapter 9). The chapter provides some stirring messages. MGNREGA is likely to have a negative impact on agriculture in the long-run. It projects that the share of agriculture in total GDP is likely to come down from 13.0% to 11.8% between 2010 and 2020. This would be due to diversion of government resources to MGNREGA from erstwhile productive sectors. Second, between 2010 and 2020, the real income in rural areas seems to come down partly due to agricultural growth and partly due to lower market wages compared to the business-as-usual (BAU) scenario of 8% GDP growth. The overall picture is that MGNREGA has contributed to the growth of the industrial sector and has provided a big fillip to the industries such as manufacturing and construction. The study reveals that under this scenario, the real income will be reallocated from rural to urban areas compared to BAU during 2010–2020. Thus, MGNREGA seems to push up the income of the urban poor and not of the rural poor in the long-run.

The study does not support the ongoing debate whether MGNREGA pushes up wages for agricultural labour and causes labour scarcity in agriculture due to diversion of labour from agriculture to works under MGNREGA. The results of annual real income per capita show lower figures for MGNREGA than the BAU scenario by 2020. Also, the increase in the market wages for unskilled labour in the rural areas is not due to MGNREGA, as MGNREGA supplements only the off-season employment and does not draw agricultural labour away from farming.

A significant message is that MGNREGA may not be sustained in the long-run given the limited resources of the government and, secondly, MGNREGA does not seem to continue to provide benefits to the rural poor, for whom the scheme was originally launched.

India's price support policies

Chapter 10 analyses the impact of global price spikes in relation to India's agricultural price policy. It may be mentioned that Indian policies aim to achieve the twin objectives of assuring remunerative prices to the producers and providing food grains to consumers at affordable prices. In order to ensure remunerative prices, the government announces minimum support prices (MSP) for major agricultural commodities and procures the produce (especially rice and wheat) for distributing under social safety-net programmes. Subsidized food grains (currently, wheat, rice, coarse grains) are distributed through the Public Distribution System (PDS), later modified as the Targeted Public Distribution System (TPDS), and now falls under the National Food Security Mission. The spread of this policy can be judged from the facts that in 2011/12, the cost of TPDS to the exchequer was almost 1% of GDP (Rs600 billion or US$11 billion) and about 26.4 Mt of rice and 23.5 Mt of wheat were earmarked for distribution across different sections of society in 2012/13 (GoI, 2012).

The Government of India could successfully shield its domestic market from the global food price spikes and volatility in 2007–2008 and 2011 through its policy of imposing bans on export of wheat and rice and lowering import duty on wheat. The shielding effect is confirmed by the Engle-Granger tests for co-integration, which does not reveal statistical significance for both rice and wheat, implying that the domestic prices of these commodities are not influenced by the world market. However, the policies did not succeed in holding the price rise in the domestic markets at a later time due to the slow release of stocks. On the other extreme, due to the ban on exports, farmers could not profit from the higher global prices and lobbied successfully for a higher MSP. This triggered an increased supply of rice and wheat. Consequently, the buffer stocks of these commodities increased substantially, which piled up, exceeding their installed storage capacity. Unfortunately, due to inefficiencies in the TPDS, the stocks were not distributed sufficiently.

The study reveals another significant effect of price support policies. It argues that with growth in GDP and population, the demand for rice and wheat will increase, but the composition of demand is likely to change and the share of food in household expenditure is expected to decrease. This has consequences for India's TPDS, which, given these developments, is expected to play a relatively small role by 2030. There is thus a risk of a future mismatch between the growing supply, with rising MSP and relatively reducing demand for the TPDS.

The adoption of the National Food Security Act (NFSA), under which 67% of the population will be given subsidized food at almost one-tenth of the economic cost, will have an impact on all agricultural policies by making them more restrictive and state-controlled besides adding significantly to the cost to the exchequer.

Input subsidies versus farm technology

The rising costs of farm inputs discourage their use and lead to a fall in the supply of agricultural commodities and farm incomes. The decline in supply of these commodities raises their market prices and adversely affects the purchasing power of consumers. To overcome the rising input prices, the Government of India is providing subsidies on key inputs. Over the years, the subsidy burden has grown, adversely affecting the government exchequer. On input subsidies, questions are often raised such as what would happen to output supply, factor demand, agricultural prices and farmers' income under the alterative subsidy and farm technology scenarios. Chapter 12 attempts to answer such questions by developing empirical unified models for wheat and rice, and analysing the role of input subsidies and farm technologies.

The study reveals that technology is the most powerful instrument for neutralizing factor price inflation and safeguarding the interest of producers as well as consumers, while input subsidy has a weak effect on output supply. The study observes that investments in irrigation, rural literacy, capacity building, research and extension and information flow are crucial to increase supply at a higher growth rate. The input subsidy has a positive effect on input use, crop supply and farm income, but technology shifters have a positive and strong influence on commodity supply and a substantial negative effect on farmers' income because of the decline in market price in the absence of MSP policy. Also, the input subsidy to farmers and price subsidy to consumers will not be feasible in the long-run as they involve a substantial share of public resources. A message from the chapter is that a viable solution can only be found with appropriate adjustments in the non-price factors.

Presenting a scenario of 10% withdrawal on one of the major farm inputs, viz. fertilizer, the chapter reveals that for its compensation, investments on agricultural research will have to be increased at the growth rate of 6.0% per annum, irrigation will have to be increased at the growth rate of 0.3–0.4% per annum and the TFP growth will have to be increased from the present level by 0.18% for rice and 0.20% for wheat by adopting appropriate measures.

Biofuel commitments in India and international trade

The recent rise in use of biofuels is driven by the concerns for energy security, climate change and rising fossil fuel prices. India implemented a pilot programme aimed at realizing 5% ethanol blending (E5) in 2001 and launched a National Mission on Biodiesel in 2003. In 2009, the Government of India approved the National Policy on Biofuels that includes an indicative 20% blending target by 2017 for both biodiesel and bioethanol (GoI, 2009). The objectives of this policy are to reduce the dependency on imports of fossil oil, reduce greenhouse gas emissions, promote rural development and generate employment. Chapter 11 in this book has addressed the consequences of the National Biofuel Policy of India and of biofuel policies in other countries on poverty, welfare, land use, trade, food security, etc., in India by 2020 using a global economic model, MAGNET (Modular Applied GeNeral Equilibrium Tool).

Since the current biofuel policies in India and other countries are based on the use of crops for producing first-generation biofuels (ethanol from sugarcane and biodiesel from vegetable oils), these would depict various effects on poverty, societal welfare, land use, trade, food security, etc. by 2020. The Indian biofuel policy seems to have a negative effect on poverty in India due to projected rise in crops' prices by 11.5% and wage rate in agriculture by just 3% by 2020. The biofuel policies outside India will also have a negative impact on poverty in India, but the effect will be less on the rural than the urban poor. As a result, the consumption of agricultural products and livestock products in India is likely to decease, although the societal welfare effects would be positive due to 'terms of trade effect' (import prices increasing less than export prices).

The chapter also presents some positive impacts of biofuel policies. In terms of production of agricultural commodities, the analysis reveals that global production of sugarcane would increase by 40%, coarse grains by 30% and vegetable oils by 50%. The increase in biofuel use would also result in increased crop yields for vegetable oils in the EU, for coarse grains in the USA and for sugarcane/sugarbeet almost everywhere.

Because sugarcane is a major crop used for biofuel production, the projected statistics for it are quite significant, particularly for India. The area under sugarcane would expand by 68%, the price of land used for sugarcane production would increase by 150% or more and the price of sugarcane would increase by 27% (and of molasses by 11%).

Discussing the welfare effects of biofuel policies in regions other than India, the chapter reveals that the allocation effect would be highly negative for the USA, as also for China. For Brazil, these policies would have a positive terms of trade effect as a result of reduced fossil fuel imports. The big loser will be Africa (including the Middle East), which is likely to pay more for food without having much benefit of lower fossil fuel prices.

The chapter, however, observes the need for a more refined regional and long-term analysis for developing biofuel policies on agricultural production, rural development and technological innovations in different countries of the world.

Emergence of Livestock Sector

Dairy sector and its impact on poverty and trade

The increasing diversification in dietary patterns towards high-value commodities such as fruits, vegetables, dairy products, meat, fish, eggs, etc., coupled with highest consumption of milk and milk products amongst all the livestock products, reveals the high relevance of the dairy sector in India. The fact that India is the highest producer of milk in the world, livestock is a source of income and employment to a large number of rural households (it was 70 million in 2002), dairy production provides opportunity for income generation to rural women, and rising awareness about quality of milk and milk products, further show the importance of this sector.

Besides surveying the supply response of milk to its high demand growth, Chapter 13 (this volume) looks into an important aspect of dairy sector, viz. its potential to act as an engine of pro-poor growth in India. It suggests that households engaged in dairy need to increase milk productivity, which is still low compared to other major dairy-producing countries in the world. India will have to overcome constraints, such as access to fodder, land and animal healthcare, before it can be confident that domestic supply will be able to keep up with domestic demand growth. Another revelation is that dairy development in India seems to have increased more on the extensive margin (more households engaged in dairy), rather than on the intensive margin (households engaged in dairy scaling-up their operations).

Food Safety and Standards

Food safety and quality issues are important for domestic and global markets. Chapter 7 (this volume) shows how these may reinforce each other. Using the dairy sector, the chapter provides evidence on food safety practices from a primacy survey of dairy producers in Andhra Pradesh (then unified Andhra Pradesh), a state in southern India (now bifurcated into Telengana and Andhra Pradesh).

Chapter 7 reveals that food safety and quality management systems are still in their infancy in India. The current food safety regulations present high risks of food hazards to many consumers and may be considered a risk to public health in India. For example, in the dairy sector, milk adulteration by adding water, oil, milk powder or soda remains common and goes widely unpunished – which means that an adulterated milk crisis may likely happen in India as it occurred in China a few years back.

The establishment of the Food Safety and Standards Act (FSSA) in 2006 and its implementation in 2014 show that concerns for food safety are on the rise, and that these have reached the policy makers. The implications of this changing mind-set are uncertain in the short-term, but once food safety standards are enforced in the way policy makers have conceived them, the potential implications for food supply chains in India will be substantial.

The agricultural supply base in India is made up of millions of small farmers, therefore the advent of food safety and quality standards will surely have a profound impact on rural livelihoods. While the expectations of such implementations may be confronted with strong opposition in India, the potential benefits should not be underestimated. Not only domestic consumers may benefit from improved food safety and quality, there may as well be important implications for international trade.

If India succeeds in aligning its own food standards with international norms as laid down in the Codex Alimentarius, it may surmount current SPS (or other NTM) barriers to exports and gain importantly from new opportunities for trade. Food standards are currently the crucial determinants of trade with high-income countries such as the EU and the USA. Those countries which are currently importing dairy products from India (mainly neighbouring countries in South Asia and countries in the Middle East) may increase their food standards over the next decades as well.

Based on a rural household survey in Andhra Pradesh, Chapter 7 (this volume) reveals that food safety practices in milk are based on traditional wisdom, vary across households, and milk farmers receive little training, guidance or inspection by milk traders or by government. However, the study also finds positive correlations between participation in dairy production, incomes per capita, land accumulation and other assets (such as mobile phone, fridge, motorbike, etc.).

Poultry sector of India and its trade potential

The Indian poultry sector grew (8–10%) rapidly during the past one-and-half decades. India is ranked third in global egg production and sixth in chicken meat production. Despite high growth in the poultry sector, and a large increase observed in the production of eggs and poultry meat, India is a very small player at the global level with some export of eggs and egg powder. In the rest of the poultry products, India's presence is negligible. Chapter 8 (this volume) analyses the price competitiveness of India in poultry products (viz. hen eggs (in shell), chicken live weight, chicken meat) with Brazil, the USA, Germany, France and The Netherlands. The findings clearly reveal that India has a definite comparative advantage in the case of egg exports, but not in poultry meat.

The assessment of the potential effects of complete elimination of tariffs (currently 87%) on poultry meat imports reveals a fall in its price, which is likely to push up its import. Such a situation would be disastrous to the domestic poultry industry and the effect will be more on small poultry producers in the country.

Rise of Modern Retail Chains

Retailing in India is changing with the spread of consolidated formats such as supermarkets. The research questions addressed in Chapter 14 (this volume) are related to customer choices over products and product attributes in organized retail vis-à-vis the traditional marketing outlets. The study assesses the valuation of non-price attributes preferred by customers in comparison with price attributes or its correlated traits.

The chapter reveals that the attributes that are dominant have in general limited pull for supply-chain coordination. The non-price attributes, specifically food safety, get a backseat in customer choices. The advent of supermarkets has only introduced an element of convenience in shopping and, for a significant majority of customers, this is an attractive feature of supermarkets.

Supermarkets seem to have played a role in diversifying the consumption portfolio of its clients, particularly with imported niche products. Further, there is a definite move in the niche space towards imported products. The erstwhile niche products, for example, broccoli or baby corn, are readily available even in the traditional markets and are no longer imported.

The message from the study is that without a strong demand for attributes like food safety and customization to customers' choice and niche products increasingly sourced as imports, incentives remain low for supermarkets to invest in the development of back-end. With the approval of Foreign Direct Investment (FDI) in the retail sector, demand pull for supply-chain coordination may remain low. A command approach by the government to enforce back-end development would work only if enforcement is strong, which may be very unlikely in India.

The Way Forward

The following are the key recommendations for accelerating agricultural growth, reforming policies for developing markets, and promoting agricultural trade for increasing farm incomes and reducing poverty.

- **Manage higher GDP growth:** The exercise on implications of economic growth under different scenario reveals that with high GDP growth (8–10%), the rural share in real income will come down. Therefore, the major policy shift should be to balance between higher economic growth and create opportunities for rural poor, especially of the farming community.
- **Increase public investments:** The supply and demand projections to 2030 depict the supply to be deficit of demand for some essential commodities including pulses, oilseeds, sugar, meat, fruits, vegetables, etc. and emphasize on strengthening the efforts at increasing production potential through public investments on agricultural research, irrigation, infrastructure, natural resources management and environment.
- **Rationalize subsidies:** The agricultural input subsidies have outlived their significance, and it is time to rationalize them to boost investments. Considering the long-term negative impacts of inputs subsidy, the public investments on agricultural research should be enhanced. There is also a need of encouraging private-sector participation in technology development.
- **Balance price support policy:** Although the price support policy helps farmers in selling their produce at a better price, it comes at a cost in terms of increased supply. This leads to piling up of stocks and creates storage problems. There is an urgent need to develop appropriate storage capacities for scientifically stocking grains. In addition, the commodity procurement needs to be made more farmer-friendly, widespread and free of reported shortcomings.
- **Develop livestock and fisheries missions:** The National Horticulture Mission (NHM)

has shown an impressive overall performance and shows the need for such programmes in other high-value sectors such as livestock and fisheries.

- **Expand export of high-value commodities:** The global trade in high-value commodities has increased over a period of time, but India is still a small player and herein lies the scope to expand further. India needs to find niche commodities and countries for promoting export of high-value commodities to take advantage of higher global prices. However, it will require compliance of sanitary and phytosanitary standards for acceptability by importing countries.
- **Compete globally in poultry sector for trade:** The exercise on finding trade opportunities for India's poultry sector, which at present is one of the fastest growing segments of agriculture, reveals that India is definitely price competitive in eggs but not in poultry meat. India should not lower import duty on poultry meat from the existing 87% to zero, because it will push up imports enormously but will prove disastrous to the domestic poultry industry.
- **Long-term trade agreements:** The exercise with trade agreements has shown that trade agreements are good for food security, with the improvements increasing over time. India, therefore, may plan for long-term trade agreements to derive longer benefits.
- **Find alternatives to MGNREGA in the long-run:** In the short-term, MGNREGA schemes have benefited the poor and contributed to food and nutritional security. However, the long-term projections reveal that MGNREGA schemes may have a negative impact on agriculture, and may push up the income of the urban poor and not of the rural poor. The policy implication is that programmes under MGNREGA may not be sustained in the long-run, given the limited resources of the government, and also MGNREGA schemes may not continue to provide benefits to the rural poor, as was intended originally.
- **Develop retail chains and supermarkets:** There is a need to institute a credible system of certification to differentiate a safe product from an unsafe product. Considering the rising demand for products that are currently being imported, there is a need to minimize the barriers to direct farm–firm linkages, and also evolving policies to diversify the set of processed items domestically might be worthwhile.
- **Develop a safer and hygienic dairy sector:** Food safety and quality management systems being in their infancy in India, they will need effective implementation of the Food Safety and Standards Act (FSSA), which will not only benefit the domestic consumers but may also have important implications for international trade.
- **Attract private sector to agriculture:** The private sector is increasingly taking ahead technology-based innovations, research and marketing in the agriculture sector. There is a need to recognize the role of private-sector participation and encourage its presence in other segments of this sector. One area may be to attract the private sector for large investments to build quality infrastructure and bring in the correct technology to make India strong in agricultural exports.

References

GoI (Government of India) (2009) *National Policy on Biofuels*. Ministry of New & Renewable Energy, New Delhi.

GoI (Government of India) (2012) Economic Survey 2011–2012: Agriculture and Food. New Delhi. Available at: http://www.indiabudget.nic.in (accessed October 2015).

Index

Note: Page numbers in **bold** type refer to **figures**
Page numbers in *italic* type refer to *tables*
Page numbers followed by 'n' refer to notes